# Orlicky's Material Requirements Planning

# About the Authors

**Carol Ptak** is currently a partner with the Demand Driven Institute, and was most recently at Pacific Lutheran University as Visiting Professor and Distinguished Executive in Residence. Previously, she was vice president and global industry executive for manufacturing and distribution industries at PeopleSoft where she developed the concept of demand-driven manufacturing (DDM). Ms. Ptak spent four years at IBM Corporation culminating in the position of global SMB segment executive.

**Chad Smith** is cofounder and managing partner of Constraints Management Group, a services and technology company specializing in pull-based manufacturing, materials, and project management systems for mid-range and large manufacturers. He has been at the forefront of developing and articulating demand-driven MRP and is also an internationally recognized expert on the theory of constraints (TOC).

Carol and Chad founded the Demand Driven Institute, an organization devoted to the proliferation and further development of demand-driven strategies and tactics in industry. Go to www.demanddriveninstitute.com to learn more.

# Orlicky's Material Requirements Planning

*Third Edition*

CAROL A. PTAK, CFPIM, CIRM
CHAD SMITH

New York   Chicago   San Francisco   Lisbon   London
Madrid   Mexico City   Milan   New Delhi   San Juan
Seoul   Singapore   Sydney   Toronto

The **McGraw·Hill** Companies

Cataloging-in-Publication Data is on file with the Library of Congress.

**Orlicky's Material Requirements Planning, Third Edition**

3 4 5 6 7 8 9 0    QVS/QVS    1 9 8 7 6 5 4 3

ISBN 978- 0-07-175563-4
MHID      0-07-175563-2

The pages within this book were printed on acid-free paper.

**Sponsoring Editor**
  Judy Bass

**Editorial Supervisor**
  David E. Fogarty

**Project Manager**
  Patricia Wallenburg

**Copy Editor**
  Jim Madru

**Proofreader**
  Claire Splan

**Indexer**
  Claire Splan

**Production Supervisor**
  Pamela A. Pelton

**Composition**
  TypeWriting

**Art Director, Cover**
  Jeff Weeks

# Contents

## PART 1

## Perspective

### CHAPTER 1

## Overview . . . . . . . . . . . . . . . . . . . . . . . . . . . . . . . . . . . . . . . . . .3

### CHAPTER 2

## MRP in the Modern World . . . . . . . . . . . . . . . . . . . . . . . .11

### CHAPTER 3

## The Four Critical Questions Answered . . . . . . . . . . . . . .25

# PART 2

# Concepts

## Lot Sizing .................................................143

## System Records and Files ....................................163

# PART 3

# Managing with the MRP System

## A New Way of Looking at Things .............................181

## Product Definition .........................................197

**CHAPTER 12**

## Master Production Schedule  . . . . . . . . . . . . . . . . . . . . . .221

**CHAPTER 13**

## More Than an Inventory Control System . . . . . . . . . . . . .245

**CHAPTER 14**

## System Effectiveness: A Function of Design and Use  . . . . . . .259

**CHAPTER 15**

## Industry Effect on MRP . . . . . . . . . . . . . . . . . . . . . . . . . . .273

**CHAPTER 16**

## Project Manufacturing

## CHAPTER 17

# Remanufacturing . . . . . . . . . . . . . . . . . . . . . . . . . . . . . . . .295

## CHAPTER 18

# Process Industry Application . . . . . . . . . . . . . . . . . . . . . . .305

## CHAPTER 19

# Repetitive Manufacturing Application . . . . . . . . . . . . . . . .315

## CHAPTER 20

# Sales and Operations Planning . . . . . . . . . . . . . . . . . . . . .329

## PART 4

# Looking Backward and Forward

**CHAPTER 21**

## Historical Context . . . . . . . . . . . . . . . . . . . . . . . . . . . . . . .365

**CHAPTER 22**

## Blueprint for the Future: Demand-Driven MRP Logic . . . . . . . . .385

**CHAPTER 23**

## Strategic Inventory Positioning . . . . . . . . . . . . . . . . . . . . . .391

**CHAPTER 24**

## Buffer Profiles and Level Determination . . . . . . . . . . . . . . . .405

## CHAPTER 25

## Dynamic Buffers . . . . . . . . . . . . . . . . . . . . . . . . . . . . . . . . . . .423

## CHAPTER 26

## Demand-Driven Planning . . . . . . . . . . . . . . . . . . . . . . . . . . . . .435

## CHAPTER 27

## Highly Visible and Collaborative Execution . . . . . . . . . . . . . . .457

## CHAPTER 28

## Demand-Driven Material Requirements Planning (DDMRP) Performance Reporting and Analytics . . . . . . . . . . . . . . . . . . . .479

## CHAPTER 29

## DDMRP Future . . . . . . . . . . . . . . . . . . . . . . . . . . . . . . . . . . . .485

# Foreword

**W**hen I migrated from engineering to manufacturing around 1980, one of the first things I heard about was a thing called MRP. I can clearly remember walking up the main street of Annapolis, Maryland with my wife Debby, a systems engineer for Hewlett Packard, and hearing her explain to me that MRP was a computer application that manufacturers use to identify what they need and when they need it through a series of calculations. "They do a forecast," she said, "then check inventory to see what they need to make and use the lead time to figure out when to start making it. Next, they use the bill of materials to multiply out the quantities of each component they'll need, check inventory again, and use the lead times to see when to start making or buying whatever they don't have."

As an engineer, I was impressed by the simple, straightforward logic and, yes, the elegance of MRP. What could be simpler and more obvious? One could do this by hand, but it was clear that this was a task ideally suited to a computer—easily defined and simply calculated but involving a lot of data and a lot of those simple multiplications, additions and subtractions. And in 1980, computing power was just becoming affordable for mid-sized enterprises through the emergence of the mini-computer as an alternative to expensive mainframes.

I wanted to learn more about MRP and, like many others, found what I needed in Orlicky's *Material Requirements Planning*. When I got involved with APICS a few years later and pursued my certification in production and inventory management, Orlicky again provided the definitive information on MRP and how it works. A whole generation of manufacturing management learned MRP from Orlicky.

Can a technique and computer application first defined and developed in the 1960s, which rose to common use in the 1980s, still be viable in the 2010s? Wasn't MRP considered passé in the mid 1990s with the emergence of APS and ERP? Yes and yes. The MRP of the 1970s became MRP II (Manufacturing Resource Planning) in the 1980s and

Enterprise Resource Planning in the 1990s as software companies added an ever-wider range of supporting applications to better coordinate all of the activities of a manufacturing enterprise and supply chain to the task of meeting product availability and customer service objectives. But MRP is still at the heart of all of these modern processes. The fundamentals of "what do we need and when do we need it" cannot be denied. Inventory availability and the bill of materials are still the basic considerations.

Sure, manufacturing has changed in the last 30 years and clever and sophisticated techniques for planning and execution continue to emerge. But nothing has replaced the basic logic and application of MRP—it is still the backbone of ERP and supply chain management and is not made unnecessary by the theory of constraints, Lean, pull, kanban, demand-flow, or any other "modern" approach to production and manufacturing.

And, yes, MRP-based applications today are certainly faster and more capable than the early products of the first Orlicky MRP era. But the fundamentals remain. This third edition presents those fundamentals and then goes on to explore both the good and the not-so-good about material requirements planning in today's fast-paced, global, Internet-connected supply chains as well as some emerging techniques that build on MRP with additional facilities to make it work better in today's world.

I've known Carol Ptak for many years and I can't think of anyone more qualified to update this classic work for today's manufacturing environment. She is both an experienced manufacturing management practitioner and a true thought leader. She practically invented the term *demand driven* and truly understands the benefits and limitations of MRP, ERP, TOC, SCM, APS and the rest of the galaxy of acronymic techniques and theories as they apply to the challenges facing manufacturing companies in the twenty-first century. Chad Smith is also a thought leader with impressive credentials in the advancement of constraint-based manufacturing theory and practice.

I fully expect that *Orlicky's Material Requirements Planning, Third Edition*, which you now hold in your hand, will once again become the definitive reference for the next generation of manufacturing practitioners and leaders, and I am extremely pleased to be associated with this effort. I believe that true innovation and advancement are built on a solid understanding of the fundamental thinking and processes that have been successful in the past. I am confident that the manufacturing industry will continue to innovate and new theories and processes will emerge as new tools (faster and cheaper computers, enhanced communications, more ubiquitous autoID technologies) become available. Yet, I know that the fundamental ideas behind MRP will endure. It is a true benefit to the industry to have this updated standard reference to keep that definition alive and to educate succeeding generations on the ABCs of MRP.

*Dave Turbide, CFPIM, CMfgE, CIRM, CSCP*

# Preface

## SCOPE FOR THIS EDITION

In addition to the comprehensive coverage of material requirements planning (MRP), this third edition will focus on the current state of the MRP application as well as identify the fundamental changes required to achieve sustainable success given the current global circumstances and technology options. While capacity requirements planning (CRP) is an integral part of the modern planning process, the scope of this book is limited primarily to materials and inventory only. This book is intended not only to describe current practice but also to articulate the next generation of MRP logic—demand-driven MRP (DDMRP).

## ABOUT THE COLLABORATION

The authors did not set out to write this book. Each was doing work in his or her respective areas and from time to time crossed paths. Chad Smith had been working primarily with midrange manufacturing companies. His firm, Constraints Management Group (CMG), saw the market opportunity for software to support the pull concepts of drum-buffer-rope (DBR) and created a product called DBR+™. CMG implemented the product in various small and midrange manufacturers. In 2005, CMG developed a product named Replenishment+®. By 2007, it was clear that the future of the company was this new product and its innovative approach to planning material. In 2008, Chad decided to take the product and some conceptual PowerPoints to an old friend named Carol Ptak because they both live in the Seattle area. Carol was the former president and CEO of the American Production and Inventory Control Society (APICS) and also was vice president and global industry executive for the manufacturing and distribution industries at PeopleSoft before acquisition by Oracle. Carol was invited to be executive in residence at Pacific Lutheran University after that acquisition. Carol immediately saw that the product and its conceptual framework made possible the vision of demand-driven manufacturing that was first developed at PeopleSoft in 2002. Chad approached another old friend who was a senior account executive for Infor and showed him what they had developed. His response was, "I don't know if you have any idea how big what you have is. I sell five different major ERP platforms and none of our MRP or SCM modules can come close to this."

Needless to say, both these reactions encouraged Chad. The immediate problem became how to get the word out because these ideas were truly breakthrough concepts. Articulating the new concepts and critical differences between the new demand-driven approach to planning and the standard and insufficient approaches was the next big obstacle. Carol and Chad wrote a white paper entitled, "Beyond MRP." On a whim, they sent it off to APICS to see if there was any interest in it. The response was almost immediate. APICS asked them to condense the article for its magazine. APICS not only put it in the magazine but also made it the cover article under the title, "Brilliant Vision" (July–August 2008). Shortly after that, APICS sponsored a Webinar in August 2008 with both Carol and Chad on the subject of the article. Over 200 companies signed up. Then, in September 2008, Carol spoke on the topic at the APICS International Conference in Kansas City, MO. The response was standing room only, with over 350 people in the room. These responses were enough to convince the authors that they had struck a chord with the mainstream world's difficulty trying to plan materials in a demand-driven environment.

With this encouragement Carol and Chad began to further articulate the solution. They described the solution with a term: *actively synchronized replenishment* (ASR). Chad spoke in November 2008 at the Theory of Constraints International Certification Organization (TOCICO) Conference in Las Vegas, NV. Chad and Carol were approached at that conference by Dr. Jim Cox to continue writing on this topic. Dr. Cox is well known in both the TOC and APICS worlds. He was to be the coeditor with John Schleier of a new book to be published by McGraw-Hill that was to be called, *The Theory of Constraints Handbook*. Dr. Cox asked the authors to contribute a chapter to the book. The chapter was submitted about nine months later. Jim and John were very enthusiastic about the chapter content and sent it to McGraw-Hill, telling the editor that there should be a whole book dedicated to this. Below is what John had to say:

> Wow! What a chapter. My head is spinning around networks of interconnected buffers pulling production from the market side of the supply chain through multilevels in a shop with other buffers protecting its supply side. This is really an exciting story about a very creative piece of work. I wrote the first MRP system for John Deere's Ottumwa, Iowa, plant in the late 1950s; automated the BOMs, routings, inventory records, MRP, shop floor scheduling, and the purchasing system. Then in the early 1960s I headed the development team that built the compliment of logistics systems for the IBM Rochester plant, later implemented at the IBM plants in Boulder and Boca Raton, with elements in IBM European plants. I only mention this to frame

my appreciation for the incredible progress reflected in your work on ASR. Congratulations! I wish we had some of these solutions back then. . . . I am really blown away by the caliber and scope of this work.

In 1975, Joe Orlicky wrote the first book on MRP, entitled, *Material Requirements Planning*. This first MRP book is still seen today as the "bible" and the genesis of standardized MRP to which every software company coded its product. In 1994, Joe's close friend, George Plossl, revised the book, and it was entitled, *Orlicky's Material Requirements Planning* (second revised edition). George was one of the thought leaders at the time in the implementation of these concepts with Joe, Oliver (Ollie) Wight, and Richard (Dick) Ling. These two editions have sold over 175,000 copies combined—clearly demonstrating this impact.

In the spring of 2010, McGraw-Hill offered Chad and Carol a contract to write the third revised edition. We are humbled to stand on the shoulders of these giants and bring MRP up-to-date and into the future. You hold in your hands the result of this journey. We understand that most readers of this book are planning personnel bound by various policy and metric restrictions; believe us, we understand. Persuade your executives to explore the alternative (and superior) metrics and rules of the demand driven methodology. The case *can* be made effectively. Focus on services levels, cash, and return on working capital employed (ROCE). Then demand better and more appropriate tools from your technology providers. Please go to www.demanddrivenmrp.com to see how changing the rules *and* tools to fit the volatile and demand driven world of the twenty-first century is not only possible but also practical. Please contact us with questions, feedback, and requests for help.

We would like to thank our colleagues and friends who have helped so much with this version. Of special note are Dick Ling, who so graciously contributed his latest breakthroughs with Andy Coldrick in S&OP; Gene Thomas, one of the original developers of the earliest precursors of MRP, who so generously shared the detailed history in Appendices A and D in addition to his review of the manuscript; David Turbide for his review and Foreword contribution, in addition to being part of the overall development of the DDMRP idea back to the concept of rapid priority management (RPM); John Ricketts of IBM for his detailed review of the manuscript; Jim Cox and John Schleier for their encouragement to write this book in the beginning; and Bruce Spurgeon for his review. We acknowledge the partners and staff of Constraints Management Group, particularly Greg Cass and Paddy Rama, for sharing their ideas, verbalizations, and experiences. Finally, we would very much like to acknowledge and thank our spouses and families for putting up with our absence during the writing process.

*Carol Ptak*
*Chad Smith*

# PART 1

## Perspective

## CHAPTER 1

# Overview

## ORLICKY'S VISION

Joe Orlicky was truly a giant. There are very good reasons why the first edition of this book sold over 140,000 copies. What he defined and articulated had a profound impact on the modern global manufacturing landscape, and much of his writing remains relevant, even visionary, to this day. This is more remarkable when you consider a rather limited set of technological concepts and tools at his disposal (the first computerized manufacturing requirements planning system was written in 8 kB of memory!). If you have not read the first edition of this book, you should. The update you hold in your hands began from Orlicky's original work rather than from the second edition because of this amazing vision.

In 1975, in the first edition of this book, Joe Orlicky's first written words were, "*Someone* had to write this book." This book represented the first significant and extensive definition of material requirements planning (MRP). Just how new was MRP at that time? Not really all that new. People such as Joe Orlicky, George Plossl, and Ollie Wight had been pioneering the installation of computer-based MRP systems before 1960.[1] Additionally, APICS was founded in 1956 in Cleveland, Ohio, to support the practitioners at the time and to share best practices and insights.

Why was there such a lag between 1960 and 1975? Orlicky explains it: "In the field of production and inventory management, literature does not lead, it follows. The techniques of modern material requirements planning have been developed not by theoreticians and researchers but by practitioners. Thus the knowledge remained, for a long time, the property of scattered MRP system users who normally have little time or inclination to write for the public." In addition, many of these early-adopter companies viewed this new technology as a competitive advantage and were reluctant to share this knowledge.

---

[1] Joseph Orlicky, *Material Requirements Planning* (New York: McGraw-Hill, 1975), Preface, p. ix.

Orlicky's preceding statement about literature still rings true today. Generally, academia continues to research and write about what is already written rather than developing new insights and thought leadership. The practitioners who use technology and techniques remain in the best position to properly and practically define the tool and techniques that actually produce sustainable positive results. However, very few will take the time to write those techniques into a book.

After the first edition of this book was written, the use of MRP in industry literally exploded. If you are interested in a detailed history, refer to Appendices A and D. Consider the following statement from the second paragraph of the Preface of the first edition:

> As this book goes into print, there are some 700 manufacturing companies or plants that have implemented, or are committed to implementing, MRP systems. Material requirements planning has become a new way of life in production and inventory management, displacing older methods in general and statistical inventory control in particular. I, for one, have no doubt whatever that it will be the way of life in the future.

MRP did become *the* way of life in production and inventory management. This planning approach is still the standard across the globe for determining what to buy and make, how much to buy and make, and when to buy and make it. The number of MRP implementations today worldwide is estimated to be in the hundreds of thousands (if not millions) of companies using it in some form. An Aberdeen Group Study showed that 79 percent of companies that bought enterprise resource planning (ERP) systems also bought and implemented the MRP module.[2]

What is important to note is that since its articulation in the first edition of this book, the definition of MRP has not really changed. The concept has evolved as technology has improved, as detailed in Figure 1-1. This evolution included closed-loop MRP, manufacturing resource planning (MRP II), advanced planning and scheduling systems (APS), and finally enterprise resource planning (ERP). An in-depth explanation of this evolution is chronicled in Chapter 21, as well as in Appendix D.

Throughout this entire evolution, the core MRP calculation kernel stayed the same. MRP fundamentally is a very big calculator using the data about what you need and what you have in order to calculate what you need to go get—and when. This has grown from the first ability to track inventory. At its very core, even the most sophisticated ERP system of the day uses these basic calculations. Typically, these calculations are implemented in a push system based on a forecast or plan with the assumption that all the input data are accurate. In the most stable of environments, this assumption may be somewhat possible, but how does the twenty-first-century global economic environment fit with this approach?

---

[2] Aberdeen Group ERP Study, 2006, Table 3, p. 17.

**FIGURE 1-1**

Planning tool
evolution.

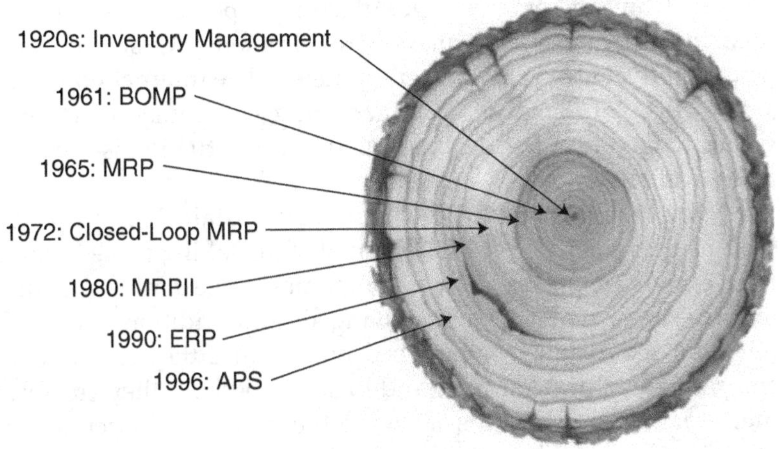

1920s: Inventory Management

1961: BOMP

1965: MRP

1972: Closed-Loop MRP

1980: MRPII

1990: ERP

1996: APS

One of the most remarkable passages in the first edition was Orlicky's description of the vast changes and removal of limitations taking place in the 1970s. These changes and more degrees of freedom allowed fundamental changes to the way companies planned, ordered, and manufactured. Consider the following passage:

> Traditional inventory management approaches, in pre-computer days, could obviously not go beyond the limits imposed by the information processing tools available at the time. Because of this, almost all of those approaches and techniques suffered from imperfection. They simply represented the best that could be done under the circumstances. They acted as a crutch and incorporated summary, shortcut, and approximation methods, often based on tenuous or quite unrealistic assumptions, sometimes force-fitting concepts to reality so as to permit the use of a technique.
>
> The breakthrough, in this area, lies in the simple fact that once a computer becomes available, the use of such methods and systems is no longer obligatory. It becomes feasible to sort out, revise, or discard previously used techniques and to institute new ones that heretofore it would have been impractical or impossible to implement. It is now a matter of record that among manufacturing companies that pioneered inventory management computer applications in the 1960s, the most significant results were achieved not by those who chose to improve, refine, and speed up existing procedures, but by those who undertook a fundamental overhaul of their systems [Orlicky, *Material Requirements Planning*, p. 4].

Almost 50 years later we are at another time of reexamination and transition. Shortly after the turn of the millennium, the world of manufacturing turned upside down. Production became more efficient in the United States. Eastern Europe was incorporated into the European Union, putting low-cost production very close to a lucrative market. China became the manufacturing powerhouse of the world. Manufacturing capacity now far exceeds market requirements. Customers have become increasingly fickle. Product life cycles have plummeted. The Internet now allows global sourcing with a few clicks of a mouse. Manufacturing companies worldwide are faced with more volatility than ever before. Exacerbating this situation is the global economic slowdown of 2008–2011.

No longer can a company achieve a sustainable competitive advantage using the old business rules. These fundamental shifts taking place in the current global manufacturing environment are forcing companies to reexamine the rules and tools that manage their business. The world has changed, and additional technology barriers have been removed. Companies will succeed not because they improve, refine, and speed up the enforcement of obsolete rules and logic but because they are able to fundamentally adapt their operating rules and systems to the new global circumstances. A new approach to planning is required.

## FOCUS AND ORGANIZATION OF THIS BOOK

This book is not meant to serve as a basic text on the general subject of production and inventory management or even inventory control. At least elementary knowledge of these subjects on the part of the reader is assumed, particularly knowledge of the fundamentals of conventional (statistical) inventory control. (See the Bibliography at the end of this chapter.) The book is written primarily for users and potential users of material requirements planning systems, that is, for manufacturing managers, materials managers, production control managers, inventory planners, systems analysts, and interested industrial engineers. It also can serve the needs of students of production and inventory management. More universities will need to include the subject of demand-driven material planning and scheduling techniques in their business curricula to better prepare their students to enter this demand-driven world.

This book's scope is limited to the system of logistics planning in a manufacturing environment (as contrasted with the pure project environment). Figure 1-2 shows the continuum of production from the extremely high variety of the project to the consistency of the flow shop. The scope of this book will include job shop, batch, assembly-line, and continuous-flow organizations. This matrix provides the framework for the industry application in Chapters 15 through 19.

This book does not extend to traditional execution subsystems, although it stresses that a high quality of the outputs generated by the planning system is a prerequisite for the effective functioning of such subsystems. The book will, however, propose an execution approach that can be used by these subsystems. A simplified chart, applicable to any manufacturing operation, of the relationships between the planning system and the execution (control) subsystems is presented in Figure 1-3. This is a closed-loop system where the execution subsystems have a direct effect on the planning system. However, the multiple linkages and feedback loops would quickly overwhelm this simple diagram.

## FIGURE 1-2

Wheelwright and Hayes product/process matrix.

| Process Structure Process Life Cycle Stage | Product Structure Product Life Cycle Stage | Low Volume Unique (One of a Kind) | Low Volume Multiple Products | Higher Volume Standardized Product | Very High Volume Commodity Product |
|---|---|---|---|---|---|
| | Project | | | | |
| Jumbled Flow (Job Shop) | | Job Shop | | | |
| Disconnected Line Flow (Batch) | | | Batch | | |
| Connected Line Flow (Assembly Line) | | | | Assembly Line | |
| Continuous Flow (Continuous) | | | | | Continuous |

## FIGURE 1-3

Planning, execution, and control systems.

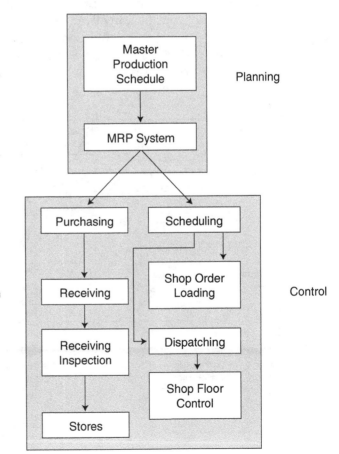

The focus of this book is on the new rules required to effectively support a manufacturing operation using material requirements planning systems in the twenty-first century. The objective is an exposition of procedural logic, function, and use of these systems rather than programming and other considerations of system implementation. All considerations that are of a purely technical data-processing nature are excluded because they are amply documented in manuals published by computer manufacturers and software providers. The software aspect of MRP is intentionally downplayed so as not to divert the reader's attention from the really important subject matter. As far as MRP is concerned, the computer's contribution lies solely in its power to execute a host of rather straightforward calculations in a very short time and display nearly instantaneous visibility to relevant information and priorities to the appropriate personnel. A comprehensive understanding of the computer aspect is not essential to an understanding of the subject in question.

The discussion of MRP concepts, principles, and processing logic is expanded to encompass system inputs and system uses reflected in functional outputs. The input-output chart depicted in Figure 1-4 can serve as a map of the topics that constitute this book. We have tried to avoid a case-study approach to the core subject so as not to obscure the general validity of the principles involved and the universal applicability of the MRP approach. However, at times in this book some real-life examples will be used. Additionally, at the end of this book, two case studies have been provided with the results achieved by these companies as a demonstration of what is now possible. Abstract

**FIGURE 1-4**

MRP system: input-output relationships.

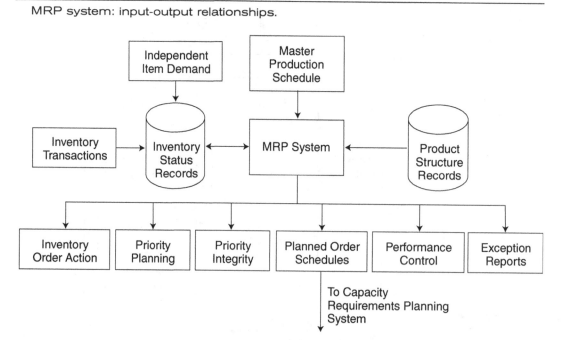

examples are used as much as possible, and there are no pictures of actual forms, documents, and computer printouts.

The book consists of four parts. Part 1 describes our overall perspective on the current manufacturing world. Part 2 examines the detailed MRP concepts as they exist in reality today. Part 3 details how these concepts are used in different industries and applications. Finally, Part 4 details the roadmap for the near and distant future for this critical manufacturing management tool.

> It is interesting to note that the number of pages written on independent-demand-type inventory systems outnumbers the pages written on material requirements planning by well over 100 to 1. The number of items in inventory that can best be controlled by material requirements planning outnumber those that can be controlled effectively by order point in about the same ratio. It is a sign of the adolescence of our field that the literature available is in inverse proportion to the applicability of the techniques.
>
> —OLIVER W. WIGHT, "DESIGNING AND IMPLEMENTING A MATERIAL REQUIREMENTS PLANNING SYSTEM," IN *PROCEEDINGS OF THE 13TH INTERNATIONAL CONFERENCE OF APICS*, 1970

## BIBLIOGRAPHY

Brown, R. G. *Decision Rules for Inventory Management*. New York: Holt, Rinehart and Winston, 1967.

Buchan, J., and E. Koenigsberg. *Scientific Inventory Management*. Englewood Cliffs, NJ: Prentice-Hall, 1963.

Hadley, G., and T. M. Whitin. *Analysis of Inventory Systems*. Englewood Cliffs, NJ: Prentice-Hall, 1963.

Magee, J. F., and D. M. Boodman. *Production Planning and Inventory Control*, 2d ed. New York: McGraw-Hill, 1967.

Melnitsky, B. *Management of Industrial Inventory*. New York: Conover-Mast Publications, 1951.

Plossl, G. W., and O. W. Wight. *Production and Inventory Control*. Englewood Cliffs, NJ: Prentice Hall, 1967.

Prichard, J. W., and R. H. Eagle. *Modern Inventory Management*. New York: Wiley, 1965.

Wagner, H. M. *Statistical Management of Inventory Systems*. New York: Wiley, 1962.

Welch, W. E. *Tested Scientific Inventory Control*. New York: Management Publishing Corp., 1956.

Whitin, T. M. *Theory of Inventory Management*. Princeton, NJ: Princeton University Press, 1957.

# CHAPTER 2

# MRP in the Modern World

**B**y definition, manufacturing is the conversion of low-value materials to higher-value products. These must have value at least equal to their price in meeting customers' needs or desires. People, capital, materials, and other resources are employed at costs that must be less than the selling price, taxes must be paid, and, hopefully, money will be left to fund research and development (R&D), expand the business, and reward the owners who provide the operating capital.

Recognition that manufacturing is a process is essential to understanding how it should work. A bewildering variety and great complexity of products, materials, technology, machines, and people skills obscure the underlying elegance and simplicity of the total process. The essence of manufacturing is flow of materials from suppliers through plants to customers and of information to all parties about what was planned, what has happened, and what should happen next.

This is true regardless of what is made, how and when it is made, and who makes it where. The first law of manufacturing is

> All benefits will be directly related to the speed of flow of materials and information.

Assuming a valid direction (not a trivial assumption), this is a universal law applying to every type of manufacturing. Difficulties in controlling manufacturing will decrease and planning will become more effective as material and information flows speed up. The best use of resources comes from eliminating problems that interrupt or slow down these flows. The obvious principle that emerges is

> Time is the most precious resource employed in the manufacturing process and the ultimate constraint.

...ounts are available to everyone, but time moves relentlessly; it cannot be ...ded, or recycled, and wealth can buy no more. Wasting time causes irre-...oss. Simple, universal logic underlies all manufacturing and can be represented ...simple questions:

1. What is to be made?
2. How many, and when are they needed?
3. What resources are required to do this?
4. How should those resources be configured and deployed?
5. Which are already available?
6. Which others will be available in time?
7. What more will be needed, and when?
8. How will this plan enable sustainable profits for the company?

Business and marketing strategies determine the answers to the first question. Internal company planning and control systems provide answers to the last four. A blend of business and marketing strategies and internal company planning, execution, and control systems provide the answers to questions 2, 3, and 4. Those questions represent the interface between strategy and tactics.

Today, many manufacturers and supply chains face a huge dilemma related to their operational strategy and tactics. The world is a much different place than it was when the first edition of this book was published in 1975. To put it bluntly, the world of "push and promote" is dead. Gone are the days when a company could use the past to predict the future, build products to that forecast, and have any hope that the market would want what they produced. Companies that continue in this mode will see continual erosion in their market share and bottom-line performance until that company simply goes out of business. Additionally, rules and tools that were developed under those conditions must be reexamined and rebuilt for the circumstances of today.

Those circumstances define a level of complexity never seen before. Over the last decade, the nature of the global manufacturing and supply-chain landscape has become much more unstable. Consider the following factors contributing to this volatility:

- Global sourcing and demand
- Increased outsourcing
- Shortened product life cycles
- Shortened customer tolerance times
- More product complexity and/or customization
- Demands for leaner inventories
- Increasing forecast error
- Material shortages
- More product variety
- Long-lead-time parts/components
- A hypersensitive global economic community
- Dramatic cutbacks in personnel and other resources across the supply chain

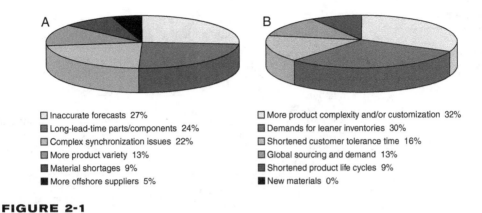

Inaccurate forecasts 27%

■ Long-lead-time parts/components 24%

Complex synchronization issues 22%

■ More product variety 13%

■ Material shortages 9%

■ More offshore suppliers 5%

More product complexity and/or customization 32%

■ Demands for leaner inventories 30%

Shortened customer tolerance time 16%

Global sourcing and demand 13%

■ Shortened product life cycles 9%

■ New materials 0%

**FIGURE 2-1**

Indentified product (A) and market (B) issues.

At the 2008 American Production and Inventory Control Society (APICS) International Conference, APICS professionals were surveyed about their number one concern about volatility. They were given two lists of six factors related to volatility and asked which were the biggest challenges to their company. Figure 2-1 represents the results of that survey.

This survey did not show any overwhelming single cause but rather showed that these challenges are widely varied across organizations. These factors combine to create more complex planning and supply scenarios for manufacturers and supply chains than ever before. A November 2010 Aberdeen Group survey noted that 48 percent of companies surveyed indicated that increased supply-chain complexity was a top priority.[1] This more volatile environment is not a temporary phenomenon. It is here to stay.

In attempting to maneuver through these complexities, companies are increasingly running into a dilemma. Figure 2-2 is a conflict diagram depicting the two conflicting operations modes. The top side of the diagram describes how in order to deal with today's complexities companies must be effective in their plans. Specifically, companies must plan in advance of actual customer orders. Customers are not willing to wait for companies to order long-lead-time materials, to incorporate sales and marketing data and plans, to plan capital and staffing levels, and to develop contingency plans for potential problems. Customers want what they want when they want it and at a price they want to pay. The successful operation is the one that can provide this at a profit to itself. In the face of this, supply chains have extended and broadened, product life cycles have shrunk, and product complexity has risen. This real need to plan drives the management team to focus on systems and approaches that enhance predictability. Some companies have developed a very sophisticated sales and operations planning process to minimize the potential for problems within the planning horizon. Other companies have invested

---

[1] Viswanathan, Nari, "Enabling Supply Chain Visibility and Collaboration in the Cloud," November 2010, p. 4.

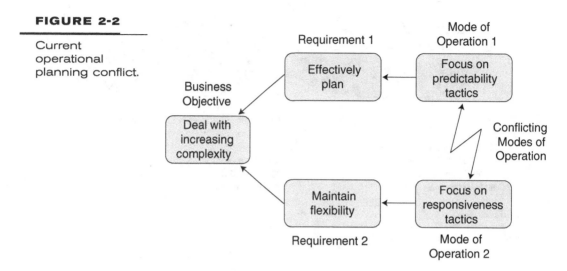

**FIGURE 2-2**

Current
operational
planning conflict.

huge amounts of money and time in advanced forecasting algorithms in the hope of gaining an insight into the future using past experience. Companies try to measure almost anything and everything that can be measured in the hope that it will tell them something that they do not already know. Most every organization generates an ocean of data, but sifting relevant information out of that ocean is quite a different matter. The reality is that worldwide companies are drowning in data and starving for accurate, actionable information.

On the bottom side of the diagram, however, the advanced commitment of capital, inventory, and capacity means that organizations are much less flexible in the short term. This forces a company into the situation of expediting, schedule deviations, and confusion. This need for flexibility has driven many managers to clamor for reduced system complexity and the implementation of highly visible and responsive pull-based strategies such as lean and drum-buffer-rope. The *APICS Dictionary* (Blackstone, 2008) provides excellent insight into these tactics:

> **pull signal:** Any signal that indicates when to produce or transport items in a pull replenishment system. For example, in Just-in-Time production control systems, a kanban card is used as the pull signal to replenish parts to the using operation. See: pull system.

> **pull system:** (1) In production, the production of items only as demanded for use or to replace those taken for use. See: pull signal. (2) In material control, the withdrawal of inventory as demanded by the using operations. Material is not issued until a signal comes from the user. (3) In distribution, a system for replenishing field warehouse inventories where replenishment decisions are made at the field warehouse itself, not at the central warehouse or plant.

> **demand chain management:** A supply chain inventory management approach that concentrates on demand pull rather than supplier push inventory models.

But do these approaches meet the required need for effective planning? In most manufacturing environments, they are grossly inadequate. This is why the conflict is so severe and the impact so significant. The things we do for planning hurt our ability to be flexible, whereas the things we do to be flexible disregard certain critical planning requirements. In many lean implementations, a stated project objective is to eliminate or disable the material requirements planning (MRP) system. There will be more on this specific conflict around the use of MRP later in Chapter 3.

In order to effectively resolve this conflict, a solution must be deployed that allows companies to plan effectively while maintaining or increasing flexibility. Most global manufacturers, enterprise resource planning (ERP) software companies and manufacturing consultants seem to ignore the obvious path to the above-stated solution requirements. Instead, the most common approach is to chase symptoms and propose incomplete and even disastrous solutions that either overcomplicate or oversimplify planning, execution, and control systems with less than desirable results.

## KEY QUESTIONS FOR PLANNING AND FLEXIBILITY

The key questions to be answered with regard to planning and flexibility are

1. *How do we minimize or eliminate shortages?* Shortages cause problems in manufacturing. Any person who spends even one single day in operations knows this fact. When those shortages are chronic and frequent, they can cripple a company's service and financial performance. Frequently, shortages result in additional spending for overtime and expedited freight, cause scheduling deviation and general chaos, and jeopardize service levels. In a time when customer tolerance times are shrinking, controlling shortages becomes even more crucial to a company's sustainable success.

2. *How do we keep production lead times as short as possible?* As mentioned earlier, customer tolerance times are shrinking. In order to stay lead-time-competitive and minimize the amount of inventory required to do so, companies are under constant pressure to compress manufacturing and purchasing lead times.

3. *How do we keep working capital (materials and manufacturing assets) synchronized with demand?* Companies are striving to minimize inventory positions while at the same time provide a high level of service. Defining this kind of strategy is a smart business move no matter what the climate. In the best of times, it means a significant return on average capital employed (RACE), and in the worst of times, it minimizes the company's exposure to downturns and recessions.

## DEALING WITH VARIABILITY

The answer to these three key questions resides in understanding and protecting against variation and volatility within both the manufacturing enterprise itself and its supply chain(s). W. Edwards Deming and Walter Shewhart clearly understood this concept with respect to the quality a company can produce. Deming and Shewhart taught process control and continuous improvement for years in the United States. Because American industry did not realize the impact that these critical tools could have on the bottom line, Deming eventually went to Japan. The rest, as they say, is history. Several years before his death, Deming came back to this country, and finally, the market was ready to listen. Reducing variation is core to the six sigma process-improvement approach. Understanding and dealing with variability are even more important today because variability and volatility are on a dramatic rise.

Recently, companies have been caught between a rock and a hard place with regard to solving the conflict inherent in the preceding three questions. In the last decade, many companies, in the attempt to embrace lean concepts, including the dramatic reduction of inventory everywhere, leaned out so much that they actually exposed themselves to increased variability and volatility. By treating inventory as a waste, they actually experienced the opposite effect and made their supply chains too brittle and less agile.

Variability can be systematically minimized and managed but not eliminated. The biggest challenge in attacking all the causes for variability and minimizing their individual impact on the system is the investment of time, effort, and money to get there and the return on that investment. The six sigma toolset provides an excellent approach to reduce variability, but even the best master black belt cannot totally eliminate variability. This is not to say that companies should not seek to use these tools that identify and dampen the variability noise. These tools, however, are complementary rather than primary to the materials planning function.

So how are variability and volatility addressed in a way that keeps a company agile while at the same time minimizing working capital considerations? There are four distinct sources of variation (two internal and two external) from an enterprise perspective. These four sources are diagrammed in Figure 2-3 and discussed below. The squiggly lines are meant to depict variations that occur within each of those four areas. The direction of the arrow is meant to depict the direction in which variation is passed.

**FIGURE 2-3**

The four sources of variation.

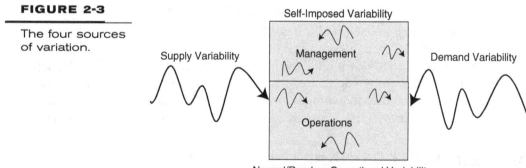

## Demand Variability

Demand variability is characterized by fluctuations and deviations experienced in demand patterns and plans. In many supply chains, demand variability is driven by MRP system nervousness of major players near the top of the chains (e.g., original equipment manufacturer [OEM]) because those systems are attempting to make adjustments in materials requirements within the demand time fence. Consequently, suppliers receive a constantly changing picture of their major customers' requirements, usually in weekly buckets.

> **demand uncertainty:** The uncertainty or variability in demand as measured by the standard deviation, mean absolute deviation (MAD), or variance of forecast errors [*APICS Dictionary* (New York: Blackstone, 2008)].

## Supply Variability

Supply variability is measured by disruptions in the supply network or deviation from requested dates and/or promised dates for supply order receipts. It is the reliability (or lack thereof) of the supply network. Remember, only one part missing can block the delivery of an end item. A company can have 99.9 percent supply reliability and still have unacceptable service levels to their customers with huge implications for cash flow. In the extreme, a 5 cent fastener can block the delivery of a multimillion-dollar assembly.

## Normal/Random Operational Variability—"Murphy"

The old adage is that what can go wrong will go wrong. This has become known as *Murphy's law*. A corollary to that law is that Murphy was an optimist. Still another corollary is that the probability that Murphy will strike is directly proportionate to the penalty. W. Edwards Deming called this *common-cause variation*. This is the normal and random variation exhibited by a system in steady state. Perfection at every point of the process is impossible. Even companies embracing the lean approach or using six sigma will acknowledge that the desired goal of perfection is impossible. Normal or random operational variability results in a process that may be within calculated control limits statistically but still varying between those limits.

## Self-Imposed Variability

Self-imposed variability is the human element. It is a direct result of decisions made within the company. This form of variability would be considered by Deming to be a type of *special-* or *assignable-cause variability*. Self-imposed variability frequently will take a process out of statistical control. According to Deming, special-cause variability is the first target for improvement. Only when the special cause of variation is addressed can the normal variation of the process be identified. This provides a steady state that is far easier to manage. We cannot remove the human element, but the damaging aspect of the human element (the assignable cause) can be addressed.

## Net Effect of Variation on Operations and Supply Chains

*MRP Nervousness*   The APICS Dictionary (New York: Blackstone, 2007, p. 86) defines nervousness as

> The characteristic in an MRP system when minor changes in higher level (e.g., level 0 or 1) records or the master production schedule cause significant timing or quantity changes in lower level (e.g., 5 or 6) schedules or orders.

APICS meetings have been held since the incorporation of APICS in 1956 to discuss MRP nervousness and how to manage it. Anyone in purchasing quickly learns the whiplash caused by even a small change at a higher level in the bill of material. Vertical dependencies are critical for effective planning and yet at the same time cause the cascade and amplification of any change. When order multiples, safety stocks, and unrealistic due dates are added into the equation, the picture quickly can become quite clouded as to what should be requested of suppliers. Figure 2-4 is meant to depict a wave of nervousness through a bill of materials (BOM). Of course, this figure depicts a single BOM. When there are shared components across BOMs, that nervousness also translates across bills of materials.

*Supply-Chain Bullwhip Effect*   Similar to system nervousness within a company, when considering the effect of variability across a supply chain, the cumulative system variation is significantly higher than the variation of any one of the parts. A typical supply chain is represented as a linear linkage from a supplier's supplier to a customer's cus-

**FIGURE 2-4**

MRP nervousness within a single bill of material within a single company.

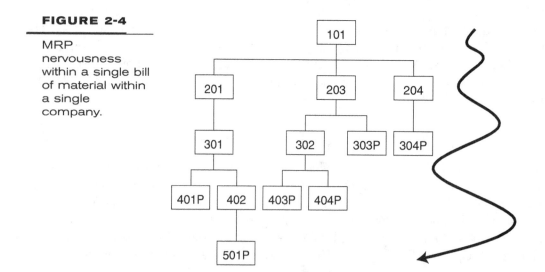

tomer. However, this is a gross oversimplification and does not represent the weblike network that more accurately reflects a company within its supply chain. Interdependencies are common. When these interdependencies are subject to variability, the effect is amplified. The more parts or dependencies there are, the worse is the cumulative effect experienced by the supply chain and the organizations comprising it. This cumulative effect is known as the *bullwhip effect*.

The *APICS Dictionary* (New York: Blackstone, 2008, p. 15) defines *bullwhip effect* as:

> An extreme change in the supply position upstream in a supply chain generated by a small change in demand downstream in the supply chain. Inventory can quickly move from being backordered to being excess. This is caused by the serial nature of communicating orders up the chain with the inherent transportation delays of moving product down the chain. The bullwhip effect can be eliminated by synchronizing the supply chain.

Figure 2-5 illustrates the bullwhip effect in both directions. The amplitude and/or frequency of the "whip" grows over the length of a particular chain of dependent events.

## Decoupling

The only way to stop nervousness and the bullwhip effect is to stop variation from being passed and amplified between the dependencies in the system. Dependencies must be decoupled from each other so that this cumulative variation is dampened or absorbed.

**FIGURE 2-5**

Bullwhip effect on both the chains of demand and supply orders.

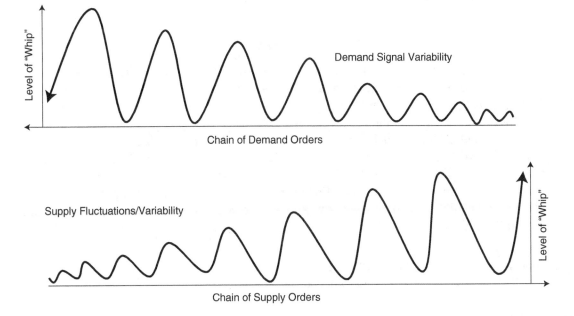

An analogy is the dampening strategy used commonly in marinas to protect boats from damage caused by waves. Marinas will build a breakwall around the moored boats to protect them. The size of the breakwall directly depends on the expected waves at a given location.

> **decoupling:** Creating independence between supply and use of material. Commonly denotes providing inventory between operations so that fluctuations in the production rate of the supplying operation do not constrain production or use rates of the next operation.[2]

> **decoupling points:** The locations in the product structure or distribution network where inventory is placed to create independence between processes or entities. Selection of decoupling points is a strategic decision that determines customer lead times and inventory investment.[2]

The APICS definition of decoupling is descriptive of one form of decoupling. However, additional decoupling points are now required to compete effectively in today's manufacturing environments. This is not to say that all dependencies should be decoupled. This is further explored in Part 4. These critical dependencies are the places where system performance will be most affected by cumulative variation. This is covered in Chapter 23, Strategic Inventory Positioning. These salient decoupling points are essential to understanding where to place inventory—the primary question to answer to effectively keep a company agile while at the same time minimizing working capital considerations.

One form of decoupling is buffering. Only certain critical dependencies should be buffered. *Buffering* is the placement of a cushion between two dependencies. There are three forms of buffering that decouple dependencies and dampen system variation—time, capacity, and stock.

## Time Buffers

According to Shri Shrikanth, a *time buffer* represents the additional lead time allowed, beyond the required setup and processing times, for materials to flow between two specified points in the product flow.[3]

Figure 2-6 shows a planned event at a discrete time. This planned event is being buffered (labeled "Buffered Event"). In front of that planned event, there is a discrete amount of time that will act as the buffer (labeled "Time Buffer"). The time buffer is determined by analyzing how much variation is typical preceding the buffered event. The string of preceding events (labeled "Pre-Buffer Sequence of Events") experiences cumulative variability that is quantifiable over time. A correct time buffer should be sized in a way that it will cover enough variation that will reliably result in acceptable service lev-

---

[2] APICS Dictionary, 12th ed. (New York: Blackstone, 2008), p. 34.

[3] Mokshagundam (Shri) Shrikanth, "DBR, Buffer Management and VATI Flow Classification," in *Theory of Constraints Handbook*, New York: McGraw-Hill, 2010, Chapter 8.

**FIGURE 2-6**

Time buffer.

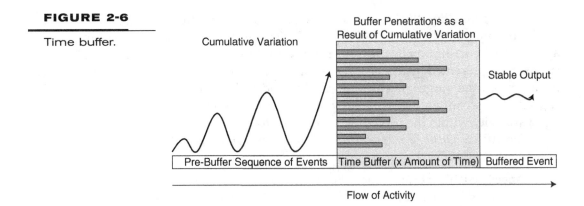

els to the buffered event. As the prebuffer sequence of events experiences disruptions, the finished work gets delayed. The length of that delay is depicted as the lines under the heading "Buffer Penetrations as a Result of Cumulative Variation." Note that in this example, no penetration was significant enough to affect the start time of the buffered event. Thus the buffered event experiences a stable/reliable output from the prebuffer sequence of events (labeled "Stable Output").

## Capacity Buffers

A *capacity buffer* is defined as the protective capacity at both constraint and nonconstraint resources that allows these resources to catch up when Murphy strikes.[4] The variability caused by Murphy is not limited to the resources required in the prebuffer sequence of events' inability to perform (Figure 2-7). Murphy also can be caused by unexpected expedited orders that pass through those same resources that represent real surges in demand and thus affect capacity requirements.

**FIGURE 2-7**

Capacity buffer.

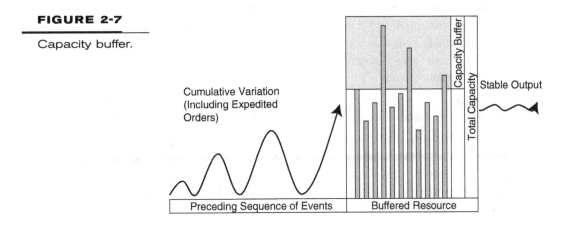

---

[4] Ibid.

In the preceding illustration, there is a resource ("Buffered Resource") that has an additional amount of capacity ("Capacity Buffer") in order to absorb surges in demand and variability passed on by previous resources ("Preceding Sequence of Events"). These surges are represented by the bars. In this case, the resource had more than enough capacity to absorb the surges, ensuring a reliable and stable output ("Stable Output"). The capacity buffer can be determined by analyzing the typical number of demand surges and variability within the preceding set of resources.

The *APICS Dictionary* (New York: Blackstone, 2008, p. 17) defines these capacity buffers as follows:

**capacity cushion:** extra capacity that is added to a system after capacity for expected demand is calculated. Syn: safety capacity. See: protective capacity.

**protective capacity:** The resource capacity needed to protect system throughput—ensuring that some capacity above the capacity required to exploit the constraint is available to catch up when disruptions inevitably occur. Nonconstraint resources need protective capacity to rebuild the bank in front of the constraint or capacity-constrained resource (CCR) and/or on the shipping dock before throughput is lost and to empty the space buffer when it fills.

**surge capacity:** The ability to meet sudden, unexpected increases in demand by expanding production with existing personnel and equipment.

## Stock Buffers

*Stock buffers* are quantities of inventory or stock that are designed to decouple demand from supply. They are commonly amounts of inventory that will provide reliable availability of the stock to consumers while at the same time allowing for the aggregation of demand orders, creating a more stable and efficient supply signal. Below are the typical and prevalent types of stock buffers used in manufacturing today and recognized by most ERP systems.

**min-max system:** A type of order point replenishment system where the "min" (minimum) is the order point and the "max" (maximum) is the "order up to" inventory level. The order quantity is variable and is the result of the max minus available and on-order inventory. An order is recommended when the sum of the available and on-order inventory is at or below the min.

**safety stock:** In general, a quantity of stock planned to be in inventory to protect against fluctuations in demand or supply [*APICS Dictionary* (New York: Blackstone, 2008)].

**supermarket:** Stores of in-process inventory used where the process cannot produce a continuous flow. Examples of supermarkets include when one operation services many value streams, when suppliers are too far away, or when processes are unstable, have long lead times, or have out-of-balance cycle times. The supplying operation controls the supermarket and its inventory. Supermarket inventory is tightly controlled.[5]

---

[5] Natalie J. Sayer and Bruce Williams, *Lean for Dummies*, Hoboken, NJ: Wiley, 2007, p. 108.

**FIGURE 2-8**

Stock buffer.

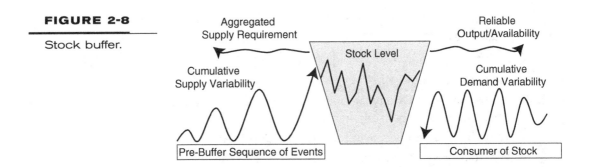

Figure 2-8 is a depiction of a stock buffer. Variability hits it from both directions. This is depicted by the "Cumulative Supply Variability" and "Cumulative Demand Variability" arrows flowing into the bucket icon. The smoother lines flowing out of the bucket icon represent the reliable availability on the consumption side and an aggregated supply requirement on the source side. Within the bucket icon, the sawtooth line represents stock levels fluctuating.

Stock buffering plays a critical role in the new rules for MRP. While min-max and supermarkets still have relevance, safety stock does not. Safety stock is an antiquated concept replaced by "strategically replenished buffers." Strategically replenished buffers are applied to strategic part positions. Determining these strategic part positions is discussed in Chapter 23. These strategically replenished buffers provide the planning, execution visibility, and priority management companies need to sustain success. Just as a supply chain has strong interrelationships, there are inherent connections between the three types of buffers (time, capacity, and stock). In most cases, the lack of one will necessitate the use of additional amounts of the others.

## MATERIALS OR CAPACITY: WHERE TO FOCUS FIRST?

The core driver behind the changes felt by every manufacturing company today is that global capacity now exceeds global demand. Where companies struggled to wring every bit of output out of scarce capacity in the late 1990s, now most companies worldwide have more internal capacity than market demand for their products. The assumptions made in the earliest days of MRP that there was infinite capacity are probably more realistic today than when MRP was first developed.

Even when a company has capacity issues, many of these issues are due to schedule deviations and expedites related to material shortages. Additionally, capacity efficiencies will be easier to drive if demand signals are better synchronized with supply signals. Most capacity improvements are rendered impotent because the materials planning system cannot synchronize these signals adequately in the more complex manufacturing and supply landscape of the twenty-first century. Thus, in most cases, materials have become the place to focus first. With this statement in mind, let's examine some critical questions about MRP:

1. Is MRP still relevant in an industrial world obsessed with supply-chain management and ERP?
2. Is MRP failing in today's environment because it is inherently flawed or because it is poorly implemented?
3. Why do lean and other pull techniques and MRP often come into conflict?
4. Why has MRP not progressed significantly in the last 30 years?

## CHAPTER 3

# The Four Critical Questions Answered

## QUESTION 1: RELEVANCE OF MRP

Is manufacturing requirements planning (MRP) still relevant in an industrial world obsessed with supply-chain management and enterprise resource planning (ERP)?

At the core of every MRP application is the calculation designed to tell companies what they have, what they need to make and buy, and when they need to make and buy it. If this were the only definition for MRP, then the answer to this question would be obvious. MRP may not be perceived as leading edge at this time, but accurate answers to the simple questions of what a company has, what it needs to make and buy, and when it needs to make and buy it can spell the difference between success and failure for any company. When taking into account the increased complexities in the twenty-first century, the failure to answer these questions accurately will cause a company to fail more quickly. In this case, "failure" is not necessarily bankruptcy. It is often a difficult middle ground where some companies perpetually teeter on the brink without actually failing, whereas others are also-rans that never can break their way into sustained growth.

In reality, MRP is critical and more relevant than ever. Unfortunately, the truth is that the core MRP concepts and rules around them have been neglected and left to stagnate for a long time. In addition, there is a complete lack of understanding that this stagnation is even a problem. Many newer operations approaches even advocate abandoning MRP completely. Some lean and theory of constraints (TOC) implementations measure their success by unplugging the formal MRP system.

In the last 50 years, there has been a revolution in logistics costs and management. Consider that the first container ship was introduced in 1955. The cost to load one ton of cargo in 1955 was $5.86.[1] Today, the cost to load one ton of cargo in a shipping container is 16 cents. At the same time, communication technology has evolved dramatically. In 1915, Bell's first transcontinental call from New York to San Francisco took 23 minutes to

---

[1] Intermodal Steel Building Units, American Association of Port Authorities.

connect using five intermediary operators. The first self-dialed long-distance call in the United States took place on November 11, 1951.[2] Today, the Internet and 4G cellular and satellite-enabled phones make communication instantaneous almost anywhere in the world. In the last 20 years, there has been significant attention and emphasis from the software companies on developing supply-chain solutions from both methodologic and technological perspectives. The development of these highly integrated systems has enabled a revolution in distribution and logistics between consumers and suppliers. Information about distribution and logistics is no longer a limitation worldwide. Now it can be well known what items were sold, when items move, and where those items are at any point in time. A logistics company can provide real-time updates as parts move around the world. However, at the heart of any supply chain is manufacturing.

In most supply chains, there are several different manufacturing sites and processes that must be coordinated and synchronized effectively to bring a finished item into the distribution pipeline.

At the heart of every supply chain is manufacturing. At the heart of manufacturing is MRP. In an attempt to understand supply chain many writers have simplified the supply chain structure into a straight line. However, supply chains are not the simple linear structures normally represented as the supplier's supplier to the customer's customer. Supply chains are web-like entities that are difficult to represent graphically. Each company is linked to several other companies depending on the product and the customers. These linkages and relationships change as market conditions change. Information flows across and throughout this web. The heartbeat for that information is the MRP system. Each node in the web is a different MRP system. An excellent graphic representation of a supply chain is the cover of this book. Therefore, a primary limitation of any supply chain will be how well MRP systems perform not just individually on each node but also collectively throughout the web. Simply put, MRP has more impact on and is more relevant to the effectiveness of today's supply chain than ever before.

A previous challenge within supply chain management was not having visibility into what is being moved and its status; now warehouse management and logistics tools have solved that problem. Now the problem is fundamentally about which specific items are actually being moved, transported, located and made. What gets put on lathes, welding jigs, assembly lines, trucks, boats and airplanes is a response to a demand or supply order generation signal.

Today, due to the increasing complexity of the global manufacturing and supply landscape the supply order generation signals that move down through our supply chains have become more and more out of alignment with actual demand.

The traditional planning rules and tools (including forecast based demand generation) employed by most manufacturers and distributors do not fit the highly volatile and variable world we live in. Those rules were constructed under a "push and promote" mentality fueled by production efficiency metrics and a market that was more tolerant of longer lead times and shortages.

---

[2] Compiled by John Loucks from Synergy Resources.

In order to get smarter and more agile supply chains, we must take a fundamentally different approach were demand is at the center of planning and not inventory. This is not just about speeding up what we already have.

## QUESTION 2: MRP—FLAWED APPROACH OR POORLY APPLIED?

Is MRP failing in today's environment because it is inherently flawed or because it is poorly implemented?

The case can be made that MRP is both a flawed approach and an approach that has been poorly applied. MRP has many well-known shortcomings, and it is commonly implemented and/or supported in an inadequate fashion. Frequently, there are finger-pointing matches within companies about which is the real problem. The truth is that both are reality. MRP has critical shortcomings and often is implemented and supported poorly. Fixing only one issue will not improve the situation dramatically. Thus companies that "reimplement" MRP do not get the fix they were expecting or leave a lot of possible benefit on the table without realizing it.

The result of MRP's shortcomings and/or poor implementation is that companies have chronic and frequent shortages at various stages of the production, procurement, and fulfillment cycles. These chronic and frequent shortages tend to lead to three main effects:

1. *Unacceptable inventory performance.* This is identified as having too much of the wrong material, too little of the right material, high obsolescence, and/or low inventory turns. Companies frequently can identify many of these problems at the same time.
2. *Unacceptable service-level performance.* Customers continue to put pressure on the company, which quickly exposes poor on-time delivery, low fill rates, and poor customer satisfaction. In addition, customers consistently attempt to drive prices down.
3. *High expedite-related expenses and waste.* In an attempt to fix the preceding two unacceptable business results, managers will commit to payment premiums and additional freight charges or increase overtime to fulfill promises. Typically, this effect is undermeasured and underappreciated in most companies.

As research for writing this book, we surveyed over 150 companies about their materials planning systems. While a minority of companies reported all three of these effects simultaneously to a severe degree, 83 percent of companies reported at least one of these effects to a severe degree over a period of multiple years. Figure 3-1 presents the results of the survey.

To address these undesirable effects, the shortcomings inherent in conventional MRP must be fixed, and the MRP system must be implemented and supported properly. Only one part of the solution is not sufficient.

MRP is defined as:

## FIGURE 3-1

Survey results.

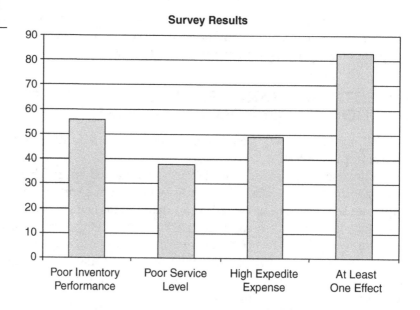

**Survey Results**

material requirements planning (MRP): A set of techniques that uses bill of material data, inventory data, and the master production schedule to calculate requirements
for materials. It makes recommendations to release replenishment orders for material.
Further, because it is time-phased, it makes recommendations to reschedule open
orders when due dates and need dates are not in phase. Time-phased MRP begins
with the items listed on the Master Production Schedule (MPS) and determines (1) the
quantity of all components and materials required to fabricate those items and (2) the
date that the components and material are required. Time-phased MRP is accomplished by exploding the bill of material, adjusting for inventory quantities on hand or
on order, and offsetting the net requirements by the appropriate lead times.[3]

Known shortcomings in conventional MRP are summarized in Figure 3-2, with the
effect resulting from the shortcomings identified.

These shortcomings are nothing new to MRP users. The rapidly changing environment of global manufacturing and the global supply chain is exacerbating these effects
and putting increased pressure on MRP systems and planning personnel.

The continued quest for forecast accuracy has led to the development of improved
forecasting algorithms and sophisticated technology dedicated to forecasting. However
the surprise is that even with all this improvement and sophistication, the overall accuracy of the forecast has not improved dramatically. Increased volatility in the market has
offset any gains in methodology. In addition, any realized improvements in forecast do
not translate to overall improvements in customer service and inventory levels. This was

---

[3] *APICS Dictionary*, 12th ed. (New York: Blackstone, 2008) p. 81.

**FIGURE 3-2**

MRP shortcomings and their undesirable effects on organizations.

| | Typical MRP Attributes | Effects to the Organization |
|---|---|---|
| **Planning Attributes** | MRP uses a forecast or master production schedule as an input to calculate parent and component level part net requirements. | Part planning becomes based on a "push" created by these forecasted demand requirements. Forecast accuracy at the individual sku and part levels is highly inaccurate. Build Plans and POs that are calculated from this forecast often are misaligned with actual market demand. This leads to excessive expediting, overtime, premium freight, increased inventory of the wrong items and missed shipments. |
| | MRP pegs down the *entire* Bill of Material to the lowest component part level whenever available stock is less than exploded demand. | Creates an overly complicated materials and scheduling picture that can totally change with one small change at a parent item. When capacity is scheduled infinitely there are massive priority conflicts and material diversions. When capacity is scheduled finitely across all resources there is massive schedule instability due to cascading slides from material shortages. |
| | Manufacturing Orders are frequently released to the shop floor without consideration of component part availability. | Manufacturing Orders are released to the floor but cannot be started due to shortages. This leads to increased WIP, constantly changing priorities and schedules, delays, lots of expediting and possibly overtime. |
| | Limited future demand qualification. Limited early warning indicators of potential stock outs or demand spikes. | Planners either have to bring in all future demand which inflates inventories and wastes capacity and materials or bring in no future demand, which makes the environment extremely vulnerable to spikes or must pour through large amounts of data in order to qualify spikes for each part. |
| | Lead time for parent part is either the manufacturing lead time (MLT) or the cumulative lead time (CLT) for the parent item. | MLT typically represents a gross underestimation of realistic lead time. When MLT is used, Manufacturing Orders are often released with dates that are impossible to achieve and/or without all component parts available. CLT typically represents a gross overestimation of lead time. When CLT is used Manufacturing Orders are typically released to far in advance, raising WIP levels and making the environment more susceptible to disrupion when order changes occur. |
| **Stock Management Attributes** | Fixed reorder quantity, order points, and safety stock that typically do not adjust to actual market demand or seasonality. | Additional exposure to forecast inaccuracies resulting in increased expediting. |
| | Past due requirements and orders to replenish safety stock are often treated as "Due Now." | There is no way to judge relative priority between stock orders. Every safety stock order looks the same, which means there is no REAL priority. To determine real priorities requires massive attention, analysis and priority changes. |
| | Priority of orders is managed by due date (if not Due Now). | There is no way to judge relative priority between stock orders. Due dates will not reflect actual priorities. To determine real priorities requires massive attention, analysis and priority changes. |
| | Once orders are launched, visibility to those orders is essentially lost until the due date of the order when it is either present or late. | There is no advanced warning or visibility to potential problems with a critical order. Critical parts are often late and disrupt parent item schedule. |

discussed in Chapter 2. The global manufacturing and supply landscape is a much more complex environment characterized by massive instability and volatility. Forecast error continues to rise despite significant investments in time and money creating more sophisticated algorithms for predicting the behavior of supply chains. There are three well-known rules of forecasting that have not changed with this investment:

1. Forecasts are always in error.
2. The more detailed the forecast, the more error will be realized.
3. The further into the future the forecast goes, the more error will be realized.

There is a legacy of make to stock manufacturing strategies that are centered on MRP and how it has been implemented traditionally. When considering today's conditions of shorter customer tolerance times with longer and more complicated supply networks, the need for holding stock at some level is a given for most manufacturers. The manner (rules and tools) in which conventional MRP plans and manages these stock positions creates the problematic undesirable effects detailed in Figure 3-2.

Most MRP systems are driven by a master schedule driven by a forecast. This plan can be developed through an interactive sales and operations process (see Chapter 19), but it is still a best guess at end-item requirements. Using the forecast to drive requirements creates, by definition, a push-based system. This is frequently in error and conflicts with the proliferation of demand-driven or pull-based approaches, methods, and tools. Companies that use forecasting to drive MRP while at the same time using demand-driven/pull-based techniques to execute often have a push-versus-pull dilemma. This "results in a constant tension between the supply side and demand side."[4] This tension creates misalignments, and the misalignments generate massive amounts of waste.

MRP is fundamentally only a planning tool. Problems tend to be identified only after they have risen to the surface. There is little or no visibility of potential problems or what constitutes the actual execution priority. This lack of visibility creates a critical gap in today's more complex supply-chain scenarios. Companies are now beginning to realize just how important this type of visibility is. A survey of 209 companies in 2009 by the Aberdeen Group concluded that:

> . . . 57 percent of respondents indicated that Supply Chain Visibility (SVC) was currently a high priority for improvement with an additional 28 percent indicating it was a medium priority. Increasing visibility is a critical strategy for enterprises aimed at reducing costs and improving operational performance in the context of their complex and multi-tiered global supply-demand networks.[5]

Most experienced planning personnel are not blind to the shortcomings of MRP. These have been discussed for years at APICS meetings. However, these shortcomings have been further exacerbated by the variability and volatility of today's hypercompetitive and hypersensitive environment. Materials and production control personnel often find themselves in a dilemma regarding their MRP system. There are powerful aspects of MRP that are still relevant and necessary. The need to be able to plan complex product structures across a complex supply chain well in advance of customer demand means that some aspects of MRP are even more relevant today than when they were conceived 40 years ago. Companies desperately need visibility within today's more complex planning scenarios.

At the same time, using MRP does not provide the flexibility to be responsive to actual market requirements and consumption. Ignoring these shortcomings has disastrous consequences in today's environment. Every person faced with materials planning under these circumstances is forced to find various, often unsatisfactory and incomplete

---

[4] Aberdeen Group, "Order-to-Delivery Excellence: Linking the Demand Chain with the Supply Chain," Boston, 2009, p. 1.

[5] Bob Heaney and Viktoriya Sadlovska, *Supply Chain Visibility Excellence; Reduce Pipeline Inventory and Landed Cost* (Boston: Aberdeen Group, 2009), p 4.

ways to sidestep these shortcomings. The typical compromises used to work around these shortcomings include

- Manual work-around proliferation
- Attempts at more efficient forecasting
- Manual reorder point systems
- Overflattening the bill of materials
- Move to a make-to-order (MTO) model
- "Dumbing down" MRP

## Manual Work-Around Proliferation

Frequently, companies try to work around the shortcomings of their MRP system by relying on stand-alone, disconnected, and highly customized data manipulation tools such as Excel spreadsheets and Access minisystems. Data are taken from a core MRP tool and then manipulated by an individual. Typically, this individual is the one who actually built the sheet or minisystem and, consequently, usually is the only one who knows how to use it. From a risk-mitigation standpoint, this is unacceptable. The company's ability to plan and execute against those plans can be crippled by the loss of this individual. These tools have serious limitations, and their proliferation makes the information technology (IT) landscape more complicated and maintenance more intensive. Their widespread use ultimately defeats the purpose behind the major investment in an integrated ERP package because the information garnered is limited to the individual user. Many have called this "Excel hell."

A recent report by the Aberdeen Group[6] showed just how pervasive this compromise is (Figure 3-3).

**FIGURE 3-3**

Companies using spreadsheets for demand management.

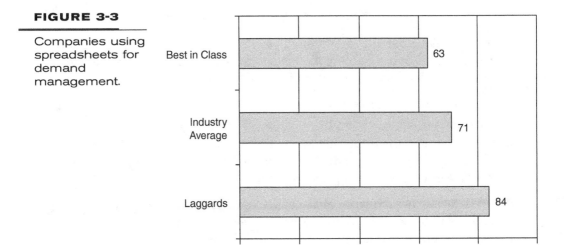

[6] Aberdeen Group, "Demand Management," Boston, November 2009.

In a survey of 135 companies,

- 63 percent of the best-in-class (top 18 percent) companies reported that they used spreadsheets for their demand management.
- 71 percent of the industry average (middle 54 percent) performers reported the use of spreadsheets for demand management.
- 84 percent of laggard companies (bottom 28 percent) used spreadsheets.

Consistent with the Aberdeen Group survey is a survey from www.beyondmrp.com regarding this strategy of MRP-related compromises (Figure 3-4). The results of these two surveys suggest that for most companies there is a basic lack of trust that MRP is providing an accurate picture of what is required and when it is required. Additionally, these reports are an indicator of the inability of ERP companies to evolve MRP rules and tools to an acceptable level or their ambivalence about such an evolution.

**FIGURE 3-4**

beyondmrp.com survey results.

Do employees develop "work-arounds" using spreadsheets and Access databases (for example) because they feel they can't work effectively within the formal planning system?

| | |
|---|---|
| Too many | 64% |
| Just a few | 28% |
| Never | 5% |
| No answer | 4% |

## More Efficient or Better Forecasting

Other companies have pursued a strategy of investing in and implementing advanced forecasting algorithms or hiring more planners in the hope of guessing/predicting better. Remember the three well-known characteristics of forecasting:

1. Forecasts are always in error.
2. The more detailed the forecast, the higher is the rate of inaccuracy.
3. The further into the future the forecast is extended, the higher is the rate of inaccuracy. This inaccuracy is even worse when the granular (stock keeping unit or part number) level is considered.

These characteristics are particularly true today, considering the increasing amount of variability, volatility, and complexity across the supply chain. In the best of times, the improvements in forecasting rarely correlated with improvements in the overall quality of the plan. When the acceleration of volatility and complexity is considered in today's supply chain, the offsetting advances in forecasting algorithms and technology are inadequate to improve the planning process. The same Aberdeen Group report[7] also noted that 57 percent of the companies it surveyed responded that they have consistently inaccurate forecasts. Additionally, the report demonstrated how forecast "accuracy" breaks down with more detail. Best-in-class companies reported 87 percent accuracy at the product family level, but at the stock keeping unit (SKU) level, accuracy dropped to 77 percent. Industry average companies reported an accuracy of 64 percent at the product family level and 54 percent at the SKU level. Finally, laggards reported 40 percent accuracy at the family and a dismal 23 percent at the SKU level. Figure 3-5 depicts this breakdown.

Even if a company succeeds in increasing signal accuracy, it does not necessarily translate well to overall effectiveness in terms of availability and fill rates. Remember, the increase of variability and volatility, especially on the supply side, easily can offset any appreciable gain in signal accuracy. Most manufacturers have multiple assembly and subassembly operations that are integral parts of their overall end-item product flow. In any type of assembly operation, it takes the lack of only one part to block a complete shipment. The more assemblies that exist, the more complex is the synchronization and execution challenge. Companies that experience this phenomenon usually compensate with additional inventory rather than risk unacceptable customer service.

Even the biggest supporters of advanced forecasting algorithms cannot argue that forecasts and any purchasing and manufacturing schedules derived from them aren't still a push-based tactic. The recent statistical advancements may make it a more educated push, but it remains a push nonetheless. For companies implementing demand-driven or pull-based manufacturing execution systems (e.g., lean or drum-buffer-rope [DBR]), this sets up conflicting modes of operation. This conflict is the push-versus-pull conflict referred to earlier.

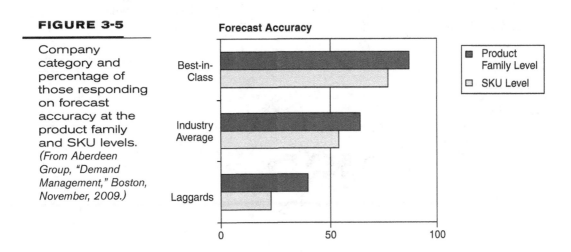

**FIGURE 3-5**

Company category and percentage of those responding on forecast accuracy at the product family and SKU levels. *(From Aberdeen Group, "Demand Management," Boston, November, 2009.)*

---

[7] Aberdeen Group, "Demand Management," November, 2009.

Forecasting assumes that the future will look like the past. Past demand is input into sophisticated forecasting algorithms to project the future. This can be compared to driving a car by only using the rear view mirror. If you are driving on a straight road where the road in front looks like the road behind, then you may be successful. However if you are driving on a road with many twists and turns, the results can be catastrophic. Business today has to face more twists and turns than ever before.

## Manual Reorder Point Systems

In order to overcome the shortcomings of MRP, some companies and many lean disciples have abandoned MRP completely. Early demand flow implementations actually had lapel buttons with the letters "MRP" and the universal "not" sign over them, as demonstrated in Figure 3-6. In its place, the pull-focused companies have implemented different forms of manual reorder point systems.

Manual reorder point systems were a common technique for managing inventory before MRP. Has industry come full circle? Is industry ready to abandon the promise of technology just because an antiquated set of rules and tools has not been updated or evolved?

Consider what Joe Orlicky himself said about reorder point systems in the first edition of this book in 1975:

> Systems based on reorder point suffer from false assumptions about the demand environment, tend to misinterpret observed demand behavior, and lack the ability to determine the specific timing of future demand. These shortcomings, inherent in all systems of this type, manifest themselves in a number of unsatisfactory performance characteristics, chief among them being an unnecessarily high overall inventory level, inventory imbalance, and stock-outs or shortages by the system itself.[8]

**FIGURE 3-6**

Lapel button
used by early
demand-flow and
lean advocates.

---

[8] Joe Orlicky, Material Requirements Planning (New York: McGraw-Hill, 1975), p. 6.

Probably the best example of a modern-day manual reorder point is the frequently used lean technique of *kanban*:

**Kanban (lean stock technique):** A method of Just-in-Time production that uses standard containers or lot sizes with a single card attached to each. It is a pull system in which work centers signal with a card that they wish to withdraw parts from feeding operations or suppliers. The Japanese word *kanban*, loosely translated, means card, billboard, or sign, but other signaling devices such as colored golf balls have also been used. The term is often used synonymously for the specific scheduling system developed and used by the Toyota Corporation in Japan.[9]

Today, many researchers and practitioners make the same comments about kanbans that Orlicky made about the manual reorder points of his day. Kanbans tend to be manually intensive, lack responsiveness to changes in the environment, limit planning visibility, and force an unnecessary and wasteful level of independence into the manufacturing environment. This is not meant to say that kanbans do not have a time and a place in modern manufacturing. The time and the place, however, are not "always" and "everywhere." Manual reorder points and kanbans are supplementary tools at best.

After years of experience in implementing kanban systems and heijunka board in a variety of industry, it is clear that as the number of parts and components grows the manual systems become increasingly difficult to manage. Kanbans are very easy to establish and begin to use. However, when the company grows to any significant size, the overhead to manage kanbans can quickly become overwhelming. The number of kanbans can quickly grow to the hundreds. Reconciling the number and location becomes a substantial task with a great deal of complexity. This is problematic for most midrange or larger discrete manufacturers. Early adopters of lean were emphatic that technology was a waste to be totally eliminated. The desired future state was one where the part would move only when pulled by the next station. However the gremlins of variability made the necessity of technology quickly apparent. John Constanza, the founder of JCIT, developed the first demand flow software to support lean. The technology was very different than the MRP systems of the day. Early adopters struggled with how to manage the sheer number of kanbans and how to trigger unique configurations. Even the first demand-flow visionaries developed software to support their approach.

Rarely are kanbans realigned with the actual demand variation owing to the overwhelming nature of completing that task. Kanbans almost always omit independent demand, including service and repair. In addition, kanbans provide no forward planning because they are a reactive pull signal assuming that the immediate future will look like the past. Few suppliers are agile enough to support a company without some forward planning. This can mean that suppliers add extra inventory for this unexpected demand or reduce the variety that can be offered to customers. Manual reorder points are even less effective in larger organizations. When an organization must plan across a large

---

[9] *APICS Dictionary*, 12th ed. (New York: Blackstone, 2008), p. 70.

enterprise, particularly a vertically integrated enterprise, it is left with few viable and satisfactory options.

Some software has been developed as an aid to reduce the complexity of managing kanbans. This software reduces the management burden and enables the company to adjust kanban sizing. However, all electronic software products lack the ability to provide the critical available inventory and netting picture that more complex manufacturing situations require.

> **available inventory:** The on-hand inventory balance minus allocations, reservations, backorders, and (usually) quantities held for quality problems. Often called beginning available balance.[10]

These are the necessary traditional MRP functions that have been abandoned in the lean world. Simply knowing the stock on hand and on order cannot provide the complete picture of what to order unless those two positions are also considered in relation to relevant demand allocations. An accurate available stock equation is not possible with manual or even electronic reorder point systems such as kanbans. The lack of visibility means that organizations of size or even moderate complexity are flying blind to the overall materials and inventory picture. The available stock equation will be discussed in depth in Part 4.

Furthermore, manual reorder points and kanbans do not consider the bill of material in its totality. The kanban is defined only at each discrete connection. This means that stock positions often must be placed at every position in the bill of material. This strategy, in turn, increases the number of stocked positions to manage and potentially raises total inventory. Often the implementation of kanbans is not limited to only each connection in the bill of material. Organizations frequently place them between each step in the routing. When there is a larger number of part numbers with sporadic demand or relatively long routings for many part numbers, inventory can rise dramatically.

*Author's note:* The preceding section on manual reorder points and kanbans is not intended to alienate the lean advocate. Its intention is to highlight accepted realities about the planning limitations of manual and even electronic reorder point positions/systems in environments of size and relative complexity. Many environments do apply kanbans successfully, especially when those kanbans are placed in the right positions.

## Overflattening the Bill of Material (BOM)

Sometimes companies try to simplify the synchronization issue by dramatically reducing the number of levels in the BOM. In many cases, reducing BOM complexity makes sense. Taken to extremes, it can close more doors than it opens. Flattening the BOM removes levels that were identified originally to define the product and the process. These dependencies provide an excellent way to stop variability from gaining momentum and disrupting the entire supply chain like a tsunami. The key to better synchronization is to

---

[10] *APICS Dictionary*, 12th ed. (New York: Blackstone, 2008), p. 8.

understand those dependencies and control them. By overflattening the BOM, companies can actually lose visibility at both the planning and execution levels. Some companies have had to reinsert a level in their BOMs in order to gain back critical leverage and decoupling points. This will be discussed at length in Chapter 23.

### Move to a Make-to-Order (MTO) Model

Some companies will attempt to address the forecasting issue by eliminating it totally and any finished end-item positions entirely. This is possible only with a pure MTO strategy. Referring back to the Wheelwright and Hayes product/process matrix in Figure 1-2, this strategy is effective only in a job-shop environment that supports high-variety, low-volume production. Attempting to use this strategy for low-variety, high-volume production will spell disaster for the company as costs spiral out of control. Companies that can employ this strategy will need carefully placed and controlled raw material and component stock positions in order to stay competitive in terms of lead time and cost. Part 4 of this book will provide more detail.

### "Dumbing Down" MRP

Finally, companies will "dumb down" the MRP system by eliminating some of the integrated nature of its planning. The power of this technology is the rapid calculation of requirements across complex interrelated BOMs. MRP at its core minimizes inventory by planning only items that are needed when they are needed. By dumbing down the system, those interrelationships and that precise timing are lost. The balance of this book will discuss the necessary conditions for an effective material planning system in a demand-driven environment.

## QUESTION 3: THE MRP CONFLICT WITH LEAN OR PULL

Why do lean and other pull methodologies often come into conflict with MRP?

In addition to the rise in volatility and complexity referred to earlier, the proliferation of lean philosophies has put additional pressure on planning personnel and MRP systems. The fundamentally different view of what is important in relation to inventory puts lean advocates and planning personnel at odds. Many lean implementations attempt to abandon MRP. This causes tremendous friction between planning personnel and those pushing for that abandonment. Lean facilitators often see MRP as an overly complex and wasteful dinosaur that simply doesn't work in the demand-driven world. Planning personnel, however, see it a completely different way. They understand that without the ability to see the total requirements picture, critical blind spots then exist in the planning process that lead to shortages and/or even excessive inventory positions. They see the lean approach as a gross oversimplification of the complex scenarios that are the "new normal." There are three conflicts that tend to occur between lean and MRP.

## Conflict 1: Planning Versus Execution

As discussed earlier, MRP is, by definition, a planning tool. MRP launches orders based on a combination of data pertaining to demand, on-hand, open-supply, BOM, and lead-time parameters. It assumes that a system of execution is in place to manage to the plan. MRP detractors cite the frequent poor quality of the plan as a reason that makes effective execution impossible and even counterintuitive. Worse, those detractors say that if organizations did in fact execute to the plan, it would be disastrous and result in shortages and service-related problems.

Thus, if the planning system cannot provide a realistic, feasible plan that delivers materials as the market needs them, many lean advocates would recommend that we abandon planning altogether. In their mind, planning often becomes relegated to some initial assumptions about the rate of demand pull, takt time, decoupling BOMs, and routings at most discrete part numbers and steps, respectively, and allowing resources (including suppliers) to "read and react" in order to fulfill materials requirements. In effect, there really is no planning after the initial setup, only execution.

## Conflict 2: Complexity Versus Simplicity

MRP advocates desire a solution that identifies and manages the inherent dependencies in manufacturing situations. In order to do this, MRP can be a highly complex system to operate. As mentioned in relation to conflict 1, the perceived need for predictability drives companies into massive complexity. Most people in manufacturing companies don't even fully understand what the planning system is or does. Furthermore, "fixing" the system seems to be a never-ending, intricate, and expensive journey. Einstein once said, "Any intelligent fool can make things bigger and more complex. It takes a touch of genius—and a lot of courage—to move in the opposite direction."

Einstein also said something else relevant to this conflict. He said, "Everything should be made as simple as possible, but not simpler." Lean advocates are seeking simplicity in a solution. The tactics tend to be manual and visible. There is nothing wrong with simple solutions—to a point. When things get oversimplified, however, the end result often can be as poor as the complex side of things. *Oversimplification* is defined as "to simplify to the point of causing misrepresentation, misconception, or error"[11] By failing to depict and account for critical dependencies and/or relationships with regard to supply, demand, on-hand inventory, and product structure, lean can oversimplify many environments. The more complex and variable these environments, the more likely it is that lean is an oversimplified approach.

MRP's complexity, combined with its critical inherent shortcomings and lean's total oversimplification, often has the same effect—chronic and frequent shortages and at the same time excessive inventory positions. The solution must provide both a level of

---

[11] www.thefreedictionary.com.

sophistication that can provide more visibility from a planning and execution perspective and at the same time simple, clear, and highly visible signals across the enterprise.

## Conflict 3: Dependence Versus Independence

This is a subtle but very important conflict to understand and is an extension of conflict 2. The solution to this conflict defines a huge innovation and leap forward for MRP logic. MRP is fundamentally about making everything dependent (see Part 2). Lean's answer to the material planning puzzle is to promote complete independence. When everything is dependent, the resulting picture can be very nervous (see Chapter 2 for a discussion of system nervousness). When everything is independent, visibility is limited to what is immediately in front of you. The solution to this conflict provides for planning and execution support for the demand-driven world in more complex operational environments. See Part 4 for a discussion of demand-driven MRP.

While these conflicts continue to rage inside manufacturers, lean advocates are beginning to understand the need for an effective technology solution. When it comes to materials and inventory planning and execution Lean advocates often get accused of being anti-technology. Do Lean advocates really want manufacturing companies to entirely abandon the promise of technology? The answer should be yes when that technology is wasteful, confusing and not reflective of reality. Unfortunately, that has been the situation for quite sometime with regard to traditional MRP and DRP systems. Point 8 of the Toyota Production System states: "Use only reliable, thoroughly tested technology that serves your people and processes."[12] Until now the prevailing materials and inventory planning and execution technology, while thoroughly tested, has been largely inappropriate to serve the people and processes in companies transforming to a demand driven approach. The proliferation and sustainability of Lean implementations has been negatively impacted by the lack of appropriate supply chain materials planning and execution technology.

Several analyst reports have concluded that there is tremendous potential for the incorporation of better planning and visibility software into Lean implementations. Manufacturing needs Lean to survive in the more complex environment of the 21st Century. Lean needs an effective demand driven materials requirements planning approach to bring that vision to reality.

What if there was an appropriate technology? What if a reliable, thoroughly tested blueprint for demand driven planning and execution of supply chain materials with high degrees of visibility could be introduced to the MRP world? Furthermore, what if that blueprint also was extremely appealing to the MRP World? Sound impossible? It's not; it is reality and it is producing significant results in short periods of time for early adopters.

---

[12] Liker, J. *The Toyota Way: 14 Management Principles from the World's Greatest Manufacturer.* New York: McGraw-Hill, 2004.

Manufacturing needs lean and the other pull-based methods to evolve in the "new normal." Lean needs an effective demand-driven planning approach in order to bring that evolution to reality.

## QUESTION 4: MRP PROGRESS IN THE LAST 30 YEARS?

Why has MRP not progressed significantly in the last 30 years?

Software providers, consultants, and the academic community have had ample opportunity to fix MRP's shortcomings. Why have they failed?

There are many knowledgeable experts in demand-driven techniques such as lean and drum-buffer-rope (DBR). Unfortunately, many of these experts do not understand the role planning technology must play to bring those techniques to full realization across a complex enterprise and supply chain. Many of them advocate the elimination of technology as the true measure of success.

Experts in variability and volatility tend to be less enterprise-focused and more specific-event-focused. Variability must be considered in relation to its impact across a holistic system. All variation does not have the same impact. Reducing variability does not necessarily improve the overall process. There are places where it must be protected against in order to keep the system stable and effective.

The generation that developed MRP is all but lost. Most people who still have an in-depth knowledge of how MRP really works are not in software companies or academia—they are seasoned planners working in private industry. Even some of the largest ERP software companies have only a small number of people (sometimes only two or three in the largest ERP providers) who truly understand what MRP is and how it works. Rarely does even the largest provider have software developers with any real-world experience using the tools they are building. The big ERP software companies cannot and will not solve a problem they cannot see. This is evidenced by the fact that none has addressed the core issues identified in this section. The proliferation of planning work-arounds proves this.

But there is light at the end of the tunnel. At the time of this writing, we know of several next-generation products that will allow for extremely flexible and unfettered configuration in alignment with the recommendations in Part 4. This is not enough, but these products do remove hard-coded restrictions that have hampered MRP users in the "new normal."

# PART 2

# Concepts

## CHAPTER 4

# Inventory in a Manufacturing Environment

In the first edition of this book, Joe Orlicky made an extensive effort to draw a distinction between manufacturing inventories and those related to distribution. Under the new manufacturing and global supply landscape, the need to treat the two separately has diminished. This chapter represents current practice. Part 4 addresses the future of inventory in a manufacturing environment.

Manufacturing inventory management is a subject in its own right. It only partly overlaps general inventory management as we know it from the literature because it represents a special problem and is governed by unique laws. This means that many of the traditional approaches to inventory management are not properly applicable to manufacturing inventories. When applied, they prove relatively ineffective. The classic inventory control theory does not adequately reflect the realities of a manufacturing environment and makes incorrect assumptions as to the function of and the demand for the individual items of which a manufacturing inventory consists.

Failure to distinguish between manufacturing and nonmanufacturing inventories accounts for a measure of the confusion or controversy observed frequently in connection with the question of applicability of a given approach or inventory control technique to a manufacturing environment. To avoid such difficulties arising from a failure of definition, this chapter is devoted to examining the attributes of manufacturing inventories and the demands to which these inventories are subject.

## MANUFACTURING INVENTORIES

A manufacturing inventory is defined as consisting of the following:

- Raw materials in stock
- Semifinished component parts in stock
- Finished component parts in stock

- Subassemblies in stock
- Component parts in process
- Subassemblies in process

Note that so-called shippable items (inventory items ready, at their stage of completion, to be delivered to a customer), such as end products and service parts, are excluded from the preceding list. They are part of a distribution inventory, as discussed later in this section. Under today's conditions, it makes more sense to treat most service parts as strategic inventory positions that consolidate internal and external consumption and supply generation. More on this can be found in Part 4.

In order to establish the attributes that set apart manufacturing inventories, let us first consider what the functions of a system of inventory management are, including that of a grocery, museum, or blood bank. Inventory management, or inventory planning and control, consists of the following functions and subfunctions:

1. *Planning.* There is normally no need for a special inventory policy pertaining to a manufacturing inventory as a whole. The least total inventory consistent with production requirements and that allows manufacturing cost to be at a minimum is always the management objective. Forecasting, within the manufacturing inventory system proper, plays a secondary role, and the type of forecasting being performed (such as the proportion of a given optional feature within a future product lot) differs from the usual forecasting of demand magnitude. Included in planning are
   - Inventory policy
   - Inventory planning
   - Forecasting

2. *Acquisition.* The order action function is expanded and exhibits several characteristics unique to manufacturing. Material in the manufacturing process is from the inventory system's point of view, being acquired and reacquired at a different configuration as it progresses through multiple stages of conversion from raw material to end product. An order for a manufactured item, once started, cannot be canceled without the penalty of scrap or rework. Neither can it normally be increased or decreased in quantity. Factors that enter into the order-quantity determination include allowances for yield or scrap, raw materials cutting considerations, and so on. The ordering function includes order suspension, that is, rescheduling the order to an indefinite future due date. Finally, the quantity and timing of an order may be affected by capacity considerations.
   - Positive order action (place or increase)
   - Negative order action (decrease or cancel)

3. *Stockkeeping.* These functions are the execution of the planning and acquisition phase.
   - Receiving
   - Physical inventory control
   - Inventory accounting (recordkeeping)

4. *Disposition.* Delivery of a manufacturing inventory item is always to an in-house demand source. Demand is represented by a dependent demand production requirement or a top-level production schedule. When an inventory item is completed (or received from a vendor), it is earmarked for consumption in the next stage of the material conversion process. If it is shippable on completion, it enters a distribution inventory. Disposition includes
   - Purging (scrap and write-off of obsolete items or lost items)
   - Disbursement (delivery to source of demand)

While any system of inventory management can be described functionally in this way, manufacturing inventory management has its own distinct characteristics and, compared with nonmanufacturing inventories, shows a difference in the content of certain key functions in every one of the four principal areas just mentioned.

The term *manufacturing inventory management* is really a misnomer. In a manufacturing environment, inventory management cannot be conceived of apart from production planning, with which it is inseparably bound up. The function of a manufacturing inventory system is to translate the overall plan of production (the master production schedule) into detailed component material requirements and orders. This system determines, item by item, what is to be procured and when, as well as what is to be manufactured and when. Its outputs "drive" the purchasing and manufacturing functions. It plans and directs purchasing and manufacturing activities because nothing will be purchased and no component parts will be manufactured without a requisition or order that generates it. The manufacturing inventory system determines (or rather should have the capability to determine) order priorities and implies the capacities required. All in all, it does considerably more than manage inventory. It is the heart of manufacturing logistics planning.

Manufacturing inventory management can be put into sharper focus by dividing business inventories into two categories based on purpose. The purpose of a manufacturing inventory is quite different from that of a distribution or marketing inventory such as is found in a supermarket, a wholesale distributor, or a manufacturer's finished goods and/or service parts warehouse.

## DISTRIBUTION INVENTORY

The purpose of a distribution inventory is to be available to meet customer demand (the term *customer* applies to any recipient of items provided from distribution inventory), which tends to be erratic and of limited predictability owing to normally expected randomness in actual demand. Total demand over a given period (termed *period demand*) typically is made up of many unit demands originating from separate sources. Period demand can be thought of as a sample drawn from a potential demand universe that is very large or infinite. The inventory investment level is governed by marketing considerations.

In contrast, the purpose of a manufacturing inventory is to satisfy production requirements. Availability can be geared to a production plan, which means that demand is calculable, that is, predictable. Period demand typically consists of a limited number of individual demands for multiple quantities of the inventory item. The production plan (including the planned production of such items as service parts) is the sole source of demand, and this demand is always finite. The inventory investment level is dictated by manufacturing (i.e., process, setup, queue, and move time) considerations. Work in process, an inventory entity unique to manufacturing, constitutes a significant part of the investment, and the level of this inventory is primarily a function of manufacturing lead times and batch sizes frequently used to improve overall resource utilization and efficiency.

In comparison with a distribution inventory, a manufacturing inventory represents a means to a different end. A manufacturing inventory, as defined previously, exists only to be converted into a shippable product. Once the product is assembled or, in the case of a service part, finished, it passes into distribution inventory. In many cases, at this point the responsibility for the inventory by manufacturing management ceases and is assumed by a marketing, distribution, or service organization. Increasingly, services are becoming an integral offering by the manufacturer as another source of profits.

The difference between distribution and manufacturing inventories is fundamental. Consequently, the respective inventory management philosophies, systems approaches, and techniques in use are (or should be) fundamentally different. In determining the desirable level of a distribution inventory, the tradeoff is between investment (and the attendant inventory carrying cost) and sales revenue realized through availability. Under the service-level concept in a distribution environment, 100 percent service theoretically requires an infinitely large inventory investment. However, this is not actually the case. In determining a manufacturing inventory level, there is no such tradeoff. The investment is dictated by production requirements, which, unlike customer demand, are given and controllable. The inventory that exceeds the minimum required brings no extra revenue. A 100 percent service level (between component items and the shippable product made from them) is a necessity, but it is feasible to achieve it with a finite inventory investment. The critical question to be answered is where the decoupling points in the bill of material are best defined to minimize inventory while at the same time compressing the overall response time to the customer. This is covered more fully later in this chapter. Part 4 of this book will discuss in depth a tradeoff inherent in manufacturing inventories with regard to shared components.

In a distribution inventory environment, demand for each inventory item must be forecast explicitly or implicitly. Uncertainty exists at the item level. The principle of distribution stock replenishment to restore availability applies, and the two principal questions are when to reorder and in what quantity. The first question cannot be answered with certainty, whereas the second is answered through computation of some form of an economic order quantity. In a manufacturing inventory environment, on the other hand, individual-item demand need not be forecast, and uncertainty exists only at the master production schedule level (will customer demand materialize to allow shipment of the product?). There is no need to forecast manufacturing inventory, only to order what is required to cover production needs. Inventory availability can be geared to the time of those needs; that is, it need not exist prior to such time. The existence of any of the in-stock categories of manufacturing inventory signifies, strictly speaking, premature avail-ability. Ideally, all manufacturing inventory would be in process, with every item imme-diately consumed (by entering into the next manufacturing conversion stage) on com-pletion or receipt. The best-managed manufacturing inventories approach this ideal.

For distribution inventory, the questions of when and in what quantity to order are being answered. The first question is answered with certainty provided by required date and lead time and the second one through lot-sizing techniques that use only known future demand (i.e., planned requirements) and take into account both its magnitude and its timing.

In practice, the question of the "correct" order quantity receives only secondary attention—and deservedly so. It is interesting to note that this question does not arise at all when the demand for an inventory item is either highly continuous (as is typical for large-volume production operations) or highly discontinuous. The Wheelwright and Hayes product/process matrix in Chapter 1 (Figure 1-2) describes the continuum for dif-ferent types of operations and the resulting inventory impact. In all cases, it can be said that it is more important to have the quantity needed at the time it is needed than to order the "correct" quantity. In the real world of manufacturing, evidence attesting to the truth of this statement abounds. Splitting lots in midproduction, double setups, teardowns caused by "hot order" expediting, and partial vendor shipments are normal occurrences. They show that it is not practical always to adhere to the calculated most economical order quantity.

The preceding description and distinction explains the structural and conceptual division between what became manufacturing requirements planning (MRP) and what became distribution requirements planning (DRP) systems, a legacy that persists at the time of writing of this third edition. It is predicated on the assumption that distribution and manufacturing inventories are fundamentally different and require fundamentally different planning tactics. Since 1975 (when the first edition of this book was published), the ramifications of treating these two inventories as completely separate and distinct has changed dramatically. Obvious differences remain. For example, distribution items (such as purchased items) have no bill of material (BOM). In the demand-driven world of the twenty-first century, however, more integration is required between these two inventory designations and, consequently, closer alignment between the two. Furthermore, as out-lined in Part 4 of this book, the two types of inventories should be planned and managed in the exact same manner when it comes to strategically managed/stocked parts (most distributed parts fall under this classification).

In general, inventories have five separate and distinct functions regardless of whether they are assigned to manufacturing or distribution. The functions are to:

1. *Decouple operations and/or stages in the system.* Different points in the system process materials at different rates and experience different rates of demand within a given time period and thus should not be linked rigidly. See Part 4 for more information on determining where to place these strategic buffers.

2. *Cushion against variability.* Demand changes and interruptions in supply have cost-ly and harmful effects. Buffers, called *fluctuation inventory*, minimize these effects.

3. *Level production.* Cyclic and seasonal changes in demand are expensive and often impossible to handle, and early production for anticipated changes is necessary. The stocks built for these reasons are called *anticipation inventory* or *stabilization stock.*

4. *Fill distribution pipelines.* Materials in transit are called *transportation inventory.*

5. *Hedge against external expected events.* Suppliers' price increases, labor strikes in suppliers' facilities or in transportation, new government regulations, and simi-lar events may make *hedge inventory* a good investment.

In general, there are four fundamental questions of management regardless of the classification of inventory:

1. Where should inventory be placed?
2. When is the proper time to reorder?
3. How much should be ordered at that time?
4. What the investment level should be appears in a different light in manufactur-ing inventory management because the problems and criteria are quite different.

These questions are covered more fully later in this chapter and are expanded on in Part 4.

## THE LOGIC OF MANUFACTURING

The fundamental logic of manufacturing includes the following questions: What will we be making? How many of each component are needed? How many do we already have? When do we need the rest, and how will we get them? This logic has been used since cave dwellers made slings, bows, arrows, and spears. In pre-MRP industry, the first question was answered using forecasts of future demand unless a large backlog of customers' orders was available and sufficient to cover the planning, acquisition, and manufacturing lead time, which is the rare exception. The next three questions required great amounts of detailed information on products, inventories, and processes so often lacking integri-ty that crude estimates and approximations were substituted.

Manufacturers of large, complex equipment (e.g., ships, trains, planes, and central station boilers and generators) had long future horizons covered with firm orders. Planning was manual, slow, and crude. Large clerical groups calculated gross require-ments for major components of their products and time-phased (albeit very roughly) these and their procurement. Revising such plans was even more tedious and was rarely done. The capability of massive data storage and manipulation required for sound inven-tory planning simply did not exist at that time.

Because of this constraint, methods of stock replenishment (order point and eco-nomic order quantity) predominated prior to the 1970s. Inventory control was attempted using paper records and electromechanical desk calculators to apply essentially simple mathematical formulas for order-quantity and safety-stock calculations.

Part of the second question above, "How many should we buy or make?" was answered in the decades preceding MRP with economic order quantity (EOQ) techniques. As mentioned in Chapter 8, Ford Harris published the first theoretical formula for EOQ in 1915. In early practice, the question of the "correct" order quantity deserved and received only secondary attention. This question does not arise at all when the demand for an inventory item is either highly continuous (typical for high-volume production operations—assembly line or continuous production in the Wheelwright and Hayes product/process matrix in Figure 1-2) or very intermittent. It is obviously more important to have the quantity needed at the time it is needed than to order an economic quantity. Evidence of this was the frequent splitting of lots in process, double, and triple setups caused by "hot order" expediting and partial vendor shipments, all of which were normal occurrences.

The first half of the last question, "When are raw materials and components needed?," received the crudest of answers prior to computer-based MRP programs. Statistical calculations of safety stocks, proposed by R. H. Wilson in 1934, gave the appearance of precision without the reality of accuracy. Calculated EOQ and safety stocks and subsequent refinements and elaborations did improve production inventory control over the guesstimates and estimates preceding them but left much to be desired.

## ORDER-POINT VERSUS MRP SYSTEMS

Under convention, there are two alternatives in fundamental approach and two corresponding sets of techniques that a manufacturing enterprise may employ for purposes of inventory management. They are order-point systems and MRP. Part 4 of this book will introduce an effective and innovative hybrid.

The first of the alternatives may be defined as a set of procedures, decision rules, and records intended to ensure continuous physical availability of all items in the face of uncertain demand. Under the order-point approach, the depletion in the supply of each inventory item is monitored, and a replenishment order is issued whenever the supply drops to a predetermined quantity—the reorder point.

This quantity is determined for each inventory item separately—parents and components—based on the forecast demand during replenishment lead time and on the probability of actual demand exceeding the forecast. The portion of the reorder-point quantity that is carried to compensate for forecast error is termed *safety stock*. It is computed on the basis of historical demand for the item in question and of the desired service level, that is, incidence of availability over the long run. The underlying belief is that higher inventory is required to provide higher levels of customer service. In an order-point system, some form of an economic order quantity computation normally determines the size of the replenishment order. The order-point system is used traditionally without any analysis to the positioning of the inventory.

An MRP system, narrowly defined, consists of a set of logically related procedures, decision rules, and records (alternatively, records may be viewed as inputs to the system) designed to translate a master production schedule into time-phased net requirements and the planned coverage of such requirements for each component inventory item needed to implement that schedule.

An MRP system replans net requirements and coverage as a result of changes in either the master production schedule, inventory status, or product composition. In the process of planning, an MRP system allocates existing on-hand quantities to item gross requirements and reevaluates the validity of the timing of any outstanding (open) orders in determining net requirements.

To cover net requirements, the system establishes a schedule of planned orders for each item, including orders, if any, to be released immediately, plus orders scheduled for release at specified future dates. Planned order quantities are computed according to one of several lot-sizing rules specified by the system user as applicable to the item in question. In its entirety, the information on item requirements and coverage that an MRP system generates is called the *material requirements plan*.

Order point is part-based without respect to any relationships of parts to each other, whereas MRP is product-oriented. Order point uses data on the historical demand behavior of an inventory item in isolation from all other items. MRP, a radically different approach, ignores history in looking toward the future, as defined by the master production schedule, and works with data specifying the relationship of components (the BOM) that make up a product.

In the face of two alternative approaches to manufacturing inventory management, the question naturally arises as to which of them is preferable. Which of them will yield better results under what circumstances, and what is the principal criterion of their applicability? This is the conventional choice available to companies. Are they really mutually exclusive? Part 4 offers an alternative.

## Dependent Versus Independent Demand

Orthodox inventory analysis and classification techniques are designed ostensibly to determine the most desirable treatment of a given inventory item or group of items. They examine various attributes of the individual items, such as cost, lead time, and past usage, but none of them takes into account the most important attribute, namely, the nature of demand. Yet it is the nature (or source) of demand that provides the real key to inventory control technique selection and applicability. The fundamental principle that should serve as a guideline to the applicability of either an order-point or MRP system is the concept of dependent versus independent demand.

Demand for a given inventory item is termed *independent* when such demand is unrelated to demand for other items—when it is not a function of demand for some other inventory item. Independent demand must be forecast unless there is sufficient order backlog to cover the planning and execution lead times.

Conversely, demand is defined as *dependent* when it is directly related to or derives from the demand for another inventory item or product. This dependency may be *vertical*, such as when a component is needed to build a subassembly or product, or *horizontal*, as in the case of an attachment or owner's manual shipped with the product. This principle was formulated originally by Joe Orlicky in 1965. In most manufacturing businesses, the bulk of the total inventory is in raw materials, component parts, and subassemblies, all largely subject to dependent demand. Such demand, of course, can be calculated. Dependent demand need not and should not be forecast because it can be precisely determined from the demand for the items that are its sole cause. These vertical and horizontal dependencies can be leveraged to shorten production times dramatically. This is further described later in this chapter.

## Order-Point Characteristics

Order-point theory makes five basic assumptions:

1. Independent demand can be forecast with reasonable accuracy.
2. Such forecasts will account for all demands.
3. Safety stocks will protect against forecast errors and unexpected events.
4. Demand will be fairly uniform in the short-term future and a small fraction of reorder quantities.
5. It is desirable to replenish inventories when they are depleted below the order-point quantity.

Forecasts of demand for components in manufacturing are most often derived from each item's past usage (intrinsic forecasts), rarely from finished product or other external demand (extrinsic forecasts). Very few people try to forecast demand separately from each product for an item common to several products. Demand forecasting determines only the average amount of demand expected in future time periods, not the specific timing of specific demands.

When computers and applicable software became available, more sophisticated applications of the order-point approach could track actual demands and compare them with forecasts (also updating these periodically) to indicate the probability of actual demands exceeding the forecast. Statistical techniques then could be applied to calculate an amount of inventory (usually called *safety stock*) that would ensure achieving a desired level of "service," meaning some minimum number of stock-outs. Again, this approach works only if the past is an accurate indicator of the future. In today's volatile climate, this is a rarity.

Forecasting is inseparable from order-point techniques. All forecasting (intrinsic as well as extrinsic) attempts to use past experience to determine the shape of the future. Forecasting succeeds only to the extent that past performance is repeatable into the future. In a manufacturing environment, however, future demand for a given part may be quite unrelated to its past demand. Forecasting therefore should be the method of last resort, used only when it is not possible to extract, determine, or derive demand from something else. In cases of dependent demand, forecasting is unnecessary because dependent demand is, by definition, derivable and calculable.

How well order-point inventory planning works depends on how closely the assumptions relate to the actual situations in the inventory. Introduced with manual calculations in the 1940s, order-point systems were an improvement over the earlier crude guesstimate approaches. The enhancements made possible by computers further improved their performance, but the fallacious basic assumptions defeated their users in getting tight control over manufactured inventories.

The order-point assumption of fairly uniform usage in small increments is invalid for manufacturing inventory. Requirements for components of products are anything but uniform; depletion occurs in discrete "lumps" caused by parent-order lot sizes. The example in Figure 4-1 shows this clearly; in it, order point is being used for all items. These could be a box wrench, the rough forging it is made from, and the forging steel. Wrenches and more complex products are not made in quantities of one piece, of course, but in reasonably sized lots. When an order is placed on the factory to produce a quantity of such an end item, it is necessary to withdraw from inventory corresponding quantities of the components; this will deplete component inventories and at some time drive them below the order points. When it does (as at the end of July in the example), the technique will act immediately to reorder the components, necessitating a large withdrawal of raw materials to produce their order quantities. If the raw materials order points are then "tripped," these materials also will be reordered immediately.

In this example, demands for the components and raw materials show marked discontinuity, causing several serious problems when an order-point approach is used for manufacturing inventory:

1. Average inventory levels are considerably higher than one-half the replenishment lot size plus safety stock, which the order-point theory commonly assumes. The shaded areas clearly demonstrate these high inventory levels.

2. The order-point system reorders prematurely, far in advance of actual need, and excess inventory will be carried for significant periods of time.

3. The schedule dates on the replenishment orders are wrong, and credibility of the system will be low.

4. Scarce capacity and materials will be applied to the wrong items.

Table 4-1 summarizes the characteristics of order-point and MRP systems.

## Mixed Independent and Dependent Demand

In a manufacturing environment, a given inventory item (subassembly, component part, or raw material) may be subject to dependent demand exclusively, or it may be subject to both dependent and independent demand. This can be further complicated with complementary products. For example, every roller chain needs at least two sprockets. But

**FIGURE 4-1**

Order point and
dependent
demand.

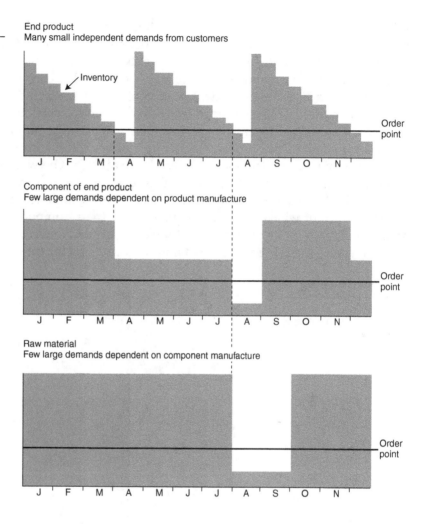

End product
Many small independent demands from customers

Component of end product
Few large demands dependent on product manufacture

Raw material
Few large demands dependent on component manufacture

**TABLE 4-1**

Comparison of Order-Point and MRP Systems

|  | Order point | MRP |
|---|---|---|
| Deals with | Parts | Products |
| Looks to the | Past | Future |
| Uses | Averages | Batches |
| Requires bills of material? | No | Yes |
| Inventory is | Maintained | Run out |
| Recommends order dates to | Start | Complete |
| Activated for an order | Once | Periodically |
| Shows future orders | None | All in horizon |
| Can be replanned? | No | Yes |

roller chain is manufactured on assembly lines, whereas sprockets are manufactured in job shops. And the length of chain is not directly related to the size of the sprockets. Furthermore, chain and sprockets can be purchased from the same or different manufacturers. Such mixed demand arises in cases of parts used in current production as well as spare-part service. The independent portion of the total demand then has to be forecast and added to the (calculated) dependent demand. Service parts no longer used in current production are subject to independent demand exclusively, and this demand is properly forecast provided that history is an accurate indicator of the future.

In manufacturing operations, the typical relationship between individual items that make up the inventory is as depicted in Figure 4-2. Material conversion stages create the relationship between raw material, semifinished part, component part, subassembly, and assembly, each of which carries a unique identity (part number) and as such represents an inventory item in its own right that must be planned and controlled. Demand for all these inventory items is being created internally as a function of scheduling the next con-

**FIGURE 4-2**

Material
conversion
stages.

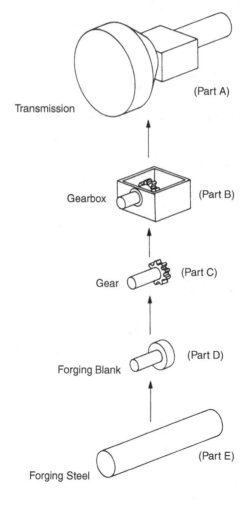

Transmission                                    (Part A)

Gearbox                                          (Part B)

Gear                                             (Part C)

Forging Blank                                    (Part D)

Forging Steel                                    (Part E)

version stage. In the example, (purchased) steel is made into a forging blank, which, in turn, is machined into a gear that then becomes one of a number of components used in assembling the gear box, a major component of a transmission. The transmission will be required for the building of some end-product vehicle, which is also an assembly.

The demand for the end product, it should be noted, may have to be forecast if insufficient backlog of orders is on hand for the end product. But none of the component items, including raw materials, need to be forecast separately. When someone manufactures wagons, for instance, he or she may have to forecast how many he or she will sell and when. Having done this, however, the manufacturer need not forecast the wheels because he or she knows that there are four wheels per wagon. This seems elementary, but the point is that the wagon wheels can be forecast independently, and the most sophisticated statistical techniques can be employed for this purpose. Some manufacturing companies do, in effect, just this. The results, of course, are bound to prove disappointing.

Min/max stock replenishment systems or kanbans are oblivious to the relationship between inventory items and to their dependence on one another. This order-point approach looks at the demand behavior of every inventory item as though it had a life of its own. This, however, is a totally false premise in a manufacturing environment.

Statistical forecasting, on which order-point systems depend, addresses only the problem of individual item demand magnitude, but for purposes of manufacturing, an added requirement is that component inventory represent matched sets. When components are forecast and ordered independent of each other, their inventories will tend not to match assembly requirements, and the cumulative service level will be significantly lower than the service levels of the parts taken individually. This is caused by the adding up of individual forecast errors of a group of components needed at one time to make an assembly.

If the probability of having one item in stock at a time of need is 90 percent, two related items needed simultaneously will have a combined probability of only 81 percent ($0.9 \times 0.9 = 0.81$). At 10 such items, the odds are against all of them being available (34.8 percent probability of all being available simultaneously). Even with the service level set slightly higher at 95 percent, the probability of simultaneous availability for 10 different component items is less than 60 percent, and at 14 items, it drops below 50 percent, as shown in Table 4-2.

These combined probabilities make it evident that when an assembly is to be built from 20 or 30 different components (a proposition not unrealistic—it would hold roughly true for the gearbox in the preceding example) ordered by an order-point system; the lack of a shortage (and the lack of a need to expedite) actually would be a fluke. Note that these shortages are not caused by some unforeseen events but are generated by the process itself.

## "Lumpy" Demand

Another dimension of demand to be considered is its relative continuity and uniformity. Order point, as already mentioned, assumes more or less uniform usage, in small incre-

**TABLE 4-2**

Probabilities of Simultaneous Availability

| Number of component items | Service level | |
|:---:|:---:|:---:|
| | 90% | 95% |
| 1 | 0.900 | 0.950 |
| 2 | 0.810 | 0.902 |
| 3 | 0.729 | 0.857 |
| 4 | 0.656 | 0.814 |
| 5 | 0.590 | 0.774 |
| 6 | 0.531 | 0.735 |
| 7 | 0.478 | 0.698 |
| 8 | 0.430 | 0.663 |
| 9 | 0.387 | 0.630 |
| 10 | 0.348 | 0.599 |
| 11 | 0.313 | 0.569 |
| 12 | 0.282 | 0.540 |
| 13 | 0.254 | 0.513 |
| 14 | 0.228 | 0.488 |
| 15 | 0.206 | 0.463 |
| 20 | 0.121 | 0.358 |
| 25 | 0.071 | 0.260 |

ments of the replenishment lot size. The underlying assumption of gradual inventory depletion at a steady rate will render the technique invalid when this basic premise is grossly unrealistic. In a manufacturing environment, where we deal with components of products, requirements typically are anything but uniform and depletion anything but steady.

Inventory depletion tends to occur in discrete "lumps" owing to lot sizing for subsequent stages of manufacture. The example in Figure 4-1 shows this clearly. Here, the end item, its component, and the raw material are all on order point. These could be a simple wrench, the rough forging that it is made of, and the forging steel. Or they could be the transmission (if it were a shippable end product), the gearbox, and the gear from the preceding example.

Wrenches (or transmissions) are not made in quantities of one. When an order is placed on the factory to produce a quantity of the end item (perhaps to replenish its stock), it is necessary to withdraw a corresponding quantity of the component. This will deplete the inventory of the component in one sudden stroke, sometimes driving it below the order point. When it does (as at the end of July in the example), the system will immediately reorder, necessitating a large withdrawal of the raw material. If its order point is thereby "tripped," this material is also reordered.

In the example, demand for the inventory items component to the end product shows marked discontinuity. Their average inventory level is considerably higher than the conventional projection of one-half the replenishment lot size plus safety stock. The order-point system reorders prematurely, way in advance of actual need, and therefore excess inventory is being carried for long periods of time when there is no real need for it.

The phenomenon of discontinuous demand illustrates the problem of timing of requirements. Inventory management literature largely concerns itself with problems of quantity, whereas in the real world of manufacturing the question of timing, rather than quantity, is of paramount importance. Order point only implies timing, based as it is on average (past) usage. But average usage data are, for all practical purposes, largely meaningless in an environment of discontinuous, dependent demand.

The example in Figure 4-1 shows graphically that order point, which essentially assumes continuity of demand subject only to random fluctuation, consequently assumes also that it is desirable to have at least some inventory on hand at all times and a need to replenish inventory as soon as it is depleted. When such inventory is subject to discontinuous demand, this is not only unnecessary but also undesirable because it causes inflation of the inventory level.

All three of the inventory items in Figure 4-1 are on order point, but this is not what causes discontinuity of component demand. It occurs even in the absence of order point because it is caused, as mentioned previously, by lot sizing at the various stages of manufacture. Where a given component item is subject to dependent demand from multiple sources (a "common" component), the demand pattern is not only discontinuous but also nonuniform; that is, the sizes of the lumps tend to be irregular. This is illustrated in the next example (Figure 4-3).

Here, no order points are involved, and short lead times allow production to be closely geared to demand but not so closely as to equal it period by period. Each item is being produced in a different lot quantity, selected for reasons of economy or convenience. In the example, item Z, a unit (pounds or feet) of steel, is used to produce two different forgings, and each of these forgings, in turn, is used to produce two different finished products, simple wrenches. In each of the simplified records of Figure 4-3, the

**FIGURE 4-3**

Causes of lumpy demand.

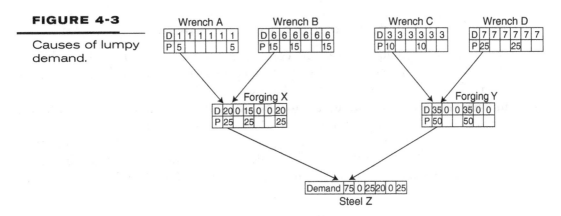

letter *D* stands for demand and *P* for production. Each record extends six periods into the future.

Because forgings will be needed every time a lot of one or the other of the various wrenches is to be produced, it is possible to construe the demand on the respective forgings by combining the wrench production lot quantities in the corresponding time periods. The forgings are produced in lot quantities of their own, and the demand for the common component material is derived in the same fashion. For simplicity's sake, the unit of measure for the steel is assumed to be equivalent to one forging.

Demand for the finished wrenches in the example is shown to be perfectly steady and level, which is hardly realistic but was chosen to highlight the fact that even with continuous and uniform demand for end products, demand for components will tend to be discontinuous and jagged. Lumpiness of demand is, of course, relative to the period chosen. By increasing the size of the period (from a week to a month, from a month to a quarter, etc.), demand measurements will show more continuity and uniformity. However, in today's climate, where lead times are being shortened and customers are more demanding, planning in larger time buckets just to improve uniformity is highly unrealistic.

As a matter of fact, in our example, even the demand for item Z, the steel, can be completely "stabilized" by increasing the size of the period 50-fold. Thus, if the period in the example represents one week, demand will turn out to be a perfectly uniform 850 units per each 50-week cycle. Such stabilization, however, has been achieved with the proverbial "mirrors," and the information is not very useful because the period spanned by the measurement is simply too large. It is the short-range lumpiness of demand that is significant because manufacturing inventory and production must be planned and controlled from day to day and from week to week.

In the example of Figure 4-3, successive period demands for item Z are 75-0-25-50-0-25. Note what may not be readily apparent, namely, that the average demand for item Z equals 17 per period. One unit of steel is consumed in the production of each wrench, and in each period, the total demand for the four wrenches is $1 + 6 + 3 + 7$. The wrenches can be forecast easily, but period demand for steel does not lend itself to statistical forecasting at all.

## Material Requirements Planning

MRP is the preferred technique to use when one item is a component of another and thus subject to dependent demand. The MRP approach does not rely on a forecast of dependent item demand and thus avoids the problems touched on in the preceding discussion. Its techniques are expressly designed for dealing with dependent, discontinuous, nonuniform demand, which is characteristic of manufacturing environments. It is forward-looking, not history-driven, like order point. MRP develops a valid inventory plan that can be replanned to keep it up to date. MRP is not without its own problems, of course. These will be understood better after the mechanics of MRP and its related techniques, covered in the balance of Part 2, are known.

When MRP was introduced, revolutionary changes occurred, and new premises were established. Orthodox practices and techniques were challenged. The existing inventory control literature, indeed an entire school of thought, had to be reexamined. It became evident that the basic tenet of old inventory control theory—that low inventory investment and high customer service were incompatible—was mistaken; successful users of MRP programs reduced inventories and improved delivery service at the same time.

Notwithstanding the difficulties experienced by many companies attempting to implement MRP-based planning, MRP has been the most successful innovation in manufacturing inventory management. It has demonstrated conclusively its operational superiority over alternatives in many companies in many industries in many countries. In addition, it has provided students of inventory management with new insights into manufacturing planning and control. MRP shows the true interrelationships among activities affecting inventories and reveals the fallacies of previous concepts, pinpointing the causes of the inadequacies of the older methods. The principles on which MRP systems are based are the subject of Chapter 5.

## THE PARADOX OF INVENTORY MANAGEMENT

A remarkable truth about inventory management is that if you cannot match inputs to and outputs from an inventory, you will never control it. The better you can match inputs and outputs, the less need there is for any inventory.

Inputs to inventory are receipts of purchased materials from suppliers and completed production of parts, subassemblies, and assemblies. Outputs are deliveries to customers, external and internal, issues of manufacturing processes, and scrap.

People concerned with inventory and its management include top-level managers and people in sales and marketing, purchasing, materials planning and control, and production activities and cost accountants. Until very recently, most of these people were involved with outputs or inputs separately or thought of them individually, not recognizing their interactions. They believed that they were victims of the environment.

Many believed that Murphy's law—"What can go wrong will go wrong at the least opportune time"—was written specifically about production. Problems abounded. Customers changed their minds about what they wanted, how many, and when they wanted them. Designs were faulty and changed frequently. Suppliers were late and often delivered defective materials. Machinery and tooling broke down. People were absent and made defective things when they worked. The consensus was that these problems were unsolvable.

This has now been proven false. Most of these problems can be solved and eliminated. Those which cannot be eliminated can be minimized. There is now little excuse for the "surprises" blamed on manufacturing problems in the past.

A major objective of this book update is to show the important role MRP can play in helping to achieve a better balance between demand and supply. The paradox stated earlier has made it clear how we should answer the question, "How much inventory is enough?" The answer is, "How much is needed because inputs and outputs don't match?" Even more important is the question, "Where can inventory be leveraged to compress overall response time?"

## Inventory: Asset or Liability?

Financial managers have always challenged amounts of manufacturing inventories. Although they are assets on the balance sheet, executives and top-level managers have viewed and treated them as liabilities. Improving inventory turnover has been the perennial goal of management. Only finished-product inventories have been accepted, albeit grudgingly, as necessary to serving customers under the belief that more inventory will yield better customer service.

This abhorrence of inventory now is seen as being eminently correct but for different reasons! In itself, some inventory is needed in manufacturing; some is even beneficial, earning an adequate return on the investment. Such inventory is an asset, but in most firms it is a very small fraction of the total.

One of the most important performance measures of the overall health of a manufacturing business is thought to be inventory turnover ratios. In the United States before 1980, these ranged from below one to as high as six. Management thought they were doing well to increase the figure by 50 percent in one year. Many who did so simply fell back to previous ratios the next year, indicating successful crisis management and actions but no permanent or sustainable improvement in performance. Many found that lowering inventories harmed customer service and caused higher costs, indicating a lack of understanding of how manufacturing should work and failure to use sound planning and control. With the advent of lean techniques, inventory turns commonly are in the double or triple digits. But is inventory turnover always an indicator of excellent or optimal inventory performance? High turns and high shortages can coincide. This happens frequently when companies lean out too much and reinforces the inherent dilemma with inventory.

In most manufacturing environments, stock in some form is a requirement. As mentioned previously, a primary reason to hold inventory is that customer tolerance times are shrinking. Customers will no longer tolerate long lead times. However, most manufacturing companies and certainly every supply chain cannot be a pure make-and-purchase-to-order system. Would you wait at the grocery store for a quart of milk if you knew the cow had not even been milked? What about at the gas station if the oil had not yet been drilled? Holding inventory is a reality in the modern world. In most cases, companies cannot address the issue effectively because they have only antiquated stock practices and tools.

Carrying inventory is a requirement. Inventory is waste only if it is located in the wrong places and in the wrong quantities. The key is to determine first where the right places are to stock and only then to determine the amounts to be stocked. Next, the process must allow those places and amounts to change as the environment and conditions change. The effective management of inventory is a dynamic closed-loop process.

This is necessary to leverage the working capital and/or capacity commitment inherent in inventory effectively to maximize the company's overall financial performance.

At the same time, it is also extremely wasteful to not carry inventory. When companies lean out too much inventory, they frequently experience shortages. When companies experience shortages, they are forced to spend additional time, effort, money, and capital to make up for it, and they can miss significant market opportunities. The question to ask is if all inventory is waste? We don't think so. Inventory can be a waste under two conditions. First, when there is not enough inventory in the right place there are disruptions and flow breaks down. Second, when there is too much inventory in the system, lead times expand as materials and capacity are tied up, expedites begin and, once again, flow breaks down. Effectively positioning and managing inventory (planning and execution) is vital for flow. Minimizing the cash and capacity we have consumed in inventory while promoting flow is vital for good return on capital.

Agility is not synonymous with zero inventories. Remember, the key to effectively leveraging the working capital and capacity commitment inherent in inventory is to find the places where that inventory can make the biggest impact and therefore provide the greatest return. Inventory can decouple otherwise dependent events so that the cumulative effects of variation are not passed and/or amplified between the dependencies. Thus inventory can be a breakwall against the variability experienced from either supply (externally and internally) or demand variability. However, as with any breakwall, it is effective if it is placed and sized properly.

Today's companies must think systematically across the supply chain and not just within their own four walls. Putting inventory everywhere is an enormous waste of company resources. Eliminating inventory everywhere puts the company and supply chain at significant risk. Strategically positioning inventory ensures the company's ability to absorb expected variability with the smallest possible investment. Unfortunately, today, most tools, training, and educational material are oriented toward determining the answer to the questions "How much?" and "When?" with little to no attention to answering "Where?" Properly determining where to place inventory is a strategic, concerted, and constant effort that should involve key personnel that represent a relevant crosssection of the company. It is primary and necessary to managing the investment in inventory effectively for the best return. There are six critical positioning factors in determining where to properly place inventory.

## The Critical Positioning Factors

- *Customer tolerance time.* The time the typical customer is willing to wait. Customer tolerance time also can be referred to as *demand lead time*. The *APICS Dictionary*[1] defines *demand lead time* as "the amount of time potential customers are willing to

---

[1] *APICS Dictionary*, 12th ed. (New York: Blackstone, 2008), p. 32.

wait for the delivery of a good or a service. Syn: customer tolerance time." Determining this lead time often will take the active involvement of sales and/or customer service.

■ *Market potential lead time.* The lead time that will allow an increase of price or the capture of additional business either through existing or new customer channels. Determining this lead time will take the active involvement of sales and/or customer service. Be aware that there could be different stratifications of market potential lead time. For example, a one-week reduction in lead time may only result in an increase in orders, whereas a two-week reduction in lead time could result in both an increase in orders and a potential price increase on some of those orders. Properly segmenting the market will prevent giving too much away at any one time.

■ *Variable rate of demand.* The potential for swings and spikes in demand that could overwhelm resources (capacity, stock, cash, etc.). This variability can be calculated by a variety of equations or determined heuristically by intuitive planning personnel. "Mathematically, demand variability or uncertainty can be calculated through standard deviation, mean absolute deviation (MAD) or variance of forecast errors."[2] If the data required for mathematical calculation do not exist, companies can use the following criteria:
  • High-demand variability = products/parts that are subject to frequent spikes
  • Medium-demand variability = products/parts that are subject to occasional spikes
  • Low-demand variability = products/parts that have little to no spike activity (their demand is stable)

■ *Variable rate of supply.* The potential for and severity of disruptions in sources of supply and/or specific suppliers. This can also be referred to as supply continuity variability. This can be calculated by examining the variance of promise dates versus actual receipt dates. When first considering the variable rate of supply, the initial variances can be caused by critical flaws in how the MRP system has been deployed. Additionally, those dates often shift owing to other shortcomings associated with the way MRP is employed rather than because of the supplier. Any critical supplier of a major manufacturer will know which day its customer regenerates its MRP. These suppliers will see a flurry of additional orders, canceled orders, and changes to orders (quantity, specification, and request date). If the data required for mathematical calculation do not exist, the following heuristics can be used:
  • High-supply variability = frequent supply disruptions
  • Medium-supply variability = occasional supply disruptions
  • Low-supply variability = reliable supply

---

[2] Ibid, p. 144.

- *Inventory leverage and flexibility.* The places in the integrated BOM structure (the matrix BOM) or the distribution network that leave a company with the most available options as well as the best lead time compression to meet the business needs. Within manufacturing, this is typically represented by key purchased materials and subassemblies and intermediate components. This becomes more important because the BOMs are deeper, more complex, and have more shared components and materials. This concept will be explored more fully in Chapter 23. The critical component of determining these points is called actively synchronized replenishment (ASR) *lead time*. Within distribution networks, this typically requires realignment of the network to a hub-and-spoke model. This is covered in more detail later in this chapter.
- *Protection of key operational areas.* It is particularly important to protect critical operational areas from disruption. In lean, these areas are called pace-setters. In theory of constraints, they are referred to as *drums*. Whatever manufacturing or operational methodology a company ascribes to, these resources typically represent control points that have a huge impact on the total flow or velocity that a particular plant, resource, or area can maintain or achieve. Similar to the BOM, the longer and more complex the routing structure and dependent chain of events (including interplant transfers), the more important it can be to protect these key areas. These types of operations include areas that have limited capacity or where quality can be compromised by disruptions.

The preceding six factors must be applied systematically across the entire BOM, routing structure, manufacturing facilities, and supply chain to determine the best positions for purchased, manufactured, and finished items (including service parts). The bigger the system these factors are applied to, the more significant the results can be. Later in this book we will examine the impact of the solution on an integrated supply chain.

Purchased parts chosen for strategic buffering tend to be critical and/or strategic parts and long-lead-time items. Typically, this will be less than 20 percent of purchased parts. Manufactured parts chosen for strategic buffering are often critical or strategic manufactured and/or service parts and at least some finished items and critical subassemblies. Typically, this will be under 10 percent of manufactured parts (for some environments with a significant number of manufactured service parts, this percentage could be higher). On the fulfillment side, most parts will be strategically buffered. This is why the distribution warehousing positions exist. This is covered more fully in Chapter 24.

Figure 4-4 is an example of a supply chain for one product called Finished Product A (FPA) after the positioning has been determined. Note that the "bucket" icon represents strategically replenished positions. Four of the ten purchased components are buffered. Three of the ten subassembly/intermediate-component positions are buffered, as well as the finished product itself. Finally, the stock positions of FPA in all three regional warehouses are buffered. Remember, the example is in relation to one end item only.

The position of the strategic buffers is accomplished through a combination of business rules and technology. The business rules are the application of critical positioning

## FIGURE 4-4

Strategic stock positions (FPA example).

## Purchasing

Critical and long lead time items.

**Supplier 1**

**Supplier 2**

**Supplier 3**

## Operations

Critical manufactured parts, sub-assemblies and finished stock.

**Bill of Materials**

Purchased Parts List

PPE
PPG
PPB
PPD
PPA
PPI
PPC
PPJ
PPF
PPH

## Fulfillment

The distribution and management of finished stock.

**Region 1**

FPA

**Region 2**

FPA

**Region 3**

FPA

factors in consideration of the overall business objectives and operating rules. This is accomplished by the people who have experience and intuition in the environment. In complex environments, technology must be leveraged to perform the necessary computations, including analyzing the product structure, qualified cumulative lead times, and shared components across the integrated BOM. The importance of carefully considering inventory positioning cannot be underestimated. Without the right strategic positioning, no inventory system can perform to its potential.

Properly positioned inventory is truly an asset, and as with any other asset, the company must expect and demand a return on that asset. This begins with sound planning and ends with tight control. MRP is the core of modern integrated planning and control systems. It can be the key to leveraging the inventory investment for the highest return.

As an example of buffer positioning, first consider a simple environment, a company that produces two products—Part 300 and Part 400. Figure 4-5 shows the BOM for these parts. Note that both end items use a common manufactured subcomponent labeled "Part 200."

Adjacent to each part is the lead time associated with each discrete part number. Manufactured parts have two lead times for each part: a *manufacturing lead time* (MLT) and a cumulative lead time (CLT). Purchased parts have a *purchasing lead time* (PLT) associated with them. The *APICS Dictionary* defines these lead times as:

**manufacturing lead time (MLT):** The total time required to manufacture an item, exclusive of lower level purchasing lead time. For make-to-order products, it is the length of time between the release of an order to the production process and shipment to the final customer. For make-to-stock products, it is the length of time between the release of an order to the production process and receipt into inventory. Included here are order preparation time, queue time, setup time, run time, move time, inspection time, and put-away time.

**FIGURE 4-5**

Bills of materials for Part 400 and Part 300.

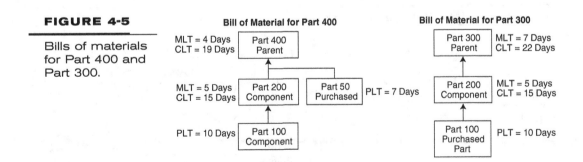

**cumulative lead time (CLT):** The longest planned length of time to accomplish the activity in question. It is found by reviewing the lead time for each bill of material path below the item; whichever path adds up to the greatest number defines cumulative lead time.

**purchasing lead time (PLT):** The total lead time required to obtain a purchased item. Included here are order preparation and release time; supplier lead time; transportation time; and receiving, inspection, and put away time.[3]

Part 400's MLT is 4 days, and its CLT is 19 days. The CLT is determined by the path Part 400 → Part 200 → Part 100. This is the longest total path in the BOM. The 19 days is calculated by summing all MLTs and the longest-lead-time purchased part (Part 100) in that CLT sequence. Part 300's MLT is 7 days, and its CLT is 22 days. Part 300 has a straight-line BOM, so determining the CLT is much easier because it is the only path in the BOM. Please note that in this example the parent-to-component usage in all BOM relationships is 1:1.

A *routing* is "information detailing the method of manufacture of a particular item. It includes the operations to be performed, their sequence, the various work centers involved, and the standards for setup and run."[4] Together, the BOM and the routing paint a relatively complete planning picture for the total requirements (materials and capacity) for all parent items and the part's contribution to the total requirements scenario (capacity and materials).

In this example, no run rates and setup times have been defined. These are not relevant for this simple example. As demonstrated in Figure 4-6, in order to make Part 400, Part 100 must be purchased and introduced into the process. This takes ten days to accomplish. The work in process passes through a sequence of three resources (A → B →

**FIGURE 4-6**

Routings for Part 300 and Part 400.

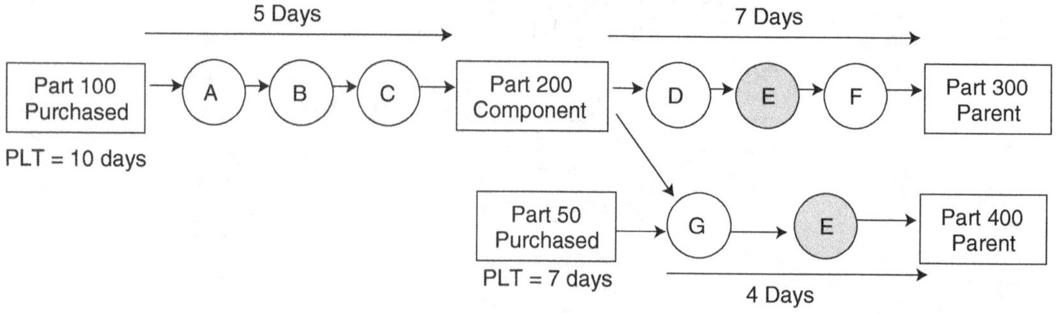

[3] Ibid, p. 112.
[4] Ibid, p. 121.

C) to transform it into the component Part 200. This can be accomplished reliably in five days. This reliability can be affected by finite capacity, quantity, quality, other orders, machine breakdown, etc. For this simple example, the assumption is made that the process can be completed reliably in five days.

Part 200 is a point of divergence. A *divergent point* means that Part 200 can be directed into different manufacturing paths. A divergent point represents a commitment that cannot be practically or cost-effectively reversed. An example would be the introduction of a log into a plywood facility, in which the first step is to peel the log into veneer. This decision precludes the log from being used or redirected into dimensional lumber. For this example, Part 200 can be directed to resource D, where it will then pass through resources E and F to be made into Part 300. This can be accomplished reliably in seven days. Part 200 can also be combined with purchased Part 50 at resource G. Resource G processes this work and forwards it to resource E to complete the process of producing parent Part 400. This can be accomplished reliably in four days.

In this example, resource E is a critical resource that should be carefully protected owing to its importance in the overall process. In this case, that protection must minimize disruption that is passed to it from preceding areas. This is a consideration for our stocking strategy. Resource E has been shaded to denote its special status.

Consider Figure 4-7 with no stock held at any level. This would be a complete make-and-purchase-to-order environment. The CLT paths for both Part 300 and Part 400 are laid over the routing. As mentioned earlier, in the case of Part 300, the CLT is 22 days. The 22-day lead time is calculated by adding the PLT of Part 100 (10 days) and the total time it takes to move from resource A through resource F (12 days). In Figure 4-8, the case of Part 400, the path is also defined by the Part 100 leg of the BOM path because that is the longer of the two legs.

Given the current competitive position, the customer tolerance time must be considered. In this example, the company can have zero inventory if the customer tolerance

## FIGURE 4-7

Cumulative production lead time for Part 300 with no stock.

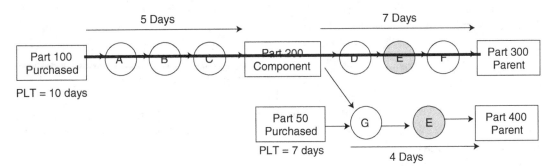

**FIGURE 4-8**

Cumulative production lead time for Part 400 with no stock.

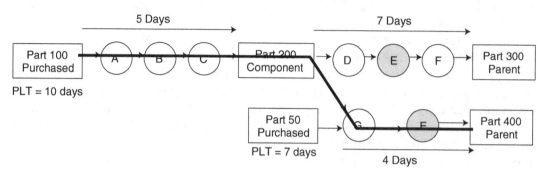

time is greater than 22 days for Part 300 and greater than 19 days for Part 400. In both cases, the assumption is made that 22 and 19 days, respectively, are reliable lead times. To have reliable lead times, the suppliers must have excellent reliability, and the internal lead time approximation must be generous enough to experience very few violations.

However, if customer tolerance is less than 22 days for Part 300 and less than 19 days for Part 400, then this company faces a choice. Either it must compress the lead times or risk losing customers. Traditionally, this has meant the attempt to forecast the independent demand and keep the finish good parts on hand or, alternatively, the processes within manufacturing must get much quicker. Unfortunately, most demand-driven staff and consultants would be quick to attack the internal processing times first. While this can be a valid point, this focus seldom results in an immediately effective significant lead time compression solution. In reality, the attempt to make major process improvements instead can bring an increase to customer-related risks in the immediate term. Instead, a well-designed inventory strategy often will stabilize the manufacturing environment so that it is easier to work on process improvement. This, in turn, reduces the required inventory commitments. In many environments, there is a proven sequence to safely, quickly, and cost-effectively compress lead times. Such a sequence starts with sensibly placed and properly sized inventories.

Small stocks of the correct inventory can have an immediate impact on lead-time compression. In Figure 4-9, both parent items can benefit by stocking purchased Part 100. Figure 4-9 shows the insertion of a stock buffer at purchased Part 100, denoted by the "bucket" symbol. In the case of Part 300, 10 days of lead time will be removed as the PLT gets subtracted from the longest sequence. Now the cumulative lead time for Part 300 is 12 days. In Figure 4-10, stocking Part 100, however, removes only eight days of lead time from Part 400 because the longest sequence shifts over to a path terminating in Part 50's purchasing lead time. Part 400's compressed lead time is now 11 days.

The benefits of the Part 100 position is that both parent item lead times are compressed. In this case, this strategy will work if the customer tolerance times are greater than 12 days for Part 300 and greater than 11 days for Part 400. In addition, possible supplier variability relative to Part 100 has been removed or minimized.

**FIGURE 4-9**

Part 300 compressed lead time—first iteration.

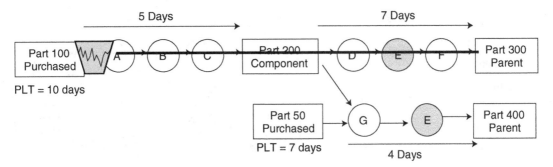

**FIGURE 4-10**

Part 400 compressed lead time—first iteration.

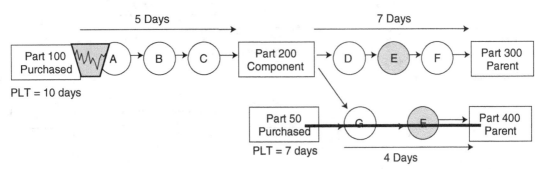

The potential risk for this strategy is that critical resource E is left relatively unprotected. Resource E can be disrupted by two different factors not related to its own operation. First, it is left susceptible to a long string of internal resource-related dependencies characterized by the path A → B → C → G. Second, it is also vulnerable to supplier issues with regard to purchased Part 50. Remember that resource G (the preceding resource to critical resource E) is a subassembly that requires both a Part 200 and a purchased Part 50.

In carrying this example forward, suppose that customer tolerance time is less than 12 days for Part 300 and less than 11 days for Part 400. This company will have to determine an inventory strategy that will compress lead times further. A possible choice would be to stock Part 200 to decouple the front end of the manufacturing processes from the back end. Figures 4-11 and 4-12 show what stocking at Part 200 will do for both finished items.

Surprisingly, by placing a stock buffer only at Part 200, there is no compression of Part 400's lead time. Part 200 does not lie on Part 400's CLT path. As mentioned earlier, if the customer tolerance time for Part 400 is within 11 days, this choice would not yield the

**FIGURE 4-11**

Part 300 compressed lead time—second iteration.

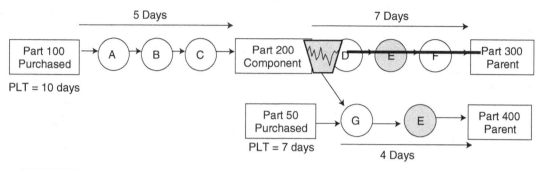

**FIGURE 4-12**

Part 400 compressed lead time—second iteration.

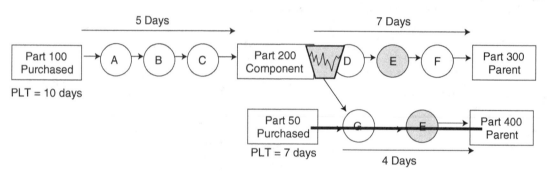

desired benefit. The only way to compress the Part 400 lead time further is to also place a stock buffer at the Part 50 position. This is depicted in Figure 4-13.

The benefits of this strategy include the following:

- Leads times are significantly compressed. In this case, this strategy will work if the customer tolerance times are greater than seven days for Part 300 and greater than four days for Part 400.
- Supplier variability relative to both Part 100 and Part 50 now has been removed from the lead-time-to-market equation.
- Critical resource E is now better protected from all upstream disruptions.
- Stocking at Part 200 allows that inventory to be leveraged against both Part 300 and Part 400 requirements.

The potential risks include the following:

- Without a buffer at Part 100, the front side of the plant is subject to supply disruption, which could affect the scheduled activity of these resources.
- Without a buffer at Part 100, the stock position at Part 200 has to be sized for a lead time of 15 days.

**FIGURE 4-13**

Part 400 compressed lead time—third iteration.

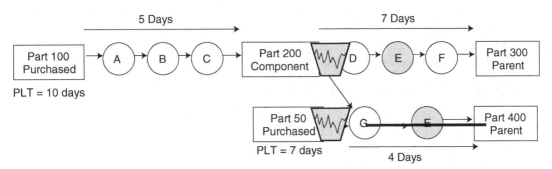

If customer tolerance time is less than seven days for Part 300 and less than four days for Part 400, then inventory positions must be implemented at the end-item level. This would move these items to a make-to-stock strategy, which then limits the product variability that can be offered affordably. Another consideration for stocking at these positions would be a potential impact on sales. If this company were able to offer dramatically shorter lead times by stocking these parts, and if dramatically shorter lead times translate directly into a competitive advantage, then is it is reasonable to assume that sales would grow on these items. A defined customer tolerance time does not mean that customers would not view it as desirable to have the part sooner. Customer tolerance time is established by market competition and only suggests that customers are unable or unwilling to wait longer. There are other suppliers in the marketplace who can deliver to meet their expectations. Market potential lead time indicates the value associated with providing products with customer tolerance time. The companies that can provide products inside the customer tolerance time frequently are rewarded with more business.

**FIGURE 4-14**

Part 300 and 400 compressed lead times—finished items stocked.

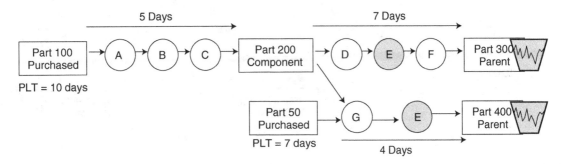

The benefits to this inventory strategy include the following:

- Leads times are fully compressed. In this simple case, this inventory strategy will work if the customer tolerance times are one or more days for Part 300 and one or more days for Part 400.
- Potential sales gains are possible for the sales of both Part 300 and Part 400 if the market will reward this company for shorter than industry standard lead times.

The potential risks include the following:

- Stocking only at the end-item level means that the level of stock needs to cover the entire cumulative lead times of both products (22 and 19 days).
- Stocking only at the end-item level leaves critical resource E exposed to both internal and external disruptions.
- The resources required to have inventory at the end-item level would necessitate lower variety offered to the market.

Another alternative would be to stock at the end items (Part 300 and Part 400) and the purchased items (Part 100 and Part 50). This represents a common picture that has its roots in a traditional MRP safety stock tactic. Safety stock is used to guard against variability in supply and demand. Therefore, the traditional approach would be to have inventory only at the finished good and purchased component level. In a traditional MRP strategy, there is reluctance to stock intermediate components because they can be manufactured internally. Figure 4-15 represents this strategy.

The benefits to this strategy include the following:

- Leads times are still fully compressed. Once again, this stocking strategy will work if the customer tolerance times are one or more days for Part 300 and one or more days for Part 400.
- All internal processes are protected from supplier disruption.

**FIGURE 4-15**

Part 300 and Part 400 cumulative lead time—purchased and end item stocked.

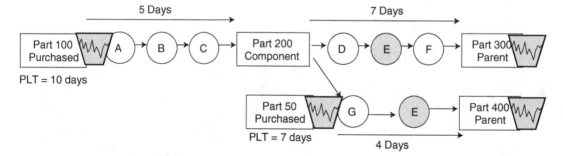

The potential risks include the following:

- End-item stocking levels are subject to the full internal lead time for both parts (12 days for Part 300 and 9 days for Part 400).
- Resource E is left unprotected from a long string of internal dependencies.
- Part 200 is not leveraged against both end-item positions.

Finally, Figure 4-16 considers stocking all part positions in this example. The benefits to this strategy include the following:

- Customer lead times are fully compressed.
- Internal processes are isolated against supplier variability.
- Resource E is best protected from internal dependencies.
- Holding stock at Part 200 allows the stock levels of parent Parts 300 and 400 to be minimized because internal lead times are minimized.
- Holding stock at Part 200 allows that stock to be directed or flowed to the different parent items as the market pulls those items.

The potential risks are that there are more stock positions to manage. This can be more difficult to manage and might be viewed as too much inventory.

Frequently, in the implementation of lean, companies will flatten their BOMs. Flattening a BOM removes a level of the bill to make it more streamlined. Consider what would happen if this company, with the purpose of flattening the BOM, removed Part 200 as a discrete part number, as in Figure 4-17. This is one of the MRP-related compromises described earlier in this book. If Part 200 does not exist as a unique item, then this company loses the ability to:

- Decouple internal lead times for both parents, resulting in higher stock positions for those end items.
- Leverage a common component against multiple end items.
- Protect resource E from a relatively long string of internal dependencies.

## FIGURE 4-16

Part 300 and Part 400 cumulative lead time—all parts stocked.

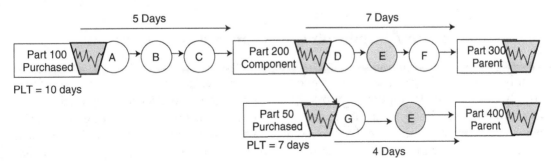

**FIGURE 4-17**

Part 300 and Part 400 cumulative lead time—without Part 200.

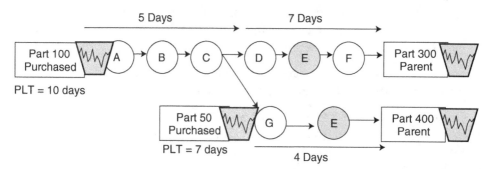

In this one very simple example, a number of different potential scenarios were considered. Obviously, as the complexity of the routings and BOM structures increases, the more potential scenarios there will be. The answer is not to simply place inventory everywhere. This is often unnecessary and can be extremely wasteful. The important thing to note is that the right answer is based on several different factors that vary from company to company and can change over time. Developing the best inventory strategy is an ongoing process with sales, operations, and financial implications. It is an ongoing closed-loop process as conditions relative to the factors named earlier change.

This example began by using manufacturing and cumulative lead times as factors for determining the right position. Within the example, however, a critical point can be realized; there is actually another type of lead time that needs to be recognized, calculated, and made visible. This lead time will be a critical factor in:

- Understanding how to best leverage inventory
- Setting buffer levels properly
- Compressing lead times
- Determining realistic due dates, if needed

This new lead time, called *ASR lead time* (ASRLT), is defined and incorporated into an extensive example in Chapter 23. It represents a dramatic leap forward for positioning and planning visibility.

## Inventory Positioning in Distribution Networks

This book is focused on manufacturing inventory. However, the current manufacturing and supply landscape dictates a discussion about the entire integrated supply chain. Part of that supply chain is the distribution network. Most distribution networks have regional and/or local warehouses holding stock. These locations are constantly attempting to balance between the critical requirement to have what the market requires within the time frame it requires it (usually instantly) and the need to turn or convert inventory into cash/profit (not have too much inventory). In most cases, the primary positioning factor in distribution networks has to do with inventory leverage and flexibility in relation to three facts.

*Fact 1:* Making the right decisions (whether predictions or actions) is inherently easier when the factors being considered are more stable and/or known. The most common factor for distribution networks is demand variability. Figure 4-18 is a depiction of a simple distribution network that has only two levels. One manufacturing plant supplies four regional warehouses. The demand variability is much higher at each discrete distribution location than at the plant for the same time period. The law of total variance means that aggregating the demand variability from the remote locations creates a natural smoothing effect for the manufacturer.

The most interesting observation in this example is to point out that common sense is not common practice. Distribution management and forecasting is typically accomplished at the distribution center level. Only net demand is passed to the manufacturer. This then is extremely lumpy and discontinuous. See the discussion about lumpy demand earlier in this chapter.

*Fact 2:* Holding inventory closer to the source actually protects the largest portion of potential consumption for the least amount of inventory. A distribution network is

## FIGURE 4-18

Simple distribution network and variability of demand.

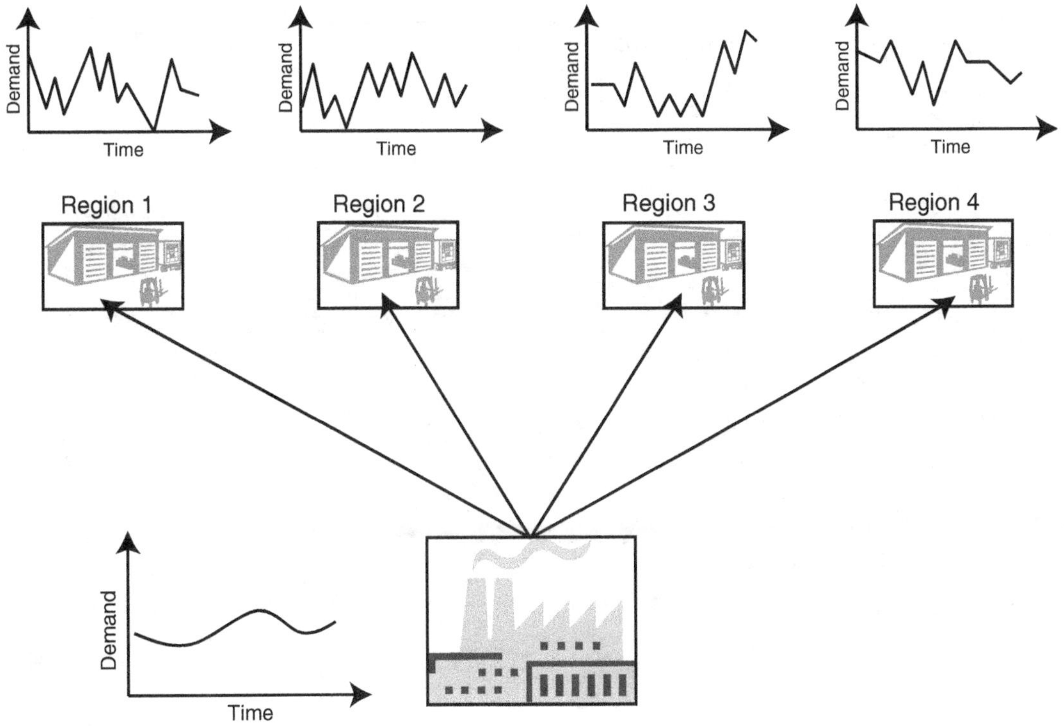

shaped like a V. In Figure 4-18, once inventory is pushed out to region 1, it is unavailable to the other regions in the immediate term. In the longer term, it can be made available, but only through costly realignment activity or cross-shipping.

As demand variability increases, it is harder to predict how much inventory will be required to protect against it. This is especially true when the reliability of the plant lead time is considered as part of the time frame a regional center must protect. If shortage penalties are high, demand variability is relatively high, and the plant lead time is a factor in the plan, then the likely situation at the regional location is either high inventory or frequent shortage. Combine this with the fact that the closer the inventory is to the point of consumption, the better it can meet the market's needs. The net result is that a significant amount of inventory is pushed out into the distribution network. Figure 4-19 shows the distribution network and inventory positions. Note that there is little inventory held at the plant.

This situation often creates circumstances in which one region inevitably does not have enough, whereas others have too much. The result is cross-shipment between the distribution centers, missed potential sales, and expedites placed in the plant's manufacturing schedule. In the aggregate, the system has enough inventory; it is just located at the wrong place. The inventory must be better aligned to:

**FIGURE 4-19**

Simple distribution network and inventory position.

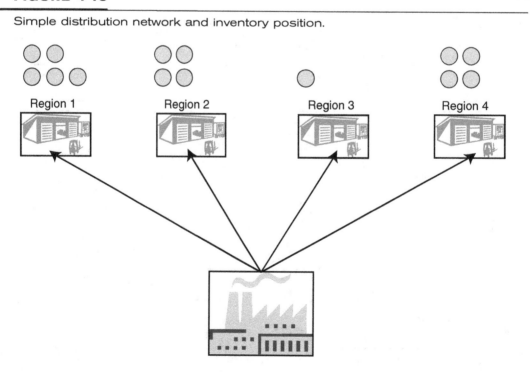

- Ensure that the alignment of inventory is better able to cover all points of consumption.
- Move inventory as few times as possible.
- Limit plant schedule disruptions that reduce capacity and complicate planning.
- Remove the plant's lead time from regional stock level consideration.

Realigning the inventory to locate the majority of it at the source will do all these things. Figure 4-20 is a realigned distribution network showing this realignment of inventory. This realignment comes with an additional bonus of better protection against variability with lower inventories.

*Fact 3:* If the order and delivery cycle is more frequent and reliable, then less inventory is needed at the destination location. The more frequently that supply orders can be launched and received means that time intervals between reliable resupply can be reduced. The common practice of batching conflicts with this more frequent replenishment. An *artificial batch* is any batch that is more than actual demand. Examples of artificial batches will include transportation multiples (frequency and quantity), ordering multiples (frequency and quantity), and any manufacturing multiples (frequency and quantity). The common belief is that the batches are more economical. Some of these artificial batching policies cannot be reduced or eliminated. However, once the realignment of inventory occurs, these policies often become the limitation of better system performance.

**FIGURE 4-20**

Simple distribution network with realigned inventory position.

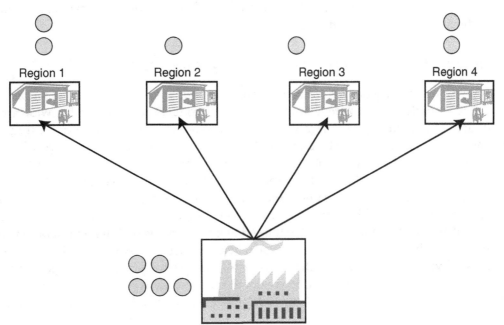

Applying these three facts transforms companies with multiple distribution centers to a type of hub-and-spoke distribution model. With regard to vertically integrated companies that own both distribution and manufacturing locations, it often will make sense to set up the hub or "central buffer" at the manufacturing plant. Figure 4-21 demonstrates how part FPA is held at the plant as a central buffer and fed to regional distribution centers as demand dictates.

This configuration is the complete opposite of how companies frequently operate. Often companies tend to push inventory out into the distribution network, assuming that the closer the majority of inventory is located to market, the better they are able to meet demand. In addition, most companies book their revenue at the point inventory is shipped to distribution. The negative result is all too common in most major distribution scenarios. On average, there is sufficient inventory to cover demand, but it is located in the wrong places. This leads to costly expediting activity, including cross-shipments and/or schedule deviations, expedited materials shipments, and overtime at the plant as the plant attempts to react to false requirements from distribution. In most large companies with both manufacturing and distribution, there is constant tension among planning, manufacturing, distribution, sales, and logistics. This solution, however, is an effective answer to that tension. The insertion of a central buffer:

- Protects regional locations by ensuring a reliable pipeline of supply defined by transportation time only as opposed plant lead time plus transportation time
- Allows consumption and corresponding resupply signals to the plant to be naturally consolidated into batches that are still sensible
- Allows the manufacturing facility to schedule close in time to actual central buffer requirements rather than to frozen schedule horizons that further limit flexibility and create frustration for sales

## A DISTRIBUTION POSITIONING EXAMPLE

This section presents an example of the power of the hub-and-spoke concept in distribution. These data were developed with a sizable consumer products company with a wide array of products ranging from personal care to food. Two product lines were chosen for study. A hub-and-spoke model was constructed, and buffers were sized and managed according to the tactics seen in Part 4 of this book. Figure 4-22 shows the simulated results over a 180-day period using actual demand for a line of cake mixes and compared against actual results.

In this case, one of the company's eight regional warehouses has been selected for the hub or "central buffer." This warehouse was chosen because of both its relative proximity to the plant and its available space. One thing that is apparent is that the central buffer carries the bulk of the inventory: $5,996,940 versus a combined $2,059,152 for all the forward locations. Obviously, the central buffer not only will supply the other seven regional warehouses but also will directly serve its regional customers. In summary, service levels increased, inventory decreased, and cross-shipments were effectively eliminated. All this occurred with no appreciable change in the overall number of plant setups.

# FIGURE 4-21

Supply-chain example of a distribution buffer.

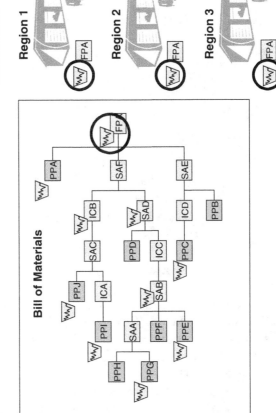

## Purchasing

Critical and long lead time items.

**Supplier 1**

**Supplier 2**

**Supplier 3**

Purchased Parts List

PPE PPG PPB PPD PPA PPI PPC PPJ PPF PPH

## Operations

Critical manufactured parts, sub-assemblies and finished stock.

**Bill of Materials**

## Fulfillment

The distribution and management of finished stock.

Region 1

Region 2

Region 3

FPA

**FIGURE 4-22**

Hub-and-spoke
results with cake
mixes.

| Whse # | Actual Inventory Turns | Service Level (Based on Stock Days) | Days Forward Covered | Average Inventory $$ |
|---|---|---|---|---|
| 1 | | 96.1% | 5 | $ 402,41 |
| 2 | | 98.7% | 9 | $ 285,25 |
| 3 | | 98.7% | 7 | $ 543,91 |
| 4 | | 98.5% | 7 | $ 324,80 |
| 5 | | 98.1% | 11 | $ 98,32 |
| 6 | | 98.3% | 6 | $ 215,85 |
| 7 | | 96.7% | 7 | $ 188,58 |
| Combined Warehouses | | 97.9% | 7 | $ 2,059,15 |
| Central Buffer | | 98.9% | 18 | $ 5,996,49 |
| In Transit Open Supply | | | | $ 1,072,76 |
| Consolidated Results | 12.5 | 97.9% | 29 | $ 9,128,40 |
| Actual Performance | 9.4 | 97.5% | 44 | $ 13,627,17 |
| Difference | 3.1 | 0.4% | 15 | $ 4,498,77 |
| % Reduction | 33% | | 34% | 33% |

Figure 4-23 shows a personal-care item. In this example, the central buffer is the actual production facility (labeled "Plant"). The plant has the space available, and it will eliminate an extra stage in transportation. The plant will not serve any regional customers except in the case of large direct customer orders. The simulation of 180 days of actual demand was put into the model and compared against actual performance. In this case, the results are even more dramatic. Service levels improve nearly a whole point, whereas inventories are cut in half. In this business, a single point of service level has significant market implications. Those implications were not factored into the identified benefit. Obviously, there would be a chance for market-share growth. Additionally, there are no cross-shipments.

## SUMMARY

The MRP approach does not rely on a forecast of item demand and thus avoids the problems touched on in this discussion. Its techniques are designed expressly for dealing with dependent, discontinuous, nonuniform demand, which is characteristic of manufacturing environments. The principles on which MRP systems are based are the subject of Chapter 5.

**FIGURE 4-23**

Personal-care
item analysis.

| Whse # | Actual Inventory Turns | Service Level | Days Forward Covered | Average Inventory $s |
|--------|------------------------|---------------|----------------------|----------------------|
| 1 | | 99.0% | 11.2 | $ 29,47 |
| 2 | | 99.3% | 5.1 | $ 59,84 |
| 3 | | 99.3% | 9.5 | $ 66,04 |
| 4 | | 99.5% | 6.3 | $ 94.28 |
| 5 | | 99.3% | 5.4 | $ 52,71 |
| 6 | | 98.1% | 18.5 | $ 26,19 |
| 7 | | 98.7% | 9.6 | $ 51,68 |
| 8 | | 97.8% | 9.3 | $ 57,14 |
| Combined Warehouses | | 98.9% | 7.1 | $ 437,38 |
| Plant (Central Buffer) | | 99.4% | 10.7 | $ 672,51 |
| Av. Daily Open Supply | | | | $ 159,71 |
| Consolidated Results | 17.8 | 98.8% | 20.6 | $ 1,269,61 |
| Actual Performance | 11.8 | 98.0% | 13.4 | $ 2,574,68 |
| Difference | 6.0 | 0.9% | 39% | 1,305,06 |
| % Reduction | 51% | | | 51% |

# BIBLIOGRAPHY

E. Goldratt, E. Schragenheim, and C. Ptak. "Necessary But Not Sufficient," 2000.

Orlicky, J. A. "Requirements Planning Systems: Cinderella's Bright Prospects for the Future," in *Proceedings of the 13th International Conference of APICS*. Great Barrington, MA: North River Press, 1970.

Wight, O. W. "To Order Point or Not to Order Point," *Production & Inventory Management* 9(3), 1968.

## CHAPTER 5

# Principles of Materials Requirements Planning

**A**s mentioned in Chapter 4, the alternative choice to statistical inventory control is material requirements planning (MRP), an approach that recognizes the realities of demand existing in a manufacturing environment. This approach is eminently suitable for the management of inventories subject to dependent demand because it does not rely on any assumptions regarding patterns of demand and inventory depletion. The MRP approach does, however, assume certain characteristics of the product and of the process used in its manufacture. This chapter reviews these and other assumptions, prerequisites, and principles that an MRP system employs.

## TIME PHASING

*Time phasing* means adding the dimension of time to inventory-status data by recording and storing the information on either specific dates or planning periods with which the respective quantities are associated. In the older era of inventory control, status of a given item normally was shown in the system's records as consisting of only the quantity on hand and the quantity on order. When physical disbursements of the item reduced the sum of these two quantities to some predetermined minimum or reorder point, it was time to place a replenishment order.

This approach was refined around 1950 with the introduction of the perpetual inventory control concept. The principal idea behind this concept was to maintain somewhat expanded status information "perpetually" up to date by posting inventory transactions as they took place. This innovation largely was made feasible by the availability of better office equipment, particularly Kardex and punched-card data-processing installations, which streamlined the clerical tasks of posting and calculation.

Inventory-status information was expanded by adding data on requirements (demand) and *availability* (the difference between the quantity required and the sum of the on-hand and on-order quantities). The classic inventory status equation was formulated and publicized as follows:

$$A + B - C = X$$

where  $A$ = quantity on hand
         $B$ = quantity on order
         $C$ = quantity required
         $X$ = quantity available (for future requirements)

Thus the status of an inventory item might appear like this:

| On hand:   | 30 | or | 30 |
|------------|----|----|-----|
| On order:  | 50 |    | 25 |
| Required:  | 65 |    | 65 |
| Available: | 15 |    | −10 |

The quantity required would be derived from customer orders, a forecast, or a calculation of dependent demand. The quantity available had to be calculated. Negative availability signified lack of coverage and a need to place a new order. The inventory system now could better answer the questions of what and how much. What remained unanswered under this approach was the crucial question of when. When is the quantity on order due to come in, and is it a single order or are there more than one? When will the demand actually have to be satisfied, and will it be at one time or is the requirement a summary figure concealing several demands spaced in time? When will the stock run out? When should the replenishment order be completed? When should it be released? The system was unable to answer these questions, and the inventory planner had to depend on his or her own estimates and guesses, on rules of thumb, or on physical manifestations of need ("We can ship only 30 against this order for 65").

Time phasing means capturing or developing the information on timing so as to provide answers to all the foregoing questions. The price of time phasing is the added cost of processing and storage of the time-phased data. The value of the additional information thus made available, however, normally more than offsets the price paid for it. For instance, if the data in the preceding example were time-phased, the status of the item, by week, might appear as shown in Figure 5-1.

Answers are now provided in the figure to every single "when" question raised previously. There is one open order for 25, due to be received in the fifth week. There are three separate demands for quantities of 20, 35, and 10, which will occur in the second, fourth, and tenth weeks, respectively. The replenishment order need not be completed

**FIGURE 5-1**

Time phasing.

| | Period | | | | | | | | | |
|---|---|---|---|---|---|---|---|---|---|---|
| On hand: 30 | 1 | 2 | 3 | 4 | 5 | 6 | 7 | 8 | 9 | 10 |
| Open orders due: | 0 | 0 | 0 | 0 | 25 | | | | | |
| Quantities required: | 0 | 20 | 0 | 35 | 0 | 0 | 0 | 0 | 0 | 10 |
| Available: | 30 | 10 | 10 | −25 | 0 | 0 | 0 | 0 | 0 | −10 |

until the tenth week. This determines the (latest) release date for that order based on the item's lead time. Time phasing, in this case, yields additional valuable information. In the fourth week, the availability is negative, which indicates that while total coverage is adequate for the first nine weeks, the timing within the nine-week span is out of phase. The inventory planner sees exactly what specific action is called for if a *stock-out*, or shortage, is to be prevented, and he or she sees it (in this case) four weeks in advance. The open order should be rescheduled to be completed one week earlier.

The difference between a time-phased and a non-time-phased inventory control system, in terms of usefulness and effectiveness, is considerable. The differences in data-handling, data-manipulation, and data-storage requirements also are considerable, as can be seen in the preceding two examples. In the first one, inventory status is expressed through four data elements. In the second example, 26 data elements are used.

The data-processing burden in maintaining time-phased status data up to date is substantial. This is so not only because there are so many more data elements involved but also because of the fact that with non-time-phased data, only changes in the quantities need to be processed, whereas time-phased status information has to reflect changes in both quantity and timing (i.e., changes in the timing even when the quantities remain unaffected). Time-phased MRP systems represent a classic computer application in the sense that here the computer is being used to do something heretofore literally impossible—handling and manipulating vast quantities of data at high speed. In a fairly typical situation where status information for 25,000 inventory items is time-phased by week over a one-year span, some 5 million data elements may be involved. Given available technology, this is no longer a problem.

Further advancement of technology and business rules has advanced the use of automated planning tools, and now, demand-driven MRP is possible. The first question to be answered is no longer "when" but "where." Chapter 4 details the information required to answer the critical "where" question.

## INVENTORY SYSTEM CATEGORIES

MRP (time phasing is implied by the term) evolved from an approach to inventory management in which the following two principles are combined:

- Calculation (versus forecast) of component-item demand
- Time phasing, that is, segmenting inventory-status data by time

The term *component item* in MRP covers all inventory items other than products or end items. Requirements for end items are stated in the master production schedule, and they are derived from forecasts, customer orders, field warehouse requirements, inter-plant orders, and so on. Requirements for all component items (including raw materials) and their timing are derived from this schedule by the MRP system.

MRP consists of a set of techniques eminently suitable for the management of inventories subject to dependent demand, and it represents a highly effective inventory control system for manufacturing environments, where the bulk of the inventory is subject to this type of demand. It should be noted that while an MRP system is oriented primarily toward dependent-demand inventory, it also easily accommodates independent-demand items such as service parts. These can be integrated into the system through the time-phased order-point technique discussed briefly below and more fully described in Chapter 7. There are, as mentioned previously, some inventory items that are subject to both dependent and independent demand, such as service parts still used in current production. In an MRP system, the service-part demand, which is forecast, is simply added to the dependent demand that has been calculated. The MRP system takes it from there.

With reference to the two above-mentioned principles, demand calculation and time phasing, any manufacturing inventory control system can be assigned to one of four categories based on combinations of these principles. This is illustrated in Figure 5-2, which shows, in matrix form, the four system categories:

- Statistical order point
- Lot requirements planning
- Time-phased order point
- MRP

*Statistical order point*, the conventional approach in the past, has already been discussed at some length in Chapter 4. It uses forecasting to determine demand and generally ignores the aspect of specific timing. In light of what is possible today thanks to computer technology, this type of system must be considered obsolete for purposes of manufacturing inventory management.

*Lot requirements planning* was developed and used by some manufacturing companies toward the end of the era of punched-card data processing, generally the 1950s and early 1960s. Some companies still use this approach, in which component-item demand is derived from a master production schedule and is calculated correctly as to quantity per lot of product or end item but in which specific timing is disregarded. Requirement and order data are summarized by (product) lot, and it is the position of the lot in the master schedule that implies timing. Specific timing of order releases, due dates, and production schedules then is established—if it is established—through procedures external to the inventory system.

**FIGURE 5-2**

Inventory system categories.

| | | Component Demand | |
|---|---|---|---|
| | | Forecast | Calculated |
| Maintenance of Status Data | Quantity Only | Statistical Order Point | Lot Requirements Planning |
| | Quantity and Timing | Time-Phased Order Point | Material Requirements Planning |

A variation of the lot requirements planning approach is the so-called single-period requirements planning system. All product lots scheduled for a given period, usually a month, are combined, in effect, into a superlot that is then treated as an individual lot in lot requirements planning. The principal output of systems of this type is an item-by-item listing (by lot or by period) or order action required without an indication of exactly when action is to be taken on each of the items.

Methods of planning by product lot had been in use at a time when the job of detailed time phasing was too big for a punched-card installation, in the sense that millions of card equivalents would have had to be processed (sorted, collated, summary-punched, etc.) at relatively slow card-handling speeds. It would have taken days or weeks to complete the job. This was state of the art at the time. Refer to Appendix D for more detail on the technological evolution occurring during this time. Lot requirements planning systems, at one time representing a significant advancement in inventory control state of the art and superior to the statistical order-point approach, became obsolete when the computer was introduced. At that point, time phasing of inventory-status data became feasible and practical.

*Time-phased order point* is a modern technique of planning and controlling inventory items subject to independent demand. It is eminently suitable for service parts, finished products in factory stock, and field warehouse items. The system processing logic is identical to MRP (see Chapter 7) except for the manner in which item demand is arrived at. Requirements for independent-demand items are forecast (using any forecasting technique the user selects) because they cannot be calculated. Any service part that is manufactured has, however, at least one component item (e.g., raw material) that is then treated the same as any other item in an MRP system.

MRP calculates item demand and time-phases all inventory-status data in time increments as fine as the user has specified. MRP represents the ultimate approach to manufacturing inventory management. Concepts of MRP, the processing logic of an MRP system, and related techniques will be discussed in the balance of this chapter and in the chapters that follow.

## PREREQUISITES AND ASSUMPTIONS OF MRP

Unlike the goddess Athena, MRP did not spring into being in full splendor and fully armed. In some rudimentary form, it has no doubt existed as long as manufacturing. The idea of planning what is really required by comparing what is a total requirement to what is on hand is as old as any shopping trip to plan meals for the week. It has been evolving gradually, moving onto successively higher plateaus with every enhancement in data-processing capability. MRP had its origin "on the firing line" of a plant. It has been painstakingly developed into its present stage of relative perfection by practicing inventory managers and inventory planners.

It never made sense to the practitioner to stock thousand-dollar castings, for instance, and to reorder them in economic lot quantities when he or she could determine exactly how many of each casting would be required and when by consulting the master production schedule. Such expensive manufacturing inventory items have, in most cases, always been treated the way an MRP system treats items under its control. Specific demand for these items has been calculated rather than forecast; existing surplus, if any, has been taken into account to determine net demand; the (usually long) lead time has served to determine order release; the items have been ordered discretely; and detailed status records have been kept. In a sense, the evolution of MRP systems is equivalent to expanding the strict and careful treatment of high-cost, long-lead-time items to items of successively lower cost and shorter lead time (as data-processing capability permitted) until all items have been covered.

Present-day MRP systems, constructed and used in what has become the standard form, imply several prerequisites and reflect certain fundamental assumptions on which these systems are based. The first prerequisite is the existence of a master production schedule, that is, an authoritative statement of how many end items are to be produced and when. An MRP system presupposes that the master production schedule can be stated in its entirety in bill of material (BOM) terms, that is, BOM (assembly) numbers. The only language an MRP system understands is *part numbers*, that is, inventory item numbers that uniquely identify specific materials, component parts, subassemblies, and end items. An MRP system cannot work with English-language product descriptions, with sales-catalog model numbers, or with BOM numbers that are ambiguous in the sense that they fail to identify a precise configuration of components for a given assembly. The ability to create a valid and complete master production schedule in BOM terms is a function of so-called BOM structure, which is discussed further in Chapter 11.

## PART NUMBERS

Another prerequisite is that each inventory item must be unambiguously identified through a unique code (part number). This requirement also extends to the identification of every manufactured item's component material (what is the item made of?) and to each item's disposition (where is it used; i.e., what is the item a component of?).

The purpose of the number is simply to provide a unique name for each individual, as Social Security numbers do. Parts having any differences in form, fit, or function so that they are not interchangeable must have different numbers. Ideal numbers should have the fewest digits, be only numeric (no alphabetic characters to speed data entry), and be assigned serially as new parts are introduced.

A trap into which many fall, greatly complicating the problem of keeping accurate records, is doing more with part numbers than giving each a name. Digits at each position are given significance, identifying some characteristic such as shape, material, or product family. This lengthens the number, shortens its useful life, and increases the likelihood of people making errors writing it or keying it into computer-based systems. Significant-digit numbers are a holdover from punched-card data processing, when the number of columns available for item data was limited.

Proponents of significant digits argue that current computer systems handle longer numbers with ease. They overlook the work of adding more new numbers (features change much more frequently than form, fit, or function), correcting more errors, and needing even longer numbers (a single digit can handle only 10 varieties of a characteristic). They also forget that computers can access subordinate files of code numbers for descriptive data without needing those codes in part numbers.

It is difficult to argue against semi-significant numbers where only two or three digits identify families of parts. Those desiring such numbers should be responsible for justifying them by showing that advantages outweigh disadvantages. It is rarely possible to replace existing part numbers in a going concern; costs and other demands on scarce resources are prohibitive. The simple solution is to adopt short, serial, nonsignificant numbers for all new items, deleting old ones as they become obsolete. Changes now proliferating in most businesses will lead them to make the conversion in a remarkably short time.

## BILLS OF MATERIAL (BOMS)

Each item in the master production schedule (MPS) must have a unique identifying number and be associated with a BOM specifying its components to be planned and controlled by MRP. BOMs identify the component(s) needed to make parent items. A parent may be as complex as a product assembled from many components or as simple as a single part made from some raw material.

Product structure data can be stored in computers using BOM processing software commonly supplied by computer manufacturers and commercial sources. These use computer storage efficiently, avoid duplication of data, and apply fast retrieval for "assembly" by the computer of BOMs in the various formats desired by different users.

Unfortunately, in practice, some parent items may have as many as five different BOMs:

- *A design parts list.* Simply listing the component items, this BOM is the last step in engineering design. It is the engineer's way of telling the rest of the organization how many of which components make up a product. Parts lists may show how engineering thinks the parent should be put together, but this is usually not the design engineer's responsibility. Engineering specifications accompany BOMs and carry other information needed to produce, inspect, and test complete, functioning products. Parts lists often do not include packaging materials and usually omit such items as glue, grease, and paint, for which it is difficult to specify a quantity needed.
- *A manufacturing BOM.* In addition to listing all components of a product, BOMs must be structured to show production people how to put the product together. Production may require subassemblies for proper welding or ease of assembly. Semi-finished parts (unpainted, not plated, incomplete machining) may reduce complexity and increase planning and production flexibility. Sales of field-replacement spare parts may be assemblies made only for this purpose. Engineering usually has no interest in these needs.

- *A material planning BOM.* Development of a valid, realistic MPS for products that offer several options to customers requires a very different BOM from those issued by engineering and those needed by production. Material planning for many varieties of the same basic product, for tooling and similar related materials, and for make-to-order products made from a few standard subassemblies requires specially structured BOMs. This subject is covered in Chapter 11.
- *A cost-accounting BOM.* This is often simplified by using one part number for many painted or plated parts and for other components with variations not affecting costs or inventory valuation.
- *A BOM describing the actual item made.* This can be different from all other BOMs for the item owing to process considerations.

Thus the existence of a BOM containing such information is also required at planning time. The BOM must not merely list all the components of a given product but must be structured so as to reflect the way the product is actually made, in steps from raw material to component part to subassembly to assembly to end item.

There is only one legitimate reason why the manufacturing BOM should differ from the way products are actually built: Last-minute design changes issued by engineering as the products were being built were not yet picked up in the computer files. Strenuous efforts should be made to keep such time delays to a minimum. BOMs showing products actually made are often different from planning BOMs. This can be caused by legitimate differences: What was planned was not what was built. Too often, however, the reason these BOM types are different is lack of data accuracy in the formal files. Each group using a BOM attempts to keep its own version; inevitably, differences creep in.

The five types of BOMs are all needed. This does not mean that five different BOM computer files are necessary; BOM processor programs can code the basic data to link components and produce a BOM for each specific purpose. The term *bill of material* is used interchangeably for that covering a single parent and its components (called a *single-level BOM*), for more complex multilevel BOMs, and for the entire BOM computer file.

Another prerequisite to MRP is the availability of inventory records for all items under the system's control that contain inventory-status data and so-called planning factors, as discussed in the next few chapters. An assumption or, rather, a precondition for effective operation of the system is file data integrity in terms of inventory status and the BOM. This is not a system assumption—an MRP system can function with faulty data and still generate outputs that are technically correct relative to the data supplied to the system—but an operational assumption. File data must be accurate, complete, and up-to-date if the MRP system is to prove successful or even useful. The requirement of file data integrity may seem self-evident, but there are two points to be made in this connection.

First, the fact is that typically the two files in question are chronically in poor shape under any system preceding the installation of MRP. And second, under an order-point system, it does not overly matter that inventory records are unreliable and that BOMs are inaccurate, incomplete, or out of date. Order point acts merely as an order-launching system (a *push* system), and it must be complemented by an expediting (pull) system in order to function at all. The assumption is that the expediters are actually putting a priority on what is really needed and not just to fill stock shelves.

Under an order-point system, the BOM is not even referenced, and the quality of its data therefore is irrelevant for purposes of inventory planning. The formal push system uses the inventory-status data, which may be (and usually are) faulty, but this is compensated for by the informal pull system that does not rely on the inventory records at all but determines specific need for inventory items, and the timing of this need, physically, in the stockroom or on the assembly line. It is the expediting action that the whole procurement and manufacturing operation then really depends on.

In contrast, under an MRP system, which provides both the push and pull functions in the formal system, there is no need for the informal system of manual shortage-list expediting, but this benefit will not be realized if the quantities and timing of orders are incorrect owing to lack of file data integrity. This integrity is vital to the MRP system, and the meticulous maintenance of the files involved calls for a special effort on the part of the system user—a novel requirement and cost.

An MRP system presupposes that lead times for all inventory items are known and can be supplied to the system, at least as estimates. See Chapter 4 for a deeper discussion on lead times. The lead time used for planning purposes normally must have a fixed value. This value can be changed at any time, but more than one value cannot be in existence simultaneously. An MRP system cannot handle indeterminate-item lead times.

An MRP system assumes that every inventory item under its control goes into and out of stock, that is, that there will be reportable receipts, following which the item will be (even if only momentarily) in an on-hand state and eventually will be disbursed to support an order for an item into which it is dispositioned. This assumption means, in essence, that the progression of the manufacturing process from one stage to the next will be monitored, usually (but not necessarily) by means of a stockroom through which the items pass physically. Lean applications of MRP can simply backflush the material that had to have been used to maintain sufficient inventory record accuracy without the components or finished goods ever passing through a stockroom.

In determining the timing of item gross requirements, the (standard) MRP procedure assumes that all components of an assembly must be available at the time an order for that assembly is to be released to the factory. Thus the basic assumption is that unit assembly lead time (the time required to produce one unit of the assembly) is short and that the several components are consumed, for all practical purposes, simultaneously. As far as subassemblies are concerned, this assumption almost always holds true. In cases of significant exceptions to this rule (e.g., where it may take several weeks to assemble a unit and expensive components are consumed successively over this period), the regular requirements computation procedure would have to be modified.

Another assumption under MRP is discrete disbursement and use of component materials. For instance, if 50 units of a component item are required for a given (fabrication or subassembly) order, the MRP logic expects that exactly 50 units can be disbursed and that 50 units will be consumed. Materials that come in continuous form (rolls of sheet metal, coils of wire, etc.) do not meet this expectation cleanly and therefore require that standard planning procedures be modified and the system adapted to handle such inventory items properly. This is also a challenge in a process manufacturing environment, where materials can be in silos or pipelines. Parts that are too small and too numerous to count also can be a challenge to this assumption. These parts may be quantified by weight rather than by counts.

An assumption implied under MRP is process independence. This means that a manufacturing order for any given inventory item can be started and completed on its own and not be contingent on the existence or progress of some other order for purposes of completing the process. Thus so-called mating-part relationships (item $A$ at operation 30 must meet item $B$ at operation 50 for the machining of a common surface) and setup dependencies (order for item Y should be set up only when a setup for item X precedes it) do not fit the scheme of things under MRP. This does not mean that MRP is inapplicable, only that it is inapplicable in its traditional form. See Chapter 18 for a complete discussion on the application of MRP to a tightly integrated environment.

To recap, the principal prerequisites and assumptions implied by a standard MRP system are as follows:

- An MPS exists and can be stated in BOM terms.
- All inventory items are uniquely identified.
- A BOM exists at planning time.
- Inventory records containing data on the status of every item are available.
- There is integrity of file data.
- Individual-item lead times are known.
- Every inventory item goes into and out of stock.
- All the components of an assembly are needed at the time of the assembly order release.
- There is discrete disbursement and use of component materials.
- Process independence of manufactured items exists.

## APPLICABILITY OF MRP METHODS

The preceding discussion of prerequisites and assumptions raises the question of the applicability of MRP to a given type of manufacturing business. Actually, all the prerequisites and assumptions listed, as such, do not represent good criteria for applicability because even where some of the required conditions do not exist, management generally can create them in order to be able to use MRP methods. Inventory items can be uniquely identified, a BOM can be created, integrity of file data can be maintained, and so forth. Whether most of the preconditions for MRP do or do not exist in a given case is usually a matter of management practice rather than an attribute of the type of business in question. The application of the MRP tool differs by the segment of manufacturing, as described in the volume/variety matrix in Figure 1-2. This application is described in Chapters 15 through 19.

It is clear that the original application of MRP methods generally is limited to discrete as compared with continuous-process manufacturing. In the past, it was thought that the use of these methods would be warranted only in manufacturing operations involving relatively complex assembled products, but developments have disproved this. Companies that manufacture some very simple products, including one-piece products, can and now do use MRP systems.

It is true that the first companies to develop and use MRP systems were manufacturers of highly engineered assembled products in the metalworking industries, typically operating machine shops (job shops) in which large numbers of orders were in process simultaneously. This type of environment represents the most severe inventory management and production planning problems, and it was to alleviate these problems that the companies in question reached for MRP methods as soon as it became feasible (with computers) actually to implement them.

Since the pioneer days of MRP, companies in so many diverse manufacturing businesses (including cable and wire, furniture, and packaged spices) have adopted this approach that applicability criteria have been obscured. What is common to all these companies (or plants), and what represents the principal criterion of MRP applicability, is the existence of an MPS (not to be confused with a final assembly schedule) to which raw materials procurement, fabrication, and subassembly activities are geared.

The MPS (discussed more fully in Chapter 12) governs component production activities. The final assembly schedule responds to external (i.e., customer, field warehouse, etc.) demand for end products, is usually stated in different terms (i.e., product models, configurations, etc.), involves a shorter lead time, and is made up later in time. In responding to the demand for end products, the final assembly schedule is constrained by the availability of components provided by the MPS. In some types of businesses, the two schedules may be identical (as in cases of small simple products or products engineered and manufactured to order), but this is a coincidence caused by the nature of the product in question.

MRP therefore is applicable to manufacturing environments that are oriented toward fabrication (the term here also covers subassembly where subassembly applies) of components. Final assembly operations normally are outside the scope of the MRP system as we know it in its present standard form. MRP can be said to be primarily a component-fabrication planning system. An MRP system can be used by any plant that has or can have an MPS.

The preceding discussion dealt with applicability by type of business. The criterion of applicability by type of inventory item (in a plant using an MRP system) is simple and straightforward: MRP is applicable to any discrete item, purchased or manufactured, that is subject to dependent demand. Other attributes of the item, such as cost, volume of usage, or continuity of demand, are irrelevant to MRP applicability. Low-cost, high-volume parts do not seem to deserve elaborate treatment, and it is a natural tendency for the first-time MRP system user to exclude such items from his or her new system. Experience has shown, however, that after several months of operation under MRP, even the lowest-cost, highest-volume items typically are incorporated into the system.

The basic reason for this is simply that it becomes evident that better results can be had under the MRP treatment. The user at this point normally also has overcome his or her initial emotional reaction to the elaborateness of MRP procedures (which, after all, are being performed by a machine with no apparent difficulty or loss of time), and the cost argument is seen for what it is—the expense of processing some additional items by an existing MRP system is trivial.

## BIBLIOGRAPHY

Everdell, Romeyn. "Time Phasing: The Most Potent Tool Yet for Slashing Inventories!" *Modern Materials Handling*, November 1968.

Wight, O. W. "Time Phasing," *Modern Materials Handling*, October 1971.

## CHAPTER 6

# The Material Requirements Planning System

The absence of an order-point approach does not a material requirements planning (MRP) system make. The term *material requirements planning* implies certain definite system attributes, such as time-phased inventory status data, the computation of net requirements, a maximum length of a planning period, a minimum planning horizon span relative to lead time, and the development of planned orders.

The reader should understand that there are genuine MRP systems and pseudo-MRP systems in industry usage. There are companies (or rather plants) that do some form of MRP without having a full-blown or real MRP system. This chapter is concerned with genuine MRP systems in one or the other of their standard forms.

There are a limited number of alternative MRP system approaches (reviewed in Chapter 15) but a number of specific techniques and special procedural features tailored to meet unusual requirements of a given system user. In the discussion that follows, the concepts and characteristics of (genuine) MRP systems that are common to such systems will be identified regardless of approach or specific technique used.

## OBJECTIVES OF THE SYSTEM

All MRP systems have a common objective, which is to determine (gross and net) requirements, that is, discrete period demands for each item of inventory, so as to be able to generate information needed for correct inventory order action. This action pertains to procurement (purchase orders) and to production (shop orders). It is either new action or a revision of previous action. New action consists of the placing (release) of an order for a quantity of an item due on some future date. The essential data elements accompanying this action are

- Item identity (part number)
- Order quantity
- Date of order release
- Date of order completion (due date)

Order action relative to purchased items takes place in two steps: a requisition placed on purchasing by inventory control and a subsequent order placed on a vendor selected by purchasing. The types of order action that effect a revision of action taken previously are limited to the following:

- Increase in order quantity
- Decrease in order quantity
- Order cancellation
- Advancement of order due date
- Deferment of order due date
- Order suspension (indefinite deferment)

To generate information for correct order action is not the only objective of an MRP system, which also serves other functions (discussed in Chapter 7), but it is the primary one. It is not much different from the objective of other (non-MRP) inventory systems in intent. The difference lies in the respective systems' ability to realize this intent. Order point systems, in particular, have difficulty in ordering the right quantity of an item (see the discussion of EOQ in Chapter 4) at the right time (see Figure 1-4), and their ability to order with a valid order due date is even more questionable. As to the ability of such systems to revise previous order action, they have virtually none.

MRP systems meet their objective by computing net requirements for each inventory item, time phasing them, and determining their proper coverage. The basic function of MRP is the conversion of gross requirements into net requirements so that the latter may be covered by (correctly timed) shop orders and purchase orders.

The *netting* process consists of a calculation of gross requirements and of allocating existing inventories (quantities on hand and on order) against these gross requirements. For example:

| | | |
|---|---|---|
| Gross requirements: | | 120 |
| On hand: | 25 | |
| On order: | 50 | 75 |
| Net requirements: | | 45 |

If safety stock is planned for the item in question—this is not usual under MRP, but the system presents no obstacle—the net requirements would be increased by the quantity of the safety stock as follows:

| | | |
|---|---|---|
| Gross requirements: | | 120 |
| On hand: | 25 | |
| On order: | 50 | |
| | 75 | |
| Safety stock: | −20 | 55 |
| Net requirements: | | 65 |

In an MRP system, the net-requirement quantities are always related to time, that is, to some date or period. The net requirements then are covered by planned orders, and the order quantities either match net requirements or are calculated by employing one of several lot-sizing techniques designed to take into account the economics of ordering. The timing of planned-order releases is also determined by the MRP system, and the information is stored for purposes of future order action.

The function of providing coverage of net requirements is served only in part through planned (i.e., future) orders. The MRP system also reevaluates the timing of open orders relative to (possibly changed) net requirements in the near future, and it signals the need for rescheduling these orders, forward or backward in time, as required, to realign coverage with net requirements.

## THE PURPOSE OF THE SYSTEM

An MRP system is capacity-insensitive in that it will call for the production of items for which capacity may not, in fact, exist. This might appear to be a shortcoming of MRP, but on a moment's reflection, it can be seen that this is not so. A system can be designed to answer either the question of what can be produced with a given capacity [i.e., what the master production schedule (MPS) should be] or the question of what need be produced (i.e., what capacity is required) to meet a given MPS, but not both simultaneously. Process industries tend to ask the capacity question first, whereas discrete manufacturing companies tend to ask the latter question first. Current MRP systems are designed to answer the both questions iteratively. The industry will determine which question is asked and answered first. See Chapters 15 through 18 for more detail by industry.

An effective MRP implementation assumes that capacity considerations have entered into the makeup of the MPS. An MRP system "believes" the MPS, and the validity of its outputs is always relative to the contents of that schedule. Another way of stating this is to say that the MPS can be invalid (vis-à-vis available capacity), but the outputs of an MRP system (assuming valid file data and correct procedures) cannot.

The output of an MRP system is not necessarily always realistic in terms of lead time, capacity, and availability of materials, particularly when the system plans requirements for an unrealistic MPS. It is then merely saying, "This is what you would have to be able to do to implement the schedule." Why the schedule is unrealistic thus is revealed in specific terms.

In any manufacturing operation, the questions of what materials and components are needed, in what quantities, and when—and the answers to these questions—are vital. An MRP system is designed to provide just these answers. MRP systems are a highly effective tool of manufacturing inventory management for the following reasons:

- Inventory investment can be held to a minimum.
- An MRP system is change-sensitive and reactive.
- The system provides a look into the future on an item-by-item basis.

- Under MRP, inventory control is action-oriented rather than clerical bookkeep-ing–oriented.
- Order quantities are related to requirements.
- The timing of requirements, coverage, and order actions is emphasized.

Because of its focus on timing, an MRP system (and only an MRP system) can generate outputs that serve as valid inputs to other systems in the area of manufacturing logistics, such as purchasing systems, shop scheduling systems, dispatching systems, shop floor control systems, supply-chain planning systems, and capacity-requirements planning systems. A sound MRP system constitutes a solid basis, a gateway, for other computer applications in production and inventory control.

The position of an inventory planning system, relative to other manufacturing logistics functions or systems, is shown in Figure 6-1. The relationships depicted in this chart exist in any manufacturing company or plant. A manufacturing operation, in essence, consists of the procurement of materials and the conversion of those materials into a shippable product. The principal outputs of the inventory system, whatever this system may be, are purchase requisitions and shop orders, each one of these calling for a specific quantity of some inventory item. Any procurement or manufacturing activity takes place only after the inventory system has generated a call for the item. The inventory system triggers all such activities. In terms of information flow, it is the upstream system.

Any and all systems along the two streams (procurement and manufacturing) of inventory system output are designed merely to execute the plan that is represented by this output. These downstream systems cannot compensate for or improve the possibly low quality (i.e., validity, accuracy, completeness, or timeliness) of the information they receive as input. Regardless of how well implemented the downstream systems themselves may be, their real effectiveness still depends on the quality of the inputs they

**FIGURE 6-1**

Upstream and
downstream
systems.

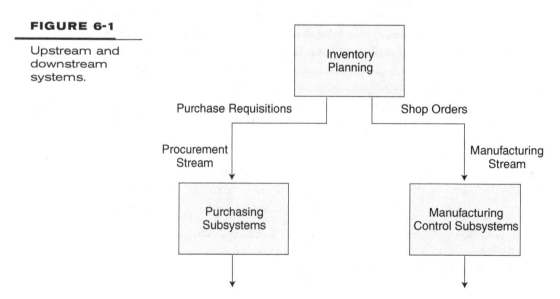

process. "Pollution" of the information, originating upstream, in the inventory system permeates all downstream functions and activities. It therefore follows that within the framework of the overall logistics system, the role of the inventory subsystem is of first importance.

When the function labeled "Inventory Planning" in Figure 6-1 is exercised by an MRP system, a reliable quarterback is directing the plays. An MRP system has the ability to generate calls for the right items in the right quantities at the right time with the right date of need for every order. The system issues its action calls according to a detailed, time-phased plan that it develops. It maintains this plan constantly up to date by reevaluating and revising it in light of ongoing changes in the environment. It also continuously monitors the validity of all open-order due dates relative to such changes. With an MRP system making the calls, the execution systems downstream can function effectively. Without it, they cannot.

## SYSTEM INPUTS AND OUTPUTS

An MRP system, properly designed and used, can provide a number of desirable outputs containing valid and timely information. The primary outputs of an MRP system are the following:

- Order-release notices calling for the placement of planned orders
- Rescheduling notices calling for changes in open-order due dates
- Cancellation notices calling for cancellation or suspension of open orders
- Item status analysis backup data
- Planned orders scheduled for release in the future

Secondary or by-product outputs come in a great variety and are being generated by the MRP system at the user's option. These outputs, discussed further in Chapter 7, include

- Exception notices reporting errors, incongruities, and out-of-bounds situations
- Inventory-level projections (inventory forecasts)
- Purchase commitment reports
- Traces to demand sources (so-called pegged requirements reports)
- Performance reports

All MRP system outputs are produced by processing inputs (relating data) from the following sources, illustrated in Figure 6-2:

- The MPS
- Orders for components originating from sources external to the plant using the system
- Forecasts for items subject to independent demand
- The inventory record (item master) file
- The bill of material (product-structure) file

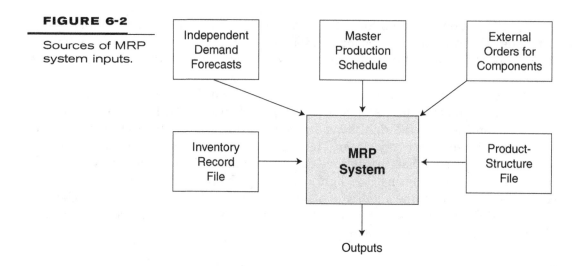

**FIGURE 6-2**

Sources of MRP system inputs.

Outputs

The MPS expresses the overall plan of production. It is stated in terms of end items, which may be either (shippable) products or highest-level assemblies from which these products are eventually built in various configurations according to a final assembly schedule. The span of time the MPS covers, termed the *planning horizon*, is related to the cumulative procurement and manufacturing lead time for components of the products in question. The planning horizon normally equals or exceeds this cumulative lead time.

The MPS serves as the main input to an MRP system in the sense that the essential purpose of this system is to translate the schedule into individual component requirements, and other inputs merely supply reference data that are required to achieve this end. In concept, the MPS defines the entire manufacturing program of a plant and therefore contains not only the products the plant will produce but also orders for components that originate from sources external to the plant, as well as forecasts for items subject to independent demand. In practice, however, such orders and forecasts are normally not incorporated into the MPS document but are fed directly to the MRP system as separate inputs.

Externally originating orders for components include service-part orders, interplant orders, original equipment manufacturer (OEM) orders by other manufacturers who use these components in their products, and any other special-purpose orders not related to the regular production plan. Components may be ordered for purposes of experimentation, destructive testing, promotion, equipment maintenance, and so on. The MRP system treats orders of this category as additions to the gross requirements for the respective component items. Beyond this, regular MRP treatment applies.

Forecasts of independent demand for component items subject to this type of demand can be made outside the MRP system, or the system can be programmed to perform this function by means of applying some statistical forecasting technique. The forecast quantities are treated as item gross requirements by the MRP system. Items subject only to independent demand (such as service parts no longer used in regular production) should be under time-phased order-point control (described in Chapter 7). Items subject to both dependent and independent demand have the forecast quantities simply added to the (computed) gross requirements. Note that service-part demand is either forecast or recorded on receipt of orders (placed by a service-part organization operating its own system) but, as a rule, not both.

The *inventory record file*, also called the *item master file*, consists of the individual item inventory records containing the status data necessary for determining net requirements. This file is kept up-to-date by the posting of inventory transactions that reflect the various inventory events taking place. Each transaction (i.e., stock receipt, disbursement, scrap, etc.) changes the status of the respective inventory item. The reporting of transactions therefore constitutes an indirect input to the MRP system. Transactions update item status, which then is consulted and modified in the course of computing requirements.

In addition to status data, the inventory records also contain so-called planning factors used principally for determining the size and timing of planned orders. Planning factors include item lead time, safety stock (if any), scrap allowances, lot-sizing algorithms, and so on. Planning-factor values are subject to change at the system user's discretion. A change in one or more planning factors normally changes inventory status.

The *bill of material* (BOM) *file*, also known as the *product-structure file*, contains information on the relationships of components and assemblies that are essential to the correct development of gross and net requirements.

All the inputs just reviewed enter into the MRP process, the principal purpose of which is to establish (reestablish) correct inventory status of each item under its control. The factors involved in establishing this status are the following:

- Requirements
- Coverage of requirements
- Product structure
- Planning factors

What sets the MRP process in motion varies depending on system implementation and system use. With so-called regenerative MRP systems, which employ batch-processing techniques, the replanning process is carried out periodically, typically in daily intervals. Here, passage of time triggers the process. With so-called net-change MRP systems, it is the inventory events (transactions) that cause replanning to take place, more or less continuously.

Changes in requirements, coverage, product structure, relevant engineering changes, or planning factors affect inventory status and therefore must be reflected in the replanning. Regenerative MRP systems, in effect, take a snapshot of these factors as they are at the time of each periodic requirements computation on the assumption that any and all changes have been incorporated during the preceding interval. These systems deal periodically with situations that are static at the time. Net-change MRP systems, on the other hand, must deal continuously with a dynamic or fluid situation. This requires that changes in any of the four factors mentioned be reported to the system as they occur.

# FACTORS AFFECTING THE
# COMPUTATION OF REQUIREMENTS

The computation of requirements is complicated by six factors:

- The structure of the product, containing several manufacturing levels of materials, component parts, and subassemblies
- Lot sizing, that is, the ordering of inventory items in quantities exceeding net requirements for reasons of economy or convenience
- The different individual lead times of inventory items that make up the product
- The timing of end-item requirements (expressed via the MPS) across a planning horizon of, typically, a year's span or longer and the recurrence of these requirements within such a time span
- Multiple requirements for an inventory item owing to its so-called commonality, that is, use in the manufacture of a number of other items
- Multiple requirements for an inventory item owing to its recurrence on several levels of a given end item

## Product Structure

Product structure imposes the principal constraint on the computation of requirements. This computation, while very simple arithmetically, requires that a rather involved procedure be followed. This is caused by the fact that a given component item can exist in its own right as a uniquely identified physical entity (e.g., a unit of raw material, a component part, or a subassembly), and it also can exist physically, but as an already-assembled component (or "consumed" material) of another inventory item, in which case it has lost its individual identity. For example, gear C in Figure 4-2 can exist as such and would be so carried on the inventory record, and it also can exist as part of the (already-assembled) gearbox B or transmission A without an identity of its own. For purposes of determining net requirements, the physically existing quantities of the item must, in effect, be accounted for irrespective of identity.

Such accounting will be the more laborious the more levels of the product are involved. Product *depth* therefore is a factor in the scope and duration of the MRP data-processing job. The concept of product level (or manufacturing level) is related to the way the product is structured, that is, manufactured. Each stage in the manufacturing process of converting material into product is equivalent to a level of product structure.

The engineering document that defines the product is the bill of material (BOM), which lists components of each assembly and subassembly. Its conventional graphic representation is shown in Figure 6-3.

The assembly in question is termed the *parent item*, and its component items, sometimes are referred to as *children parts*, are listed by identity code (part number) with the quantity per (one unit of) parent item. The term *bill of material* is used interchangeably for a single-item bill (such as shown in Figure 6-3) for all such bills, collectively, pertaining to a given product and for the entire BOM file. (BOMs are discussed further in Chapter 11.)

**FIGURE 6-3**

Bill of material.

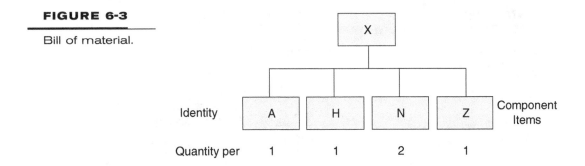

**FIGURE 6-4**

Hierarchy of bills of material.

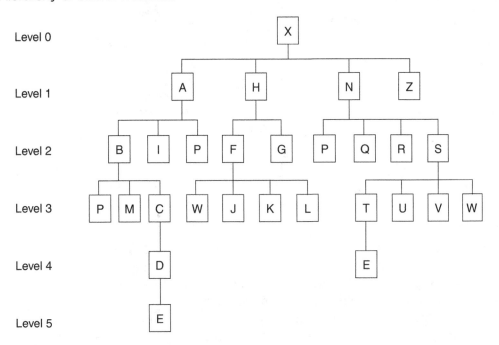

When the individual BOMs defining a product are linked together graphically, as in Figure 6-4, they form a hierarchical, pyramid-like structure, and the levels come into view. By convention the levels are numbered from top to bottom, beginning with level 0 for the end product. The structure may be likened to a Christmas tree, and vertical lines of progression (such as X-A-B-C-D-E in Figure 6-5) are called *branches*. The items on the highest level in a product-structure file are termed *end items*.

**FIGURE 6-5**

Parent-
component
relationship.

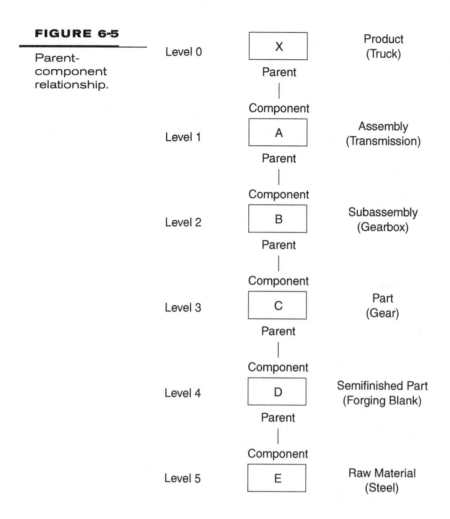

The concept of the product level is usually associated with relatively complex assembled products that contain many (typically 6 to 10) levels. But any manufactured product, no matter how simple, involves at least two and probably more levels. A one-piece wrench has at least three levels, that is, steel, forging blank, and finished wrench. A nuts-and-bolts manufacturer purchases steel rod (one level) that is drawn into wire (another level) from which a screw (a third level) is formed. Even the simplest products assembled from manufactured components have at least three and most likely more levels.

In determining net requirements for a low-level (note that the lowest level carries the highest number) inventory item, the quantity that exists under its own identity, as well as any quantities existing as (consumed) components of parent items, parents of parent items, and so on, must be accounted for. The basic logic of "netting" requirements is best demonstrated through an example. Let us assume that 100 trucks $X$ are to be produced and that the following are in inventory (on hand and on order):

Transmission A:              2
Gearbox B:                  15
Gear C:                      7
Forging blank D:            46

The task is to determine net requirements for these items. When first confronted with this type of problem, most people will apply the following logic:

Item A:      100 –  2 = 98
Item B:      100 – 15 = 85
Item C:      100 –  7 = 93
Item D:      100 – 46 = 54

This, however, is incorrect. The true net requirement for item $D$, for instance, is not 54 but 30, and the correct netting logic that leads to it is as follows:

| | |
|---|---:|
| Quantity of trucks to be produced: | 100 |
| Transmissions required (gross): | 100 |
| Transmissions in inventory: | 2 |
| Net requirements, transmission A: | 98 |
| Gearboxes required for 98 transmissions (gross): | 98 |
| Gearboxes in inventory: | 15 |
| Net requirements, gearbox B: | 83 |
| Gears required for 83 gearboxes (gross): | 83 |
| Gears in inventory: | 7 |
| Net requirements, gear C: | 76 |
| Forgings required for 76 gears (gross): | 76 |
| Forgings in inventory: | 46 |
| Net requirements, forging D: | 30 |

The net requirement quantity for item $D$ can be verified as follows:

| | | |
|---|---:|---:|
| Quantity of trucks to be produced: | | 100 |
| Quantity of item D that will be consumed: | 100 | |
| Inventory of D: | 46 | |
| Inventory of C containing D: | 7 | |
| Inventory of B containing C: | 15 | |
| Inventory of A containing B: | 2 | |
| | 70 | |
| Net requirements for item D: | 30 | |
| Totals: | 100 | 100 |

The computation of net requirements proceeded in the direction from top to bottom of the product structure in a level-by-level fashion. This procedure accounts for, or flushes out, component item $D$ in its consumed state "hiding" in higher-level items $A$, $B$, and $C$ that will be used in the manufacture of product $X$. If the computation proceeded in the other direction, the where-used traces might lead into other branches of the BOM that do not apply to product $X$. For example, an additional quantity of item $D$ might be found hidden in parent item $Y$. If it entered into the netting process, the net requirements for item $D$ would be understated because item $Y$ is not used in the manufacture of truck $X$.

Net requirements are developed by allocating (reallocating) quantities in inventory to the quantities of gross requirements in a level-by-level process. The level-by-level netting procedure is laborious, but it cannot be circumvented or shortcut. The net requirement on the parent level must be determined before the net requirement on the component-item level can be determined.

The downward progression from one product level to another is called an *explosion*. In executing the explosion, the task is to identify the components of a given parent item and to ascertain the location (address) of their inventory records in computer storage so that they may be retrieved and processed.

The BOM file (or product-structure file) guides the explosion process. Product-structure data are not operated on but merely consulted by the system to determine component identities and quantities "per" (previously illustrated in Figure 6-3). The generic name of the computer program (software) that organizes and maintains the product-structure file is the *BOM processor*. The program also handles the retrieval of individual BOMs as required during the explosion process.

## Lot Sizing

Lot sizing (as defined earlier) is also a factor in the requirements computation, and it is another reason why the top-to-bottom, level-by-level procedure must be followed. In the preceding example, a tacit assumption was that parent items $A$, $B$, and $C$ will be ordered in quantities equal to the respective net requirements for those items. In reality, though, lot sizing, where employed, would invalidate this assumption. This is so because the gross requirement for a component derives directly from the (planned) order quantity of its parent(s).

If we modify the preceding example by stipulating that for gear $C$, production order quantities must be multiples of five (because of some consideration in the gear machining process), the net requirement of 76 will have to be covered by a planned order for 80. This will increase the gross requirement for forging blank $D$ correspondingly, as illustrated in Figure 6-7. When the planned order for 80 gears (parent item) is released, 80 forging blanks (component item) will have to be issued.

Lot sizing, that is, the particular technique used to determine order quantities for a given inventory item, therefore affects the requirements for its components. For an MRP system to be able to carry out a complete explosion, lot sizing must be part of the procedure, and the respective lot-sizing rules (algorithms) must be incorporated into the com-

**FIGURE 6-7**

Timing of a gross
requirement.

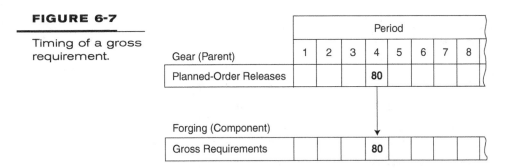

puter program that controls the requirements computation. The subject of lot sizing will
be reviewed in Chapter 8.

The general rule of MRP logic can be stated as follows: The mutual parent-component relationship of items on contiguous product levels dictates that the net requirement
on the parent level, as well as its coverage by a planned order, be computed before the
gross requirement on the component level can be determined correctly. The timing of a
gross requirement for a component item coincides with the timing of an order release
planned for its parent, as shown in Figure 6-7.

## Item Lead Times

Lead times of the individual inventory items are a complicating factor in the computation of material requirements. The preceding example of net requirements determination
for forgings, gears, and so on was oversimplified in ignoring, among other things, item
lead times. It is these lead times that will determine the timing of releases and scheduled
completions for the orders in question. Because a component-item order must be completed before the parent-item order that will consume it can be started, the back-to-back
lead times of the four items in the example make up the cumulative lead time, a sort of
critical path that determines the earliest time that the end products could be built or,
given the end-product schedule date, the latest time for the start of the lowest-level item
order. Cumulative lead time is represented graphically in Figure 6-8. See also the discussion of lead time in Chapter 4.

If the manufacturing lead times for the four items in the example were

| | |
|---|---|
| Forging blank D: | 3 weeks |
| Gear C: | 6 weeks |
| Gearbox B: | 2 weeks |
| Transmission A: | 1 week |
| | 12 weeks (cumulative lead time) |

and assembly of the end product, truck X, were scheduled for a date arbitrarily designated as week 50, component-order release dates and completion dates could be calculated by successively subtracting the lead-time values from 50:

**FIGURE 6-8**

Cumulative lead
time.

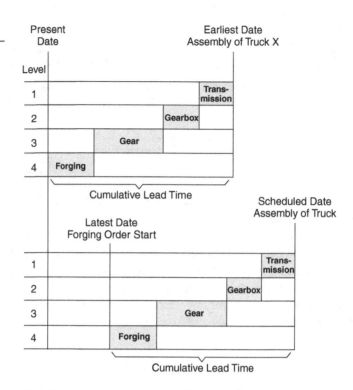

| Complete order for item A: | week 50 |
| Minus lead time of item A: | 1 week |
| Release order for item A: | week 49 |
| Complete order for item B: | week 49 |
| Minus lead time of item B: | 2 weeks |
| Release order for item B: | week 47 |
| Complete order for item C: | week 47 |
| Minus lead time of item C: | 6 weeks |
| Release order for item C: | week 41 |
| Complete order for item D: | week 41 |
| Minus lead time of item D: | 3 weeks |
| Release order for item D: | week 38 |

The preceding release dates of parent orders would establish the timing of gross requirements for the respective component items, as illustrated in Figure 6-7. Because, in the example, none of the gross requirements can be fully satisfied from inventory, the timing of net requirements for each of the items coincides with that of gross requirements. The timing of the orders providing coverage of net requirements is geared to the latter (for completion dates) and is a function of the respective lead times (for release dates). In the preceding example, it is assumed that lot sizing does not affect the lead time, that is, that item lead time does not vary with the quantity being ordered.

Time phasing solves the problem of the effect of item lead time on the timing of requirements. Lead-time values (or procedures for determining these values based on order quantity) must be supplied to the MRP system, which stores them for use in establishing a proper alignment of requirements and planned-order data in the course of the requirements explosion. Subtracting the lead time from the date of the net requirement, that is, positioning the planned-order release forward of the timing of the net requirement it covers, is called *offsetting for lead time*.

## Recurrence of Requirements Within the Planning Horizon

The planning horizon of the MPS usually covers a time span large enough to contain multiple (i.e., recurring) requirements for a given end item. This represents another complication in the computation of component requirements. The truck $X$ example assumes that the inventory quantities of all the component items involved are available for netting against the gross requirements generated by the lot of 100 trucks. But there may be another lot (or several) of the same end product (or of a different end product using the same transmission) that precedes the one for 100 in the MPS. If this is the case, its net requirements must be accounted for before the net requirements for the lot of 100 can be determined.

If we assume that there is a preceding lot of 12 trucks $X$ (lot number 1), the net requirements for the respective lots (retaining the lot-sizing rule for gear C) would be calculated as follows:

|  | Lot no. 1 | Lot no. 2 |
|---|---|---|
| Transmission A |  |  |
| Gross requirements: | 12 | 100 |
| Inventory: | 2 | 0 |
| Net requirements: | 10 | 100 |
| Gearbox B |  |  |
| Gross requirements: | 10 | 100 |
| Inventory: | 15 | 5 |
| Net requirements: | −5 | 95 |
| (available for lot no. 2) |  |  |
| Gear C |  |  |
| Gross requirements: | 0 | 95 |
| Inventory: |  | 7 |
| Net requirements: | 0 | 88 |
| Planned-order quantities: | 0 | 90 |

Forging blank D

| Gross requirements: | 0 | 90 |
|---|---|---|
| Inventory: | | 46 |
| Net requirements: | 0 | 44 |

As can be seen, the existence of a preceding lot of 12 has changed the net requirements for the lot of 100. The net requirements for the forging blank, for instance, have increased from 34 to 44. Note that this increase of 10 does not match the increase of 12 in the end-product requirements. This difference is caused by the lot-sizing rule (order in multiples of 5) applied at the item C level.

The requirements-calculation procedure followed in this example is identical to the one used in lot requirements planning, which was discussed briefly in Chapter 5. Under this approach, requirements are determined for one end-item or product lot at a time, and to the extent that the various end items use common components, the chronological sequence of the lots (for all products) must be strictly observed for the requirements data to be valid. These data, once developed, remain valid as long as the chronological sequence of the lots in the MPS continues unchanged.

In the development of lot requirements, component inventories have been allocated (applied against gross requirements) according to the sequence of (all) the end-item lots. If this sequence changes subsequently—as it often will in practice—component inventories must be reallocated and requirements redetermined. A change in end-item lot sequence affects not only the timing but also the quantities of requirements. This can be seen in the preceding example of the forging blank. Had lot number 2 preceded lot number 1 (this will become more realistic if we think of the lots as representing two different models of trucks that use the same transmission), the requirements developed in the earlier example would have been correct—the net requirements for forging D would have been 34 for lot number 2 and 15 for lot number 1 for a total of 49, but when the sequence of the lots is reversed, the net requirements for the same item are (as per the latest example) 0 for lot number 1 and 44 for lot number 2 for a total of 44.

This phenomenon constitutes a severe problem in lot requirements planning. Because the sequence of the lots (of all products in the MPS) normally keeps changing, either the individual lot requirements are being reprocessed repeatedly or they are not—the usual case—and the plant operates with invalid requirements information and orders that may be wrong in both quantity and timing.

The sequence of end-item lots also affects both the quantities and the timing of requirements in an MRP system, but because this system computes requirements by product level rather than by lot, and because time phasing follows the correct chronological order of requirements as per the current MPS, no distortions occur. For this reason, a changing lot sequence is of little concern under the time-phased MRP approach, and it has virtually no effect on the data-processing burden of the MRP system.

## Common Usage of Components

Common usage of a component item by several parent items is another complicating factor in the computation of requirements by an MRP system. In the examples used thus far, all the component items had a single parent, that is, a single source of demand. In the real situation, many components will be found to have multiple parents. The lower the level of the component item, the more parent items it tends to have. In our example, forging blank *D* is used to make gear *C*. But the same blank might very well be used for the manufacture of half a dozen different gears. The lower-level steel, from which this forging is made, almost certainly would be used for a large number of different parts made from that particular grade and diameter of steel bar. This also would make this part a possible candidate for a strategic buffer to compress the overall manufacturing lead time. See Chapter 23 for the discussion of strategic buffers and actively synchronized replenishment lead time (ASRLT).

In order to determine the net requirements for such a common-usage item correctly, its gross requirements stemming from all its parent items for which there are planned orders must be determined first. (*Note:* The mere existence of a parent item is not tantamount to a gross requirement.) In addition to dependent demand generated by its parent items, a component also may be subject to independent demand if it is used as a service part. Such demands, if any, also must be added to the gross requirements for the item in question. The gross-requirements schedule of an inventory item represents a summary of demands originating from one or more sources and applicable to various points in time. This is illustrated in Figure 6-9.

The level-by-level (rather than lot-by-lot) approach to the computation of requirements under an MRP system minimizes the problem of multiple parent demands. All the parents of a given component item tend to be on the same, next-higher level. Because that entire level (i.e., all the items on that level) is processed first, all the parent planned orders

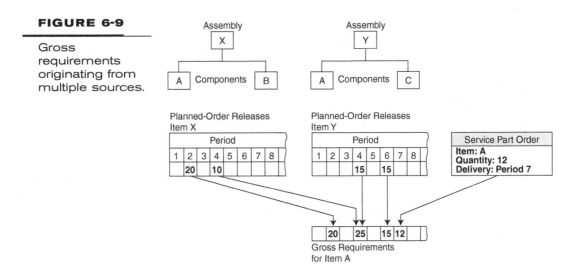

**FIGURE 6-9**

Gross requirements originating from multiple sources.

are developed before the next-lower level is netted. This means that the component's requirements need not be reprocessed and renetted subsequently owing to emergence of another gross requirement generated by another parent item. The foregoing applies provided the component item in question appears on only one level in the product structure. Some items, however, may appear on two or more different levels. This leads to the next complicating factor.

## Multilevel Items

Recurrence of gross requirements for a given item during the requirements-computation process may be caused by the fact that the item's several parents are on different BOM levels. Because a component item, by definition, is always on the next level below its parent item, this means that whenever an item's parents appear on more than one level, the item itself has multiple levels associated with it. There are two cases: An item may exist on different levels in the structure of different end items that use it in common (Figure 6-10), or it may exist on different levels in the structure of one end item (Figure 6-11). In a given case, both these conditions may exist at the same time. The problem here is one of having to reprocess and renet at every recurrence of gross requirements stemming from parent items that appear on multiple levels. This would mean multiple retrievals of

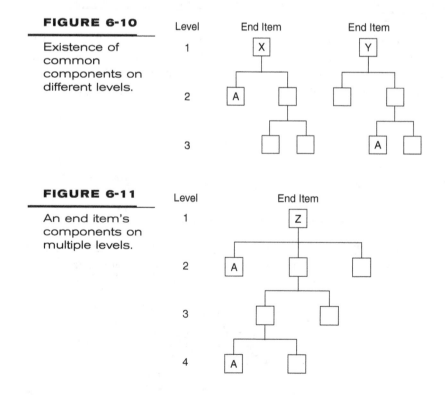

**FIGURE 6-10**

Existence of common components on different levels.

**FIGURE 6-11**

An end item's components on multiple levels.

the component-item record from storage during the requirements explosion and a reduction in data-processing efficiency.

This problem is solved by employing a low-level coding technique. The lowest level at which any inventory item appears is identified through an analysis of the BOM material file, and this information—the low-level code—is added either to the BOM or to the item's inventory record. In the level-by-level requirements-computation process, the processing of the item then is delayed until the lowest level on which it appears is reached. At that point, all the possible gross requirements stemming from parent items at any of the higher levels have been established, and the need for multiple retrievals and processing has been forestalled. The concept of low-level coding is illustrated in Figure 6-12.

All the factors just discussed that affect and complicate the MRP process must be taken into account by the procedural approach used by any MRP system. These approaches have been perfected and streamlined gradually, and they are by now largely standardized in what is called the *processing logic* of an MRP system. Processing logic will be the subject of discussion in Chapter 7.

**FIGURE 6-12**

Low level coding.

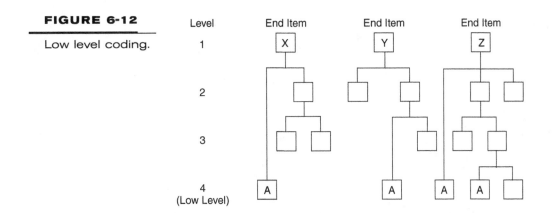

# Processing Logic

The term *logic*, when used in the context of material requirements planning (MRP), pertains to the reasoning behind a procedure or a system of procedures rather than to specific procedural steps. The validity of the results a procedure or system will yield is a function of the soundness of this reasoning. This chapter describes the inner workings of an MRP system at a level of detail required to make this description reasonably comprehensive. The problem in doing so, however, is that there are alternative approaches to MRP system implementation and that the individual systems installed in different industry applications cover a spectrum of special features, functions, and procedures. How an MRP system works therefore must be described largely in terms of processing logic rather than in terms of specific procedure. Chapters 15 through 18 provide additional information on the specific application of MRP to a variety of industries.

## INVENTORY STATUS

The status of an inventory item must be known before it can be determined what, if any, inventory management action is to be taken on that item. Inventory status (or stock status) is expressed by means of data that define an item's current position. Status information is intended either to answer or to help answer (depending on how complete the information is) the essential questions of:

- What do we have?
- What do we need?
- What do we do?

The answer to the last question follows from an evaluation of the status, which can be performed either by an inventory planner or by a computer executing an evaluation procedure.

The most primitive expression of inventory status is limited to data on quantities on hand and on order. Action then is determined by comparing need (demand) with that availability status or on depletion of inventory with some predetermined minimum, in which case the action taken reflects anticipated future need. A more elaborate expression of inventory status is provided by the classic "perpetual inventory control" equation introduced in Chapter 6 and repeated below, which states the elements of expanded inventory status and their relationship. This equation, while valid, is still somewhat primitive:

$$A + B - C = X$$

where $A$ is the quantity on hand, $B$ is the quantity on order, $C$ is the quantity required, and $X$ is the quantity available.

If $X$ has a positive value, it indicates the quantity available for future requirements. If it is negative, it is an indication of an impending shortage, that is, inadequate coverage. The idea behind this venerable approach to inventory control is for the value of $X$ to equal or exceed zero at all times. This is accomplished by increasing the value of $B$ by placing a new order whenever $X$ approaches zero or turns negative. This policy would appear to preclude shortages, but it does not because the expression of inventory status is too crude on three counts:

1. Information on timing of the demand and supply is lacking.
2. The data on $B$ and $C$ represent summaries.
3. The status formula does not provide for planned (future) coverage.

For example, status might be indicated as follows:

On hand:      100
On order:      120          $100 + 120 - 200 = 20$
Required:      200

The technique signals that all is well and that no action is called for, but in fact, there will be a shortage, as becomes evident when information on timing is associated with the status data:

On hand:      100
On order:      120, due June 1
Required:      200, May 15

Coverage is adequate in terms of quantity but not in terms of timing. To illustrate the opposite case, let the status be as follows:

On hand:        20
On order:      100          $20 + 100 - 200 = -80$
Required:      200

The technique signals that an order for 80 or more pieces should be placed to provide coverage. Let us assume that the status is examined by an inventory planner on March 1 and that 80 units of the item, which has a lead time of four weeks, are ordered on that date. This action proves incorrect in light of more detailed information on inventory status, such as

| | |
|---|---|
| On hand: | 20 |
| On order: | 100, due March 10 |
| Required: | 110, March 15 |
| | 90, June 15 |

This technique of expressing inventory status does not make the adequacy of coverage sufficiently clear. In both these examples, the action signals are false. The message in the first case should have been, "Reschedule the open order to May 15," and in the second, "An order for at least 80 will have to be placed, but not for another 10 weeks." The technique in question does not provide for orders planned for future release (as the second of the examples illustrates) and in assessing status depends on open orders only.

In an MRP system, all the preceding shortcomings in expressing inventory status are overcome by expanding the number of status elements and by time-phasing status data. The classic equation is, in effect, expanded to

$$A + B + D - C = X$$

where $D$ quantity is planned for future order release.

The MRP system evaluates the status of each inventory item, automatically establishes planned-order coverage, and signals for consistently correct action. Under MRP, the elements of inventory status (all of them associated with timing information) are these:

- Quantity on hand
- Quantities on order
- Gross requirement quantities
- Net requirement quantities
- Planned-order quantities

These status data can be divided into two categories by type:

1. Inventory data (replenishment)
2. Requirements data

Inventory data consist of on-hand and on-order quantities and their timing. These data are reported to the system and can be verified by inspection. Requirements data consist of the quantities and timing of gross requirements, net requirements, and planned-order releases. These data are computed by the system and can be verified only through recomputing. This can be done in a regenerative mode (triggered by passage of time) or net-change mode (triggered by an event).

The three principal functions of MRP are to

1. Plan and control inventories
2. Plan and replan released-order priorities
3. Provide data for capacity requirements planning

Design of MRP programs should meet these objectives:

1. Planning inventories:
   - Order the right part
   - Order in the right quantity
   - Order at the right time
2. Planning priorities:
   - Order with the right due date
   - Keep the due date valid
3. Planning capacity:
   - A complete load
   - An accurate (valid) load
   - An adequate time span for visibility of future load

## TECHNIQUES OF TIME PHASING

One of the oldest inventions of human civilization is the division of time's continuous flow into increments suitable for measuring its passage and the construction of calendars to provide a frame of reference. Our Gregorian calendar serves satisfactorily for most purposes, but when inventory-planning and production-scheduling procedures are to be automated (i.e., their execution transferred from a human being to a machine), certain characteristics of this calendar present difficulties. The Gregorian calendar does not employ a decimal base, months have an uneven number of days, and the pattern of holidays is irregular.

### Scheduling Calendars

Since these factors would complicate time-related computing procedures unnecessarily, it is common to devise special decimal calendars that are used for this purpose. There are a variety of so-called scheduling calendars or shop calendars, but what all have in common is the principle of consecutively numbering weeks and/or days. This functionality is now embedded in all commercially available MRP systems and is invisible to the user. Commonly, there is one calendar for the company's own work and another for supplier lead times. Recommended action dates are adjusted by the declaration of working versus nonworking days in the calendar.

In the use of this calendar, one week is normally equated with five working days. Thus, if an item had a five-week manufacturing lead time, 25 would be subtracted from its due date to arrive at the order release date. The actual span of time might be more than five weeks if holidays or a plant vacation happened to intervene.

## Dates and Time Buckets

In an MRP system, inventory status data are time phased by associating them with either days (relatively short time periods) or planning periods such as weeks or months (relatively long time periods). While the specific method of time phasing that is being employed will determine the terms in which the internal arithmetic is carried out by the system, the method of display of time-phased data can be selected independently. A given day can be converted to its respective planning period for purposes of display, and a planning period can be expressed in terms of one of its constituent days, usually its starting day. Figure 7-1 demonstrates how these data can be displayed by different time buckets.

**FIGURE 7-1**

Row of different time values display.

| | 1 | 2 | | 23 | 24 | 25 | 26 | Jul | Aug | Sep | Oct | Nov | Dec | I | II | III | IV |
|---|---|---|---|---|---|---|---|---|---|---|---|---|---|---|---|---|---|
| | | | | | | | | | | | | | | Period | | | |
| Gross Requirements | 10 | 15 | | 5 | 20 | 10 | 10 | 40 | 30 | 50 | 50 | 50 | 30 | 120 | 150 | 150 | 100 |

MRP systems are designed in such a way that requirements and inventory data can be entered, stored, and processed internally by date/quantity but displayed in time-bucket format. Technological developments over the past 30 years have allowed MRP systems to develop into "bucketless" systems. The development of time buckets that correspond to periods of time normally spanning more than a single day is no longer required. This represents a rather coarse division of time. For this reason, the precise meaning of the timing of data assigned to time buckets was fixed by convention. Currently, MRP systems schedule to the day. Therefore, the logic of time buckets is no longer necessary. In the following discussion of gross and net requirements, although time periods can be a day, a week, a month, or a quarter, the common deployment in MRP is that a period is a single day.

## Planning Horizon

To ensure that MRP provides data on items at all levels in bills of material (BOMs), the planning horizon at least should equal the largest total of item lead times (cumulative lead time) in the critical (longest) path leading from raw material to the end item appearing in the master production schedule (MPS). If planning horizons are too short, the process of successively offsetting for lead time in level-by-level planning will run into past periods when it reaches items on the lowest level. To ensure some *forward visibility* of data on purchased items, planning horizons should be significantly longer than the critical-path lead time.

With multilevel product structures, because of successive lead-time offsetting, there is a partial loss of horizon at each lower level. The effective planning horizon at each level is successively diminished as MRP progresses from one level to the next. One consequence of very short horizons is the inability to apply some lot-sizing techniques effectively because of lack of sufficient net requirements data. This is discussed in Chapter 8. Another, more serious consequence is lack of data for capacity requirements planning. Short horizons limit capacity requirements planning on the low-level, usually fabricated, parts where it is most desirable. However, as is discussed in depth in Chapter 9, long horizons result in plans with low validity because of greater potential changes in requirements, design details, and processing methods and more upsets. The resolution of this dilemma also is covered in Chapter 9.

## GROSS AND NET REQUIREMENTS

The concept of gross versus net requirements was reviewed briefly in Chapter 6. The gross requirement for an inventory item equals the quantity of demand for that item in that time period. Its net requirement is arrived at by allocating the available inventory on hand and on order to (i.e., subtracting from) the gross requirement in that time period. The term *gross requirement* has a specific meaning in this context, however. It is the quantity of the item that will have to be disbursed, that is, issued to support a parent order (or orders), rather than the total quantity that will be consumed by the end product. These two quantities may or may not be identical.

To illustrate, in the Chapter 6 example of truck $X$ and component item $D$, 100 trucks are to be produced, and each unit of the end product contains one unit of item $D$. The gross requirement for $D$ therefore could be said to equal 100. This figure, while meaningful in product costing and so on, is quite meaningless for purposes of MRP, where the question is not what quantity of a component will go out the door with the product but what (minimum) quantity will have to be procured or manufactured (i.e., the net requirement) to allow the product to be built and shipped in the quantity required. In the example mentioned, the gross requirement for item $D$ is computed to be 76 and the net requirement 30. The gross requirement could have been 100, however, had there been no inventories (e.g., gears, gearboxes, and transmissions) at higher levels. In an MRP environment, the gross requirement is equivalent to demand at item level rather than to demand at product or MPS level.

There may be multiple sources of demand, and therefore of gross requirements, for a given component item. This also was brought out previously. An item may be subject to dependent demand from several parent items that use it in common, and it also may be subject to independent demand generated from sources external to the plant. These gross requirements for the item are combined and summarized, by planning period, in the gross requirements schedule, that is, the respective row of time buckets, as illustrated in Chapter 6. This is reproduced in Figure 7-2. Using these data to develop subsequent examples, we must now introduce some inventory data so that net requirements can be determined and time phased correctly.

**FIGURE 7-2**

Gross requirements schedule.

| | Period | | | | | | | | Total |
|---|---|---|---|---|---|---|---|---|---|
| | 1 | 2 | 3 | 4 | 5 | 6 | 7 | 8 | |
| Gross Requirements | | 20 | | 25 | | 15 | 12 | | 72 |

## Net Requirements

Assume that there are 23 units of this item on hand and 30 on order and due in period 3. This is displayed in time-bucket format in Figure 7-3.

**FIGURE 7-3**

Time-phased status before net requirements computation.

| | Period | | | | | | | | Total |
|---|---|---|---|---|---|---|---|---|---|
| | 1 | 2 | 3 | 4 | 5 | 6 | 7 | 8 | |
| Gross Requirements | | 20 | | 25 | | 15 | 12 | | 72 |
| Scheduled Receipts | | | 30 | | | | | | 30 |
| On Hand | 23 | | | | | | | | |

The logic of the net requirements computation is simple:

| | Gross requirements |
|---|---|
| *minus* | Scheduled receipts |
| *minus* | On hand |
| *equals* | Net requirements |

In using this formulation, it is understood that if the result is a negative value, that is, if the sum of quantities on hand and on order exceeds the gross requirement, the net requirement is zero. With data in time-phased format, the calculation is performed successively for each period, and unapplied inventory is carried forward. This is shown in Figure 7-4.

**FIGURE 7-4**

Calculation of net requirements.

| Period | Gross requirements | Scheduled receipts | On hand | | Result | Net requirements |
|---|---|---|---|---|---|---|
| 1 | 0 | −0 | −23 | = | −23 | 0 |
| 2 | 20 | −0 | −23 | = | −3 | 0 |
| 3 | 0 | −30 | −3 | = | −33 | 0 |
| 4 | 25 | −0 | −33 | = | −8 | 0 |
| 5 | 0 | −0 | −8 | = | −8 | 0 |
| 6 | 15 | −0 | −8 | = | 7 | 7 |
| 7 | 12 | −0 | −0 | = | 12 | 12 |
| 8 | 0 | −0 | −0 | = | −0 | 0 |
| Totals | 72 | 30 | | | | 19 |

The correctness of the computation is verified by a check of totals as follows:

|         | Total gross requirements: | 72 |
|---------|---------------------------|----|
| *minus* | Total scheduled receipts: | 30 |
| *minus* | On hand:                  | −23 |
| *equals* | Total net requirements:  | 19 |

This is illustrated in Figure 7-5 in time-phased record format. Under an alternative method of calculating net requirements, the quantity on hand is time-phased, that is, projected into the future, period by period, and the first negative value then represents the first net requirement. Differences between successive negative values equal the net requirements in the respective later periods.

The logic of the on-hand/net-requirements computation is as follows:

|         | Balance on hand at the end of a given period |
|---------|-----------------------------------------------|
| *plus*  | Quantity on order and due in the succeeding period |
| *minus* | Gross requirements of the succeeding period   |
| *equals* | Balance on hand at the end of the succeeding period |

Under this method, the "negative on hand" is understood to equal a net requirement, recorded cumulatively. The calculation is performed in Figure 7-6.

This method provides a more efficient display in that it combines the projected on hand and the net requirements information in a single row of time buckets, as shown in Figure 7-7. Although it may appear that the data, when presented this way, are more dif-

## FIGURE 7-5

Time-phased status after net requirements computation.

|                    |     | Period | | | | | | | | Total |
|--------------------|-----|---|---|----|---|---|----|----|---|-------|
|                    |     | 1 | 2 | 3  | 4 | 5 | 6  | 7  | 8 |       |
| Gross Requirements |     |   | 20 |   | 25 |   | 15 | 12 |   | 72    |
| Scheduled Receipts |     |   |   | 30 |   |   |    |    |   | 30    |
| On Hand            | 23  |   |   |    |   |   |    |    |   |       |
| Net Requirements   |     |   |   |    |   |   | 7  | 12 |   | 19    |

## FIGURE 7-6

Alternative calculation of net requirements.

| Period | On hand at beginning of period | Scheduled receipts | Gross requirements |   | On hand at end of period |
|--------|--------------------------------|--------------------|--------------------|---|--------------------------|
| 1 | 23  | +0  | −0  | = | 23  |
| 2 | 23  | +0  | −20 | = | 3   |
| 3 | 3   | +30 | −0  | = | 33  |
| 4 | 33  | +0  | −25 | = | 8   |
| 5 | 8   | +0  | −0  | = | 8   |
| 6 | 8   | +0  | −15 | = | −7  |
| 7 | −7  | +0  | −12 | = | −19 |
| 8 | −19 | +0  | −0  | = | −19 |

**FIGURE 7-7**

Alternative
method of net
requirements
display.

| | | | | Period | | | | | | Total |
|---|---|---|---|---|---|---|---|---|---|---|
| | | 1 | 2 | 3 | 4 | 5 | 6 | 7 | 8 | |
| Gross Requirements | | | 20 | | 25 | | 15 | 12 | | 72 |
| Scheduled Receipts | | | | 30 | | | | | | 30 |
| On Hand | 23 | 23 | 3 | 33 | 8 | 8 | −7 | −19 | −19 | −19 |

ficult to interpret, little additional skill is actually required for interpretation. The discrete, period-by-period net requirements need not be mentally calculated from the cumulative figures but can, in fact, be read directly from the record. This is due to the fact that in all periods that have net requirements, these net requirements equal gross requirements except for those periods during which inventory either runs out or is added to by a new receipt. Figure 7-7 provides an example. The first net requirement of seven in period 6 is shown as such, and the subsequent net requirements equal gross requirements, that is, 12 and 0. Note also that the last (cumulative) net requirements figure, −19, equals the total net requirements for the eight periods covered by the record.

The timing of component gross requirements is linked directly to time-phased parent planned-order due dates. Planned-order quantities are determined by ordering policies that select lot-sizing techniques specified by users. Different techniques can be applied to different items or item classes.

MRP covers net requirements with planned orders for future release. Planning horizons should extend far enough into the future so that all components of the MPS end items have at least one net requirement and one planned order. Chapter 8 gives details of lead-time–planning-horizon relationships.

To generate a planned order correctly, MRP must determine

1. The timing of order completion (due) date
2. The timing of order release (start) date
3. The order quantity

The timing of order completion must coincide with the timing of the net requirements being covered if shortages are to be prevented.

## Safety Stock and Net Requirements

Earlier there was mention of the fact that the planning of safety stock on the item level affects the calculation of net requirements. For purposes of this calculation, the quantity of safety stock is either subtracted from the on-hand quantity or added to the gross requirement. Either alternative produces the same effect, namely, a corresponding increase in net requirements and sometimes a shift of the first net requirement one period forward. In the Figure 7-7 example, had there been a safety stock of two units, the projected on-hand and net requirements quantities would have been as shown in Figure 7-8.

## FIGURE 7-8

Net requirements after safety-stock deduction.

| | | | | | Period | | | | | Total |
|---|---|---|---|---|---|---|---|---|---|---|
| | | 1 | 2 | 3 | 4 | 5 | 6 | 7 | 8 | |
| Gross Requirements | | | 20 | | 25 | | 15 | 12 | | 72 |
| Scheduled Receipts | | | | 30 | | | | | | 30 |
| On Hand | 23 | 23 | 3 | 33 | 8 | 8 | −7 | −19 | −19 | −19 |
| Net Requirements | | | | | | | 9 | 12 | | 21 |

When safety stock is planned at the item level, the material requirements planning logic attempts to conserve its quantity and to "protect" it from being used up so that this quantity always might be on hand. To the extent that the system succeeds in thus safeguarding safety stock, it creates "dead" inventory that is carried along but never used—a distinctly undesirable condition. Item safety stock forces the MRP system to overstate requirements, which in itself is undesirable. The system either tells the truth or does not. An overstated requirement sometimes leads to distorted timing, when the safety stock causes the net requirement to be pulled forward in time. Treatment of safety stock is usually considered as a current period demand causing the components to always be past due.

Overstated requirements and false timing (order due dates) tend to cause confusion, unnecessary expense, and most important, loss of credibility by the MRP system. Factory personnel will quickly discover whether or not they can rely on the integrity of information being generated by the system. Shop supervisors using common sense almost certainly will disregard the due date indicated on a given shop order and delay its completion if they know that a quantity of the item is available in stock or that the order is scheduled for completion before actually needed. In time, vendors of purchased items will learn that missing a due date has no serious consequences, and they will tend to miss more due dates thereafter.

Safety stock at the item level is part of the stock-replenishment concept and as such has no legitimate place in an MRP system despite the fact that it easily can be incorporated into such a system. The primary purpose of safety stock is to compensate for fluctuations in uncertain demand, that is, for forecast error. In an MRP system, however, demand for the individual component items is not being forecast and therefore is not subject to forecast error. Component-item demand is certain relative to the MPS.

This schedule may be based on a forecast of demand for products or end items. Variability of demand is at the MPS level, not at the component-item level. Strategic buffers are placed at the necessary leverage points to ensure that sufficient compression of the reaction lead time is less than the customer tolerance time. In some cases, safety stock, where required, therefore should be provided through the MPS; that is, it should be planned in terms of end items. The MRP system, which explodes the contents of the MPS into detailed component-item requirements, need not and should not duplicate safety-stock inventory at the component-item level. Any variability at the lower levels in the BOMs is protected by the strategic buffers.

Be aware that when safety stock is planned through the MPS, it has the added impact of planning additional components in matched sets, allowing the safety-stock quantities of end items actually to be built. This will increase overall inventory investment. The preceding arguments regarding the proper place of safety stock can be summed up by stating that safety stock is applied properly only to inventory items subject to independent demand when the actively synchronized replenishment (ASR) lead time is greater than the customer tolerance time. The resupply (completion) of manufactured items need not be erratic because performance to schedule is controllable, particularly in an MRP environment. Establishment of a replenishment buffer for component items can compress the recovery time dramatically when volatility is experienced. This is covered in Part 4.

## COVERAGE OF NET REQUIREMENTS

The quantities and timing of net requirements for a given inventory item can be thought of as indicating impending shortages caused by lack of coverage. Assuming an adequate planning horizon, an MRP system detects such shortages sufficiently in advance to allow their coverage to be planned in an orderly manner. An MRP system detects future potential shortages and plans their coverage so that actual shortages will not occur.

### Planned Orders

In an MRP system, net requirements are covered by *planned orders*, that is, new orders for the respective items scheduled for release in the future. Depending on the planning horizon, the level of the item in the product structure, and the applicable lot-sizing rule, an item with net requirements will have one or more planned orders indicated. The timing of the first (earliest) planned order is governed by the timing of the first net requirement. The order quantity must equal or exceed the net requirement. If this quantity exceeds the net requirement, the timing of the next (second) planned order may be affected. A planned order may cover net requirements occurring in one or more planning periods.

To generate a planned order correctly, the system must determine the following:

1. The timing of required order completion (due date)
2. The timing of order release
3. The order quantity

The timing of order completion derives, of course, from the timing of the net requirement being covered. As mentioned previously, the timing of the planned-order release is arrived at by offsetting for lead time, that is, by subtracting the value of the lead time (expressed in shop calendar units) from the shop calendar date of order completion.

The individual-item lead times used by the MRP system in determining planned-order releases are, by necessity, estimates in most cases. The lead-time value of the four periods used in the preceding example represents the amount of time that could be expected to elapse between order release and order completion—if everything else went according to plan. This is the *planned lead time*, not to be confused with the actual lead time. The latter generally can be determined only in retrospect because it is the amount of time it actually took to complete the order in light of possible changed requirements and unplanned events. The actual lead time of a given item may and often does vary widely from order to order.

## Lead-Time Contents

Planned, or normal, lead times must be used by the MRP system for purposes of planning, but their *accuracy* is not crucial. These lead times are, after all, used merely to determine order release dates, which are considerably less important than the completion dates related to actual lead times. The lead time of a manufactured item is made up of a number of elements, listed here in descending order of significance:

- Queue (waiting to be worked on) time
- Running (machining, fabrication, assembly, etc.) time
- Setup time
- Waiting (for transportation) time
- Inspection time
- Move time
- Other elements

In a general machine shop (job shop) environment, the first of the elements listed normally accounts for roughly 90 percent of average total elapsed time. In an individual case, queue time is a function of a job's relative priority as it finds itself in contention for a given productive facility with other jobs. The queue time and, consequently, the actual lead time of the respective order will increase or decrease as its priority is changed—the "hottest" orders spend little time in queue. Actual lead time is usually quite flexible and in emergency situations can be compressed to a small fraction of the planned lead time.

An MRP system has an inherent ability, discussed in Chapter 6, to reevaluate all open order due dates and to indicate changes in work priorities required for the orders to finish on the dates of actual need irrespective of the originally assigned due dates. The disparity between planned and actual lead times therefore is of no concern. Planned lead times serve to time order releases and no more than that.

In establishing planned lead-time values, it is possible to compute them through more or less elaborate procedures and formulas based on work standards, in-plant travel distances, average or planned queue times, and so on, but the precision thus achieved is spurious. Lead-time accuracy is indeterminate—the concept is elusive and devoid of meaning. This is why an empirically derived manufacturing lead time or any reasonable estimate will do for purposes of MRP. Most methods used, including straight estimates, yield a fixed lead time that will not vary with the quantity of the order. In cases where machining time per operation per piece is significant (e.g., turning large shafts, planing and milling heavy castings, etc.), a lead-time computation procedure can be devised that takes lot size into account.

Planned lead time is sometimes artificially inflated by the inclusion of an element called *safety lead time* or *safety time*. This element is inserted at the end of the normal lead time for the purpose of completing an order in advance of its real date of need. Where safety lead times are used, the MRP system, in offsetting for lead time, will plan both order release and order completion for earlier dates than it would otherwise. Order due dates will be advanced from dates of real need by the amount of safety lead time.

The concept of safety lead time is actually quite similar to that of safety stock. The primary purpose of both is to compensate for the vagaries of item demand. The effect of safety lead time is to create an inventory excess that then can be used to meet unanticipated demand. In practice, however, this inventory tends to remain in work-in-process, and the extra time serves to facilitate expediting completion of the order by the date of real need, that is, the date that would have been the order due date in the absence of safety lead time.

## Timing and Size of Planned Orders

The ability to generate planned orders, that is, to plan for coverage of all future net requirements, is one of the most significant characteristics of an MRP system. For every inventory item with net requirements, the system develops a planned-order schedule consisting of quantities and timing of as many planned-order releases as may be required to cover net requirements throughout the planning horizon. This schedule details inventory order action that will have to be taken in the future.

Planned-order schedules constitute one of the most valuable outputs of an MRP system despite the fact that the bulk of the planned-order data is not related to current order action. The main value of planned orders lies in providing the basis for a correct determination of their component-item requirements (a component gross requirement derives directly from a parent planned order, as pointed out previously) in terms of both quantity and timing. Planned orders provide "visibility" into the future and form the basis for various projections, including projected on-hand inventory, future purchase negotiations and commitments, and most important, production capacity requirements.

Once the net requirements for a given inventory item are determined and time-phased, the timing of any covering planned order can be established in a straightforward manner, as described earlier in this chapter. The answer to the question of planned-order quantity, however, is not equally clear-cut. At this point in the MRP process, one of a number of possible ordering policies or ordering rules is applied. Thus the planned-order quantities or lot sizes are a function of the lot-sizing rule specified for the item in question. Different lot-sizing rules can—and usually do—apply to different items or item classes within one MRP system.

A number of approaches to lot sizing in an environment of discrete period demands are possible, and several new techniques (lot-sizing algorithms) have been developed since the late 1950s when the first MRP systems were implemented. Because there are so many different lot-sizing techniques to be described, an entire chapter (Chapter 8) will be devoted to the subject.

## EXPLOSION OF REQUIREMENTS

The key to the entire MRP (replanning) process is the linkage between parent and component-item records. There is only one logical link between items on contiguous levels of the product structure, and that is the parent planned-order release and the component gross requirement. These coincide in time because the component item(s) must be planned to be available at the time the parent order is released for production, at which time the component item(s) will be consumed. The linkage of parent and component inventory records is shown in Figure 7-9.

The MRP process, that is, the so-called explosion of requirements from the MPS down into the various component-material levels, is guided by the logical linkage of inventory records. Gross requirements for the high-level items are processed against inventory (on hand and on order in each time period) to determine net requirements, which then are covered by planned orders. The quantity and timing of planned-order releases determine, in turn, the quantity and timing of component gross requirements. This procedure is carried out repetitively for items on successively lower levels until a purchased item is reached, at which point the explosion progression terminates. The MRP process stops when all the explosion paths that follow the branches of the BOM have reached purchased items (component parts or raw materials). The results of a requirements explosion for three items on contiguous levels are illustrated in Figure 7-10.

**FIGURE 7-9**

Linkage of parent and component records.

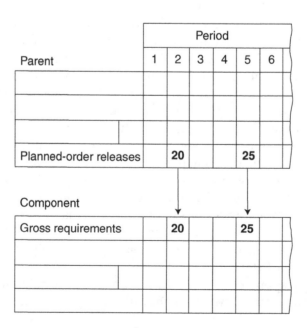

**FIGURE 7-10**

Explosion of
requirements.

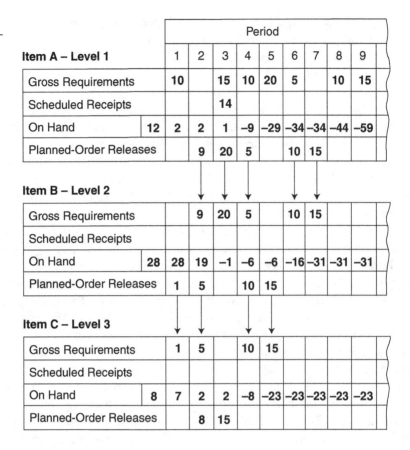

| Item A – Level 1 | | Period | | | | | | | | |
|---|---|---|---|---|---|---|---|---|---|---|
| | | 1 | 2 | 3 | 4 | 5 | 6 | 7 | 8 | 9 |
| Gross Requirements | | 10 | | 15 | 10 | 20 | 5 | | 10 | 15 |
| Scheduled Receipts | | | | 14 | | | | | | |
| On Hand | 12 | 2 | 2 | 1 | –9 | –29 | –34 | –34 | –44 | –59 |
| Planned-Order Releases | | | 9 | 20 | 5 | | 10 | 15 | | |

| Item B – Level 2 | | | | | | | | | | |
|---|---|---|---|---|---|---|---|---|---|---|
| Gross Requirements | | | 9 | 20 | 5 | | 10 | 15 | | |
| Scheduled Receipts | | | | | | | | | | |
| On Hand | 28 | 28 | 19 | –1 | –6 | –6 | –16 | –31 | –31 | –31 |
| Planned-Order Releases | | 1 | 5 | | 10 | 15 | | | | |

| Item C – Level 3 | | | | | | | | | | |
|---|---|---|---|---|---|---|---|---|---|---|
| Gross Requirements | | 1 | 5 | | 10 | 15 | | | | |
| Scheduled Receipts | | | | | | | | | | |
| On Hand | 8 | 7 | 2 | 2 | –8 | –23 | –23 | –23 | –23 | –23 |
| Planned-Order Releases | | | 8 | 15 | | | | | | |

In this example, component items *B* and *C* do not have multiple parents; that is, they are not common-usage items. In the real situation, however, multiple parents can be expected to exist, particularly for low-level items. If the explosion progressed in the manner shown in Figure 7-10, that is, straight down from item *A* to its immediate components and from those to their components, the results would have to be recomputed as subsequent explosions of high-level items revealed common usage of items *B* and *C* (and other components of *A*). While the final result would be the same, the data-processing efficiency would be unnecessarily low.

The downward progression from one product level to another just described is called an *explosion*. Net requirements are developed by applying quantities of each item on hand and on order to meet gross requirements for the item at each level. This netting process may seem laborious, but it cannot be circumvented or shortcut; net requirements on the parent level must be determined before the correct gross and net requirements on the component level can be determined.

To cover net requirements, the MRP program develops a time-phased schedule of planned orders for each item, including orders, if any, to be released immediately plus orders scheduled for release in specified future periods. Planned-order schedules will be revised by the computer when the MRP program is rerun if there have been changes in any of the parameters involved (e.g., product structure, MPS, on-hand inventory, etc.). Planned-order quantities are computed using lot-sizing rules (see Chapter 8) specified by the system user for each item.

MRP netting applies existing on-hand quantities to meet item gross requirements. MRP also reevaluates the validity of the timing of any released (open) orders that may now be needed earlier or later than previously scheduled. Good business practice leaves to planners the decision about rescheduling open orders; computers simply send a message recommending this. In its entirety, the information on item requirements and coverage that MRP generates is called the *material requirements plan*. MRP works with three types of orders—planned, released (scheduled receipts), and firm planned—and handles them quite differently.

The standard technique for maximizing processing efficiency is to process all the items on a given level before addressing their components on the next-lower level. This is known as *level-by-level processing*. Under this approach, planned orders are developed for all the items on level 1 so that they may be consolidated for purposes of determining gross requirements for any common components on level 2. This means that each item inventory record is retrieved and processed only once. When a given item appears on more than one level of the product structure, low-level coding (discussed in Chapter 5) will cause the processing of its record to be delayed until the lowest level on which the item finds itself is processed. There is no logical requirement for low-level codes in MRP; their absence will not prevent the program from arriving at correct results. Use of low-level codes yields higher computer-processing efficiency, which usually outweighs the cost of developing and maintaining the coding. Practically every BOM processor software program does this automatically. The level-by-level approach to MRP is illustrated in Figure 7-11.

Level-by-level computation of requirements by MRP minimizes the problem of handling multiple-parent demands. If a component appears on only one level in the product structure, all its parents will be on the next-higher level. Because all items on each level

**FIGURE 7-11**

Level-by-level processing.

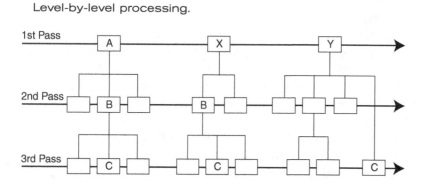

are processed before the next level below, all parent planned orders are developed before gross requirements are placed on components.

Product structure is the principal constraint on the computation of requirements. The number of levels in product structures, discussed in Chapter 11, is determined partially by the sequence of steps in which the product is made from raw material to end item, combining components into parents that then become components of a higher-level parent, and so on to the top level. The commonality of parts and the inventory buffer strategy also can affect the product structure. Product-structure depth (number of levels in its structure) is a major factor in the scope and duration of the MRP data-processing job and in the work of storing and maintaining the product structure.

Each individual item exists as a uniquely identified physical entity (e.g., a unit of raw material, a component part, or a subassembly) and also exists physically built into inventories of one or more parents. In the parents, it has lost its individual identity.

MRP assumes that each parent item in inventory is complete (with all components). MRP also assumes that released orders to make more of a parent will be accompanied by the correct quantity of each component. Additional components therefore will be needed only when more parents have to be made. Hence planned orders for each parent determine gross requirements for its components. The product structure for parent items in inventory or work-in-process makes it possible to determine the number of components in them. Having this traceability is necessary in the event of product recalls or mandatory design changes.

The general rule of MRP planning logic is that net requirements for parent items expressed as lot-sized planned orders become gross requirements for components (taking quantity of component per parent into account) in the same time period that parent orders are to be released. Determining net requirements for a low-level item (the lowest level carries the highest level number) must include consideration of the quantity present under its own identity, as well as all quantities of the item now present in all its parent and higher-level items.

The previously mentioned link between item inventory records is standard, but the link between the hierarchy of records and the MPS, the so-called master schedule interface, is susceptible to optional treatment. There are three options to choose from, and for any given MRP system, the option chosen must be clearly defined if the system is to function satisfactorily. Figure 7-12 represents a sample MPS in the usual matrix format. This would be the key input to the MRP system, but the question remains as to the meaning of the data contained in this schedule. Does the quantity of 100 of end item *A* in period 1 represent a gross requirement, a production requirement, or a planned order? Its treatment by the MRP system will vary depending on the answer to this question.

If the MPS represents gross requirements, its contents simply would be entered into the gross requirements schedules of the respective end-item records, and the processing would be standard (Figure 7-13).

The MRP system will net against both on-hand and on-order quantities, with the result that only an additional 120 sets of item *A* components would need to be produced

**FIGURE 7-12**

Master
production
schedule.

| End Item | Period | | | | | |
|---|---|---|---|---|---|---|
| | 1 | 2 | 3 | 4 | 5 | |
| A | 100 | | 100 | | 100 | |
| B | 15 | 20 | 25 | 20 | 15 | |
| C | 50 | 60 | | 60 | | |

**FIGURE 7-13**

MPS interface:
gross
requirements.

| Item A Lead time: 1 | Period | | | | |
|---|---|---|---|---|---|
| | 1 | 2 | 3 | 4 | 5 |
| Gross Requirements | 100 | | 100 | | 100 |
| Schedule Receipts | 100 | | | | |
| On Hand | 80 | 80 | 80 | −20 | −20 | −120 |
| Planned-Order Releases | | 20 | | 100 | |

                                              20          100

in the first five periods. Under this option, the MPS does not reflect a production plan but a requirements plan. This is not a recommended treatment—confusion will arise if management views the schedule as a plan of production, but the MRP system does not treat it that way.

The second alternative treats the MPS as reflecting production requirements; that is, 300 units of item *A* are to be finished in the first five periods. In this case, the system must be programmed to exclude any on-hand quantities (but not on-order quantities) from the netting process for highest-level items. This requires a modification of the regular processing logic, applicable to these items only. This procedure is sound, and it presupposes that end-item demand has been netted against on-hand inventories during the preparation of the MPS. An example of this alternative is shown in Figure 7-14.

Under each of the two alternative treatments just discussed, the MPS, in conjunction with the MRP system, will "produce" item *A* in the quantities indicated—the system will order correct quantities to be assembled or completed, and barring some difficulty, item *A* will be available according to what the schedule calls for.

The third option treats the MPS as a schedule of planned-order releases, which means that the schedule will not "produce" the end items but only their components (Figure 7-15). The assembly of end items then would have to be ordered apart from the MRP system, most likely via a final assembly schedule. This may or may not be desirable depending on the type of product being manufactured. If the end items in the MPS are

**FIGURE 7-14**

MPS schedule
interface:
production
requirements.

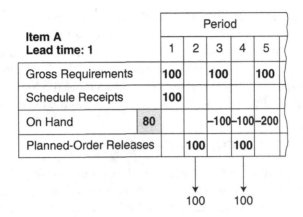

**FIGURE 7-15**

MPS interface:
planned-order
releases.

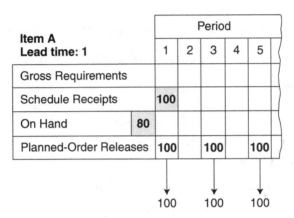

not shippable products but merely major assemblies of products, this treatment will prove suitable only for items that are actually produced in conjunction with the final assembly of the product rather than in advance of final assembly.

Where the MPS items are described by planning BOMs and are never actually built, this obviously is the correct option. If the end items in the MPS are shippable products or major assemblies of such products, this option will provide flexibility to change final assembly operations to respond to the latest customer product needs.

Locking such execution to the planned MPS much earlier blurs the distinction between planning and execution. This leads to *fences* in the MPS to limit near-term replanning for differences between customer orders and master schedules. The whole frozen-time fence approach is a fallacy. This is clear when the implications of fences are considered. Implicitly, two naive and untrue statements apply. First, "If we haven't planned it, we can't make it," and second, "If we planned it, we must make it even if we don't need it now." The only true statement is, "When enough customers' orders arrive to fill the front end of the MPS, stop replanning and execute."

Another variation of this treatment of the MPS interface uses firm planned orders (described in Chapter 9) for the quantities in question, forcing the MRP system to "produce" the end items by releasing each such order in due time and scheduling its completion. In a make-to-stock manufacturing operation, where the time-phased order-point approach is used to control finished-goods inventory, the planned-order schedules of the stocked items constitute, collectively, the MPS. Here, the firm planned-order technique would be used to space the planned orders so as to level the load on the plant. The time-phased order point and its uses will be reviewed in the next section.

## TIME-PHASED ORDER POINT

Time-phased order point is an approach that allows time-phasing techniques to be used for the planning and control of independent-demand items. The demand for such items has to be forecast, and their supply normally would be controlled by means of order points.

### Conventional Order Point Versus Time-Phased Order Point

For an example, a service part supplied from factory stock might have the following planning factors:

> Lead time $L = 2$
> Safety stock $S = 100$
> Period-demand forecast $F = 17$
> Order quantity $Q = 50$

The order point then would be determined as follows:

$$S + (F \times L) = 100 + (17 \times 2) = 134$$

The same example, in time-phased format, is shown in Figure 7-16. The forecast is projected over the entire planning horizon and represents gross requirements. The current quantity on hand of 170 will have dropped below safety stock of 100 in period 5, and

**FIGURE 7-16**

Time-phased order point.

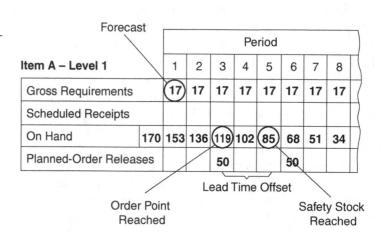

a replenishment order of 50 is planned to arrive at that time. Offsetting for lead time, the planned-order release is scheduled for period 3. In period 8, the quantity on hand (34 + 50) once again will be less than 100, and another planned order, scheduled for release in period 6, will cover this. The results are identical to those obtained under statistical order point (note that the previously computed order point of 134 is reached some time in period 3, at which time the replenishment order is planned to be released), except that under time-phased order point an entire schedule of planned replenishment orders is developed instead of one order at a time, as is done under the conventional statistical order-point approach.

Actually, the term *time-phased order point* is a misnomer because no order point as such is used for any purpose, and its value (134 in our case) is not even computed by the system. The MRP logic takes over and provides for ordering at the right time. The value of the order point need not be known because the replenishment order is not being triggered by it but rather by the on-hand quantity dropping down to the safety-stock level and the lead-time offset timing the order correctly.

The MRP system requires no modification for purposes of implementing time-phased order point. The system treats the independent-demand item exactly the same way it handles any item for which safety stock is specified (see Figure 7-8). The time-phased order point does everything its conventional counterpart does, and in addition, it provides the capabilities to:

1. Keep open-order due dates valid
2. Furnish the capacity system with a planned-order schedule
3. Plan gross requirements for the item's component material(s)

This means that both dependent- and independent-demand manufactured items can be covered by the same priority planning system, and the relative priorities of all shop orders can be kept valid. The MRP logic keeps reevaluating the validity of open-order due dates, and such orders are rescheduled in accordance with the dates of actual need. This is especially significant for items under time-phased order-point control because their gross and net requirements are expected to keep changing owing to forecast error.

When time-phased order point is used properly, open orders for items under its control are being continually rescheduled backward and forward so as to make their due dates coincide with dates of (projected) actual need and to keep their relative priorities valid. If this is done, it can be seen that the system will maintain the correct inventory status and correct coverage irrespective of the vagaries of actual demand. A close monitoring of the efficacy of the forecasting model, tracking signals, and so on therefore loses its importance. The time-phased order point works well even when forecasting is poor, as will be explored further in Chapter 10. It also has been suggested that close monitoring of forecasts via tracking signals is less important. This ignores the effects of nervousness on good execution by suppliers and production plants. MRP programs cope easily with a myriad of changes; production sometimes cannot react with the same agility.

There is one more advantage to planning and controlling independent-demand items by means of time-phased order point. In a manufacturing environment, when such items are under the control of this technique, the capacity requirements report can reflect a complete load with good visibility because planned-order schedules spanning the planning horizon are available to the capacity requirements planning system for all items. This will be reviewed in more detail in Chapter 14.

## Known Future Demand

Time-phased order point allows a known future lump requirement to be entered correctly and processed, in contrast to the conventional order point, which is quite unequipped to handle a situation of this type. For example, a service part that is forecast, based on past usage, at 25 units per period may be ordered by a customer in a quantity of 200, to be delivered at some specified future date. With conventional order point, the question is whether an order for 200 should be released immediately even though the lead time is less than the delivery time or it should be released at the proper (later) time. In the latter case, the problem is what special procedure to institute so as not to miss the (future) order release date.

The statistical order point technique itself will, of course, reorder when the inventory drops down to the order-point quantity, whenever that may happen. Moreover, once the order for 200 is released, it will increase the (on-hand plus on-order) inventory correspondingly, and this will tend to act to prevent another replenishment order from being released until the order in question is shipped. By then it may be impossible to meet regular service-part demand if stock on hand has been depleted. With time-phased order point, none of these problems arises. The order for 200 is simply entered as a gross requirement in the appropriate bucket of the inventory record and is added to whatever quantity has been forecast for that period to cover regular demand. No more need be done because the system will automatically order correctly, at the proper time, to cover both forecast and known future demand.

## Effect on Management of Parts Service

Time-phased order point is an outgrowth and extension of the MRP system, and it is interesting to note what effect on the service-part function such a system tends to have. Historically, this function in a manufacturing company has been evolving toward organizational separation and independence from the manufacturing function in general and production and inventory control in particular. This has been caused primarily by the marketing nature of a service-part inventory and by the marketing orientation of the people in charge of this inventory. In this evolutionary process, service-part stock is first segregated from production-part stock physically to prevent manufacturing personnel from borrowing service parts for purposes of production. Such borrowing is done to cover shortages caused primarily by deficiencies in the inventory control system. Next comes a separate service-part department, independent of the production control manager. A separate, different inventory control system evolves for service parts. Finally, the service-part organization operates its own warehousing and distribution facilities geographically remote from the plant that manufactures the parts.

The installation of an MRP system by the plant tends to reverse this trend. First, the service-part organization comes under pressure to adopt the time-phased order-point approach so that the plant might have better visibility into future service-part requirements. The two previously disparate inventory systems become compatible and then, in effect, two parts of one (MRP) system. With good planning and control of inventories, the former justification for physically segregating the inventories no longer exists. A consolidation of these inventories results in a lower level of total inventory investment. Stock is freely "borrowed" between production and part service, but under control by the MRP system, which simply (and impartially) allocates existing stock to production and service-part requirements in time sequence of actual need. The result is improved customer parts service and fewer production shortages.

## ENTRY OF EXTERNAL-ITEM DEMAND

In the typical manufacturing environment in which an MRP system would be used, the bulk of the component-item demand derives from the MPS and is generated internally through the requirements planning (explosion) process. At least some of the demand for components, however, normally also comes from sources external to the plant (e.g., service-part and interplant requirements) or nonproduction sources within the plant (e.g., experimental, quality-control, and plant maintenance requirements). The demand, if any, in the latter category is usually minimal and sporadic, not warranting separate planning or forecasting. Service-part and interplant demand, on the other hand, may be both significant and recurring. This demand is conveyed to the MRP system in one or more of the following forms:

1. Entry of orders placed by a service warehouse
2. Entry of orders placed by another plant
3. Forecasting of service-part demand
4. Processing of planned-order schedules of a service warehouse time-phased order-point system
5. Processing of planned-order schedules of another plant's MRP system

### Entry of Orders

Service-part and/or interplant orders are entered into a plant's MRP system by means of transactions that increase gross requirements of the item in question in the period corresponding to the order due date. Gross requirements stemming from parent planned orders and those generated by external-demand sources thus are consolidated, and the

MRP process proceeds in regular fashion. Orders for service parts may be generated by a service warehouse or by an organizationally autonomous service-part department that maintains an inventory control system of its own.

When external (this term also covers business units of the company other than the plant using the MRP system under discussion) component-item demand is expressed in the form of orders, the supplying plant specifies delivery lead times, which means that the organization requesting the parts normally must make commitments well in advance of actual need. Under this arrangement, there tend to be never-ending complaints about orders being received with less than the agreed-on lead time. The supplying plant typically regards these orders as firm and noncancellable. Generally speaking, this method of treating component-item demand between different organizations of a company is never entirely satisfactory, and delivery service tends to be poor.

### Forecasting of Service-Part Demand

In cases where a plant using an MRP system has a service-part responsibility (rather than a separate service-part organization ordering from it), the logical and most effective way to treat service-part demand is to use the time-phased order-point approach. This permits service parts to be integrated into the MRP system without any need to modify its processing logic.

As mentioned earlier, service-part demand for items in current production has to be (explicitly or implicitly) forecast, and the forecast quantities, by period, are added to the gross requirements of the items in question. Time-phased order point simply extends this principle to cover parts no longer used in current production. The forecast quantities are recorded in the gross requirements buckets of the inventory records of the respective items, even though there are no other (parent planned-order) requirements. The particular statistical techniques used in the past to determine safety stock and to forecast demand for these items can be retained when they are put under time-phased order point.

Time-phased order point provides the way for smoothly integrating independent-demand items into an MRP system. The processing logic of this system profitably applies to service parts that are no longer used in current production despite the fact that the system is intended primarily to plan production items. Note that many service parts are subassemblies and that manufactured service parts entail at least one lower (raw material) level. Time-phased order point permits, as no other technique does, requirements for component items of service parts to be determined and timed correctly.

### Planned-Order Schedules of "Customer" Systems

Demand for component items that comes from sources external to the plant (other units of the company are viewed as "customers" for these components) can be entered into the plant's MRP system by means of planned-order schedules generated by the customer's system. This presupposes that a service-part warehouse uses time-phased order point.

Instead of conveying their requirements by means of actual orders, the customer organizations simply submit schedules of planned orders for processing by the supplying plant's MRP system. This represents the most desirable and most effective method of registering external demand with the supplying plant because it provides maximum visibility of future requirements.

The principle of feeding the output of a customer system directly to the supplying plant's MRP system also extends to the company's product warehouses or distributorships and to actual customers who operate MRP systems of their own. In both cases, the precondition for thus interfacing the systems is a product that appears as an end item in the supplying plant's MPS and has an item inventory record in the plant's system or a method to link the two records. The use of this method of conveying demand has increased dramatically because of its many advantages to both trading partners.

When a warehouse's time-phased order-point system or another plant's MRP system interfaces directly with the supplying plant's system, two time-phased inventory records of the same item are logically linked in a pseudo–parent-component relationship. Let us take an imaginary service part X that has a record in a time-phased order-point system and another record in the supplying plant's MRP system, as illustrated in Figure 7-17.

At the warehouse-system level, item $X$ acts as a pseudoparent item, linked by its planned-order release schedule to the gross requirements schedule of pseudo–component item $X$ at the plant system level. The two records of item $X$ are, as it were, put one on top of the other. When this is done, independent demand for the item at the warehouse level is translated by the system's processing logic into dependent demand at the plant level.

Because the item in question has two time-phased inventory records, its normal lead time must not be duplicated but rather divided between the records. At the parent level,

**FIGURE 7-17**

Pseudo–parent-component relationship.

| Item X Safety Stock: 20 Lead Time: 1 | | Period | | | | | |
|---|---|---|---|---|---|---|---|
| | | 1 | 2 | 3 | 4 | 5 | 6 |
| Gross Requirements | | 10 | 10 | 10 | 10 | 10 | 10 |
| Scheduled Receipts | | | | | | | |
| On Hand | 48 | 38 | 28 | 18 | 8 | -2 | -12 |
| Planned-Order Releases | | | 30◄┐ | | | 30◄┐ | |

| Item X Lead Time: 4 | | | | | | | |
|---|---|---|---|---|---|---|---|
| Gross Requirements | | | 30 | | | 30 | |
| Scheduled Receipts | | | 30 | | | | |
| On Hand | 5 | 5 | 5 | 5 | 5 | -25 | -25 |
| Planned-Order Releases | 25◄ | | | | | | |

planned lead time should equal the transportation or delivery-time portion of the total, for example, one week; at the component level, planned lead time should equal the manufacturing portion of the total, for example, several weeks or months. In Figure 7-17, these lead times are shown as one and four weeks, respectively. The products can be linked between different entities in a vertically integrated company or across different companies of the supply chain.

## SYSTEM NERVOUSNESS

A very interesting attribute of material requirements planning is the system's apparent *hypersensitivity* or *nervousness*. Since file updating is equivalent to replanning, it may appear that the system calls for continual revision of user action taken previously. This is of concern especially where due dates of open purchase orders are concerned, when it is not considered practical to subject these due dates to constant revision. In evaluating this aspect, it is important to draw a distinction between

- The system being informed, up to the minute
- The frequency of action taken on the basis of the information

The latter obviously can be decided on (based on practical considerations) independent of the former. A deliberate withholding of user action in the full knowledge of current facts is preferable to a lack of action caused by ignorance of those facts. The critic of a nervous system argues, in effect, that it is better for an inventory and production-planning system to be out of date. Such an argument cannot be considered seriously. *Nervousness* on the level of planning is a virtue, not a drawback, of a net-change MRP system. Hypersensitivity on the level of reaction can and should be dampened. Not every change in inventory status calls for reaction. Many minor changes of the type that otherwise would require action are absorbed by inventory surpluses that exist as a result of previous changes and/or inventory management decisions. These surpluses are created by safety stock, scrap allowances, and temporary excesses in inventory owing to lot sizing, engineering changes, reduced requirements, overshipments, overruns, and premature deliveries by suppliers. The system constantly strives to use up such temporarily excessive inventories as early as possible through the net requirements planning process. Inventory excesses thus are automatically prevented from accumulating, but under normal conditions, they exist in some measure at any point in time.

Prompt reaction to changes in requirements or other elements of inventory status generally is called for when requirements increase or when the timing of planned performance advances. For the opposite type of change, a delay in reaction can be tolerated. Changes can occur every minute of the day. Inventory status is not affected significantly by most of the updating entries, but certain transactions, for example, unscheduled stock disbursements, scrap, physical inventory adjustments (short counts), and miscellaneous demand exceeding forecast, do cause replanning/rebalancing of inventory status.

## Action Cycles

Many changes may occur in the same inventory record on the same day, in which case the timing of open orders appears to require revision several times that day, even though the changes may have a mutually canceling effect. The inventory planner's reaction to change, however, can be decoupled from the rate at which individual changes occur and are processed by the system. The most common method of dampening reaction to change is simply to delay such reaction. In practice, this takes the form of periodic action cycles on the part of the inventory planner. He or she need not react to the continuous stream of individual changes but can let them accumulate for some period of time.

The system can provide output of action requests on a cyclic basis. Some action messages typically would be generated in a batch once a day. Most requests for normal order action (e.g., release of shop orders and purchase requisitions) belong in this category. Different action cycles apply to various types of actions depending on their purpose. Thus due dates for all open shop orders may be reevaluated once per shift so as to maintain the validity of shop priorities. For certain types of messages (e.g., premature supplier deliveries), a weekly cycle would be sufficient. Other types of messages, however, should be generated without any delay because corrective action time is critical. For example, an open purchase order may become a candidate for cancellation as a result of changed requirements. A 24-hour delay in reacting to the new situation can make the difference between being or not being able to cancel. Other examples of situations that call for reaction without delay are excessive scrap and a significant downward adjustment of on-hand inventory following a physical count. When major changes in the MPS are being processed or following regular periodic issues of the MPS, all action-request output should be suppressed until the entire change has been processed completely by the system. This type of change may affect thousands of records, and the status of an inventory item may change several times during the processing of such a change.

Planning cycles and action cycles are established on a more or less arbitrary basis. Delaying action on available information does dampen reaction to change, but delay obviously cannot be prolonged indefinitely. Under any action cycle, once delay is terminated, subsequent changes still can invalidate the action taken. As a general rule, it is better to act with less delay under a system capable of frequent—or continuous—replanning, reevaluation, and revision of previous action than to tolerate unresponsiveness by operating on long planning and action cycles. An MRP system offers a range of responses from zero delay to weekly and monthly cycles. The relative promptness of reaction to change should be a function of the type of change in question. This means that it can be transformed whenever its user is ready into an online communications-oriented inventory management system without a change in approach, reeducation, and fundamental system overhaul.

# BIBLIOGRAPHY

"Designing and Implementing a Material Requirements Planning System," in G. W. Plossl and O. W. Wight (eds.), *Proceedings of the 13th International Conference of APICS*. New York: American Production and Inventory Control Society, 1970.

"Inventory Management," in *Communications Oriented Production Information and Control System (COPICS)*, Vol. 4 (Form No. G320-1977). New York: International Business Machines Corp., 1972, Chapter 5.

"Manufacturing Control at American Bosch." Division on the IBM RAMAC 305, Application Manual Form No. E20-2053. International Business Machines Corp., 1960.

*Material Requirements Planning by Computer*. New York: American Production and Inventory Control Society, 1971.

Orlicky, J. A. "Net Change Material Requirements Planning." *Production & Inventory Management* 13(1), 1972.

Orlicky, J. A. "Net Change Material Requirements Planning," *IBM Systems Journal* 12(1), 1973.

## CHAPTER 8

# Lot Sizing

The traditional interest in the classic problem of the *economic order quantity* (EOQ) has shifted to lot sizing in an environment of discrete period demands. This development has been stimulated by the emergence of material requirements planning (MRP) systems, which expresses demand for inventory items in discrete time-series fashion by computing time-phased gross and net requirements.

A significant portion of the literature related to MRP, including virtually all the scientifically oriented writing on the subject (see the Bibliography at the end of this chapter), is devoted to discrete-demand, time-series lot sizing. This is, without a doubt, the best-researched aspect of MRP. A number of distinct techniques have been developed, the most important of which are described and evaluated in this chapter, along with the more traditional approaches to lot sizing.

## COSTS IN LOT SIZING

Two categories of costs enter into decisions of how much of an item should be purchased or made. These are:

1. *Ordering costs.* A composite of all costs related to placing purchase orders or preparing work orders, including
   - Processing paperwork—preparing requisitions, purchase orders, receiving documents for purchased materials, and shop packets for manufactured items
   - Changing machine and workstation setups
   - Inspection, scrap, and rework associated with setup
   - Recordkeeping for work-in-process
2. *Inventory carrying costs.* The total of costs related to carrying the resulting inventory, including
   - Obsolescence caused by market, design, or competitors' product changes

- Deterioration from long-term storage and handling
- Recordkeeping
- Taxes and insurance on inventory
- Storage costs for equipment, space, heat, light, and people
- Cost of capital invested in inventory, or foregone earnings of alternate investments

Ordering and inventory carrying costs rarely can be determined from traditional cost-accounting data; they have to be "engineered" specifically for each company's operations. While ordering costs can be estimated fairly accurately, they should be actual out-of-pocket costs and only costs that are affected by the decision of how many to buy or make.

Carrying cost, expressed usually as a decimal fraction of inventory value, may appear precise but in reality will be only very approximate. Estimates of several factors obviously will be little more than educated guesses at best, particularly the last one listed earlier. Which of the two choices is used for this factor depends on company policy.

In practice, carrying costs vary from as low as 15 percent to as high as 80 percent per year and can change during a year. Higher values are used by companies that must procure outside capital rather than use retained earnings and by those who believe that lot-sizing decisions should be charged at the same rates the business expects other capital investments to earn. Many professionals in inventory management believe that detailed studies to estimate carrying costs are unwarranted. They prefer to view these as "management policy variables" to achieve management's objectives in inventory investment. Increasing the carrying cost used in EOQ computations will result in smaller lot sizes, and vice versa. Thus the inventory carrying cost in use at any given time reflects the premium that management is putting on the conservation of capital.

Order sizing creates cycle-stock or lot-size inventory in both order-point and MRP approaches. In reality, the average amount of such inventories is not equal to the theoretical one-half of the quantities being ordered, as assumed in traditional EOQ calculations. In MRP, the lack of validity of such an approximation is clear. Order quantities determined by such techniques for a given inventory item will equal net requirements for one or more planning periods, causing the quantity ordered and the inventory to vary significantly from one order to the next.

The number of periods covered by an order quantity will be affected by the relative continuity of demand for the item. In cases of very intermittent demand, the order quantity often will equal the requirement for only one period. This usually also will be true for all assembled items because of typically minor assembly setup considerations.

## LOT-SIZING TECHNIQUES

The most widely recognized approaches to lot sizing are as follows:

1. Fixed order quantity (FOQ)

**2.** Economic order quantity (EOQ)

**3.** Lot for lot (LFL)

**4.** Fixed-period requirements—sometimes referred to as *period of supply* (POS)

**5.** Period order quantity (POQ)

**6.** Least unit cost (LUC)

**7.** Least total cost (LTC)

**8.** Part-period balancing (PPB)

**9.** Wagner-Whitin algorithm

The first two are *demand-rate-oriented*; the others are called *discrete lot-sizing techniques* because they generate order quantities that equal the net requirements in an integral number of consecutive planning periods. Discrete lot sizing does not create *remnants*, that is, quantities that would be carried in inventory for some length of time without being sufficient to cover a future period's requirements in full.

Lot-sizing techniques can be categorized into those which generate fixed, that is, repetitively ordered, quantities and those which generate varying order quantities. This distinction between fixed and variable is not to be confused with that between static and dynamic order quantities. A *static order quantity* is defined as one that, once computed, continues unchanged in the planned-order schedule. A *dynamic order quantity* is subject to continuous recomputation as and if required by changes in net requirements data. A given lot-sizing technique can generate either static or dynamic order quantities depending on how it is used.

Of the nine techniques listed earlier, only the first one is always static, and the third one is, by definition, dynamic. The rest, including the EOQ, can be used for dynamic replanning at the user's option. The last four are expressly intended for such replanning. It must be pointed out that dynamic order quantities are a mixed blessing in an MRP environment. While they always reflect the most up-to-date version of the materials plan, they affect the requirements (and thus also the planned coverage) for their component items. A recomputation of a parent planned-order quantity often will mean that component-item open orders have to be rescheduled in addition to recomputing and/or retiming planned orders.

Upsetting previous plans on component-item levels sometimes can cause severe problems, and while such problems inevitably arise in the course of operations, some of them could be avoided to the extent that they are caused internally by the system recomputing previously planned orders. There is merit in the recommendation made by some users of MRP systems that a planned order, once established, be "frozen" as to its quantity (if at all possible) and that only its timing be changed subsequently as required by changing net requirements. This practice is especially recommended for planned orders that are timed within the span of the cumulative product lead time (as opposed to orders planned for the longer-term future) because only those orders create gross requirements on lower levels that are likely to be covered by open orders. A review of the nine lot-sizing techniques enumerated earlier follows. These techniques usually are discussed in connection with manufactured inventory items, and the term *setup* covers all (fixed) costs of ordering. The reader should understand, however, that the logic on which these techniques are based is not limited to manufactured items. Where the cost of ordering purchased items is significant and/or where quantity discounts apply, any of the economics-oriented lot-sizing techniques can be used after appropriate modification.[1]

## Fixed Order Quantity (FOQ)

An FOQ policy may be specified for any item under an MRP system, but in practice, it would be limited to selected items only, if used at all. This policy would be applicable to items with ordering cost sufficiently high to rule out ordering in net requirements quantities period by period. The FOQ specified for a given inventory item may be determined arbitrarily, or it can be based on intuitive/empirical factors. The quantity may reflect extraneous considerations, that is, facts not taken into account by any of the available lot-sizing algorithms. Such facts may be related to the capacities of certain facilities or processes, die life, packaging, storage, and so on. It is understood (and the MRP system would be programmed accordingly) that when using this lot-sizing rule, the order quantity will be increased if necessary to equal an unexpectedly high net requirement in the period the order is intended to cover. For example, if the FOQ were 60 and the earliest net requirement were 75, the planned-order quantity normally would be increased to 75 because it would make little sense to generate two orders of 60 each for the same period. Note that this also applies to the EOQ, particularly where it is used as a fixed quantity repetitively ordered over a period of time (typically a year). An example of an FOQ of 60 is provided in Figure 8-1. Note that in this and subsequent examples, the order quantities are not offset for lead time; that is, each quantity is shown under (keyed to) the earliest period it is intended to cover.

**FIGURE 8-1**

Fixed order quantity (60).

| Period | 1 | 2 | 3 | 4 | 5 | 6 | 7 | 8 | 9 | Total |
|---|---|---|---|---|---|---|---|---|---|---|
| New Requirements | 35 | 10 | | 40 | | 20 | 5 | 10 | 30 | 150 |
| Planned-Order Coverage | 60 | | | 60 | | | | | 60 | 180 |

## Economic Order Quantity (EOQ)

The EOQ formula was first derived by Ford W. Harris in 1915. It is the oldest technique in the field and is still an aid to sound planning when its limitations are recognized. Although it predated and was not intended for an MRP environment, it can be incorporated easily into an MRP system if the user so wishes. Figure 8-2 shows EOQ coverage of the same net requirements as used in the preceding example.

---

[1] *Purchasing* (General Information Manual Form No. GH20-1149-1). New York: International Business Machines Corp., 1973, pp. 140–164.

**FIGURE 8-2**

Economic order
quantity.

| Period | 1 | 2 | 3 | 4 | 5 | 6 | 7 | 8 | 9 | Total |
|---|---|---|---|---|---|---|---|---|---|---|
| New Requirements | 35 | 10 | | 40 | | 20 | 5 | 10 | 30 | 150 |
| Planned-Order Coverage | 58 | | | 58 | | | | | 58 | 174 |

These net requirements data will be carried over into subsequent examples of lot sizing to point up the differences in the performance of the various techniques. The periods will be assumed to represent months, and the following cost data will be used throughout:

Setup $S$              $100
Unit cost $C$          $ 50
Carrying cost $I$      0.24 per annum, $0.02 per period

These cost data will facilitate calculations required in the use of some of the discrete lot-sizing techniques because the cost of carrying one unit of the inventory item for one period is $1. The EOQ calculation is as follows:

$$Q = \sqrt{\frac{2US}{IC}} = \sqrt{\frac{2 \times 200 \times 100}{0.24 \times 50}} = \sqrt{3{,}333} = 58$$

where $Q$ is the economic order quantity and $U$ is the annual usage (in units). The value of $U$ in this calculation was obtained by annualizing the nine-month demand (net requirements) of 150:

$$9{:}150 = 12{:}X$$

$$X = 150 \times 12/9 = 200$$

In this case, the known future demand, rather than historical demand, was used as a basis for estimating annual usage. The example illustrates a problem all forward-looking lot-sizing techniques face, namely, a finite, or limited, planning horizon. In our example, an EOQ based on future demand would require a year's demand data, but the system provides only nine months' visibility. Most of the discrete lot-sizing techniques are not based on annual usage, but they assume a certain minimum visibility for each lot in the planned-order schedule, including the last one. In most cases, however, the quantity of the last lot is truncated by the proximity of the far edge of the planning horizon, as will be seen in subsequent examples.

As to the effectiveness of the EOQ in a discrete-demand environment, a look at Figure 8-2 reveals that the first order quantity of 58 includes a "remnant" of 13 pieces that are carried in inventory in periods 1 through 3 to no purpose. Similarly, 6 pieces are carried unnecessarily in periods 4 through 7 owing to the size of the second lot. The ordering strategy provided by the EOQ approach (of ordering three times in quantities of 58) will be seen to be relatively poor in comparison with some of the other examples that follow.

The EOQ is based on an assumption of continuous, steady-rate demand, and it will perform well only where the actual demand approximates this assumption. In our example, the demand is both discontinuous and nonuniform. The more discontinuous and nonuniform the demand, the less effective the EOQ will prove to be.

## Lot-for-Lot Ordering (L4L)

This technique, sometimes also referred to as *discrete ordering*, is the simplest and most straightforward of all. It provides period-by-period coverage of net requirements, and the planned-order quantity always equals the quantity of the net requirements being covered. These order quantities are, by necessity, dynamic; that is, they must be recomputed whenever the respective net requirements change. Use of this technique minimizes inventory carrying cost. It is often used for expensive purchased items and for any items, purchased or manufactured, that have highly discontinuous demand. Conversely, items in high-volume production and items that pass through specialized facilities geared to continuous production (equivalent to permanent setup) normally are also ordered lot for lot. Figure 8-3 provides an example of this method of ordering.

**FIGURE 8-3**

Lot-for-lot approach.

| Period | 1 | 2 | 3 | 4 | 5 | 6 | 7 | 8 | 9 | Total |
|---|---|---|---|---|---|---|---|---|---|---|
| New Requirements | 35 | 10 | | 40 | | 20 | 5 | 10 | 30 | 150 |
| Planned-Order Coverage | 35 | 10 | | 40 | | 20 | 5 | 10 | 30 | 150 |

## Fixed-Period Requirements

This technique is equivalent to the primitive rule of ordering "X months' supply" used in some stock-replenishment systems, except that here the supply is determined not by forecasting but by adding up discrete future net requirements. In its rationale, it is similar to the fixed order quantity (FOQ) approach—the span of coverage may be determined arbitrarily or intuitively. This lot-sizing rule is sometimes referred to as *period of supply* (POS).

Under this technique, the user specifies how many periods of coverage every planned order should provide. Whereas under the FOQ approach the quantity is constant and the ordering intervals vary, under POS, the ordering interval is constant and the quantities are allowed to vary.

For example, if two periods' requirements were specified, this technique would order every other period, except when zero requirements in a given period would extend the ordering interval. This method is illustrated in Figure 8-4.

**FIGURE 8-4**

Fixed-period requirements.

| Period | 1 | 2 | 3 | 4 | 5 | 6 | 7 | 8 | 9 | Total |
|---|---|---|---|---|---|---|---|---|---|---|
| New Requirements | 35 | 10 | | 40 | | 20 | 5 | 10 | 30 | 150 |
| Planned-Order Coverage | 40 | | | 40 | | 25 | | 40 | | 150 |

## Period Order Quantity (POQ)

This technique, sometimes called *economic time cycle*, is based on the logic of the classic EOQ modified for use in an environment of discrete period demand. Using known future demand as represented by the net requirements schedule of a given inventory item, the EOQ is computed through the standard formula to determine the number of orders per year that should be placed. The number of planning periods constituting a year then is divided by this quantity to determine the ordering interval. The POQ technique is identical to the one just discussed except that the ordering interval is computed.

Both these fixed-interval techniques avoid *remnants* in an effort to reduce inventory carrying cost. For this reason, the POQ approach is more effective than the EOQ approach because setup cost per year is the same but carrying cost will tend to be lower under POQ. A potential difficulty with this approach, however, lies in the possibility that discontinuous net requirements will be distributed in such a way that the predetermined ordering interval will prove inoperative. This will happen when several of the periods coinciding with the ordering interval show zero requirements, thus forcing the POQ technique to order fewer times per year than intended.

Using the previous EOQ example and the annualized demand data, the POQ is determined as follows:

EOQ = 58
Number of periods in a year = 12
Annual demand = 200
250/58 = 3.4 (orders per year)
12/3.4 = 3.5 (ordering interval)

The application of these results (assuming the interval alternates between 4 and 3) appears in Figure 8-5. Note that the third order covers only one period's requirements because of insufficient horizon and will have to be recomputed (probably three times) in the future. In comparison with some of the other discrete lot-sizing techniques described below, the effectiveness of POQ, like that of the classic EOQ from which it springs, proves relatively low in the face of discontinuous, nonuniform demand.

**FIGURE 8-5**

Period order quantity.

| Period | 1 | 2 | 3 | 4 | 5 | 6 | 7 | 8 | 9 | Total |
|---|---|---|---|---|---|---|---|---|---|---|
| New Requirements | 35 | 10 | | 40 | | 20 | 5 | 10 | 30 | 150 |
| Planned-Order Coverage | 85 | | | 35 | | | | | 30 | 150 |

## Least Unit Cost (LUC)

This technique and the three that follow have certain things in common. All of them allow both the lot size and the ordering interval to vary. They share a common assumption of discrete inventory depletions at the beginning of each period, which means that a portion of each order, equal to the quantity of net requirements in the first period covered by the order, is consumed immediately on arrival in stock and thus incurs no inventory carrying charge. Inventory carrying cost, under all four of these lot-sizing methods, is computed on the basis of this assumption rather than on average inventories in each period. All four of the techniques share the EOQ objective of minimizing the sum of setup and inventory carrying costs, but each of them employs a somewhat different attack.

The LUC technique is best explained in terms of trial and error, and this approach is used here, although less primitive methods of computation do exist. In determining the order quantity, the LUC technique asks, in effect, whether this quantity should equal the first period's net requirements or should be increased to cover the next period's requirements or the one after that, and so on. The decision is based on the *unit cost* (i.e., setup plus inventory carrying cost per unit) computed for each of the successive order quantities. The one with the least unit cost is chosen to be the lot size. Figure 8-6 shows the computation of the first lot. The next one is computed in identical fashion starting with period 4.

The LUC is found at lot quantity 45, which will cover periods 1 and 2. The next order of 60 will cover periods 4 through 6, and the third order of 45 will cover periods 7 through 9 (Figure 8-7). The limitation of the LUC approach lies in the fact that the technique considers only one lot at a time. The unit cost varies, sometimes widely, from one lot to the next.

Tradeoffs between consecutive lots sometimes could be made that would reduce the total cost of two or more lots. Our example contains such a situation: If the requirement in period 7 were added to the quantity of the second lot, its inventory carrying cost would increase by $15, but that of the next lot would decrease by $40. The lot-sizing technique described next attempts to overcome this flaw in LUC logic.

**FIGURE 8-6**

Computation of least unit cost.

| | | | | Carry cost, $ | | | |
|---|---|---|---|---|---|---|---|
| Period | Net require-ments | Carried in inventory (periods) | Prospec-tive lot size | For lot | Per unit | Setup per unit, $ | Unit cost, $ |
| 1 | 35 | 0 | 35 | 0.00 | 0 | 2.86 | 2.86 |
| 2 | 10 | 1 | 45 | 10.00 | .22 | 2.22 | 2.44 |
| 3 | 0 | 2 | | 0 | | | |
| 4 | 40 | 3 | 85 | 130.00 | 1.53 | 1.18 | 2.71 |

Setup: $100
Inventory carrying cost: $1 per unit per period

**FIGURE 8-7**

Least unit cost.

| Period | 1 | 2 | 3 | 4 | 5 | 6 | 7 | 8 | 9 | Total |
|---|---|---|---|---|---|---|---|---|---|---|
| New Requirements | 35 | 10 | | 40 | | 20 | 5 | 10 | 30 | 150 |
| Planned-Order Coverage | 45 | | | 60 | | | 45 | | | 150 |

## Least Total Cost (LTC)

The LTC technique is based on the rationale that the sum of setup and inventory carrying costs (total cost) for all lots within the planning horizon will be minimized if these costs are as nearly equal as possible, the same as under the classic EOQ approach. The LTC technique attempts to reach this objective by ordering in quantities at which the setup cost per unit and the carrying cost per unit are most nearly equal. A second look at Figure 8-6 will show that the LUC technique chose a quantity at which setup cost per unit ($2.22) significantly exceeded carrying cost per unit (22 cents).

Because it seeks the equality of these costs, the LTC technique is able to avoid the relatively laborious computation procedure of the LUC approach and can proceed toward its goal in a more direct fashion. The vehicle for this is computation of the so-called economic part-period (EPP). The part-period measure is analogous to a man-year or a passenger-mile. It is one unit of the item carried in inventory for one period. The part-period is a convenient expression of inventory carrying cost for purposes of comparison and tradeoff; that is, it can be said that to carry a quantity of an item in inventory for a certain period of time will "cost" times the number of part-periods. The EPP is defined as that quantity of the inventory item which, if carried in inventory for one period, would result in a carrying cost equal to the cost of setup. It is computed simply by dividing the inventory carrying charge per unit per period $IpC$ into setup cost $S$. In our example, this is EPP $= S/IpC = 100/0.02 \times 50 = 100$.

The LTC technique selects the order quantity at which the part-period cost most nearly equals the EPP. An example of LTC computation appears in Figure 8-8. The quantity chosen for the first lot would be 85 because the 130 part-periods that it would cost most nearly approximate the EPP of 100. This order would cover requirements of periods 1 through 5, and the second order of 65 would cover requirements of periods 6 through 9. This is shown in Figure 8-9.

The LTC approach to lot sizing generally is favored over the LUC approach, but the arguments its advocates put forward, as well as results of limited simulations, are not entirely convincing. The LTC logic has a flaw of its own in the premise that "the least total

**FIGURE 8-8**

Computation of least total cost.

| Period | Net Requirements | Carried in inventory (periods) | Prospective lot size | Part-periods (cumulative) |
|--------|------------------|-------------------------------|----------------------|---------------------------|
| 1 | 35 | 0 | 35 | 0 |
| 2 | 10 | 1 | 45 | 10 |
| 3 | 0 | 2 | | |
| 4 | 40 | 3 | 85 | 130 |

**FIGURE 8-9**

Least total cost.

| Period | 1 | 2 | 3 | 4 | 5 | 6 | 7 | 8 | 9 | Total |
|--------|----|----|---|----|---|----|---|----|----|-------|
| New Requirements | 35 | 10 | | 40 | | 20 | 5 | 10 | 30 | 150 |
| Planned-Order Coverage | 85 | | | | | 65 | | | | 150 |

cost is at the point where the inventory (carrying) cost and setup cost are equal."[2] This holds true for the EOQ approach but not for the discrete lot-sizing approach, which assumes that inventory depletions occur at the beginning of each period, as pointed out previously.

In the graphic model of the general relationship between setup and carrying costs, the total cost is at a minimum at the point of intersection of the carrying-cost line and the setup curve only when the line passes through origin (point 0 on the X and Y axes). In the discrete lot-sizing model, however, if a line were fitted to the carrying-cost points, it would have a negative intercept, caused by the assumption that the quantity equal to the demand in the first period incurs no carrying cost. This point can best be illustrated on a model of a series of uniform discrete demands, such as

| Period: | 1 | 2 | 3 | 4 | 5 | 6 | 7 |
|---|---|---|---|---|---|---|---|
| Demand | 20 | 0 | 20 | 0 | 20 | 0 | 20 |

Figure 8-10 shows a graph of cost relationships, with setup of $100, for such a series. In a graph of nonuniform demand, there would, of course, be not one line but several connected shorter lines, each with a different slope. This is why simulations of LUC reveal that the ratio of the setup and carrying-cost elements in the LUC is lopsided some-times one way and sometimes the other. In most cases, however, the setup-cost element

**FIGURE 8-10**

Cost relationships for discrete-demand series.

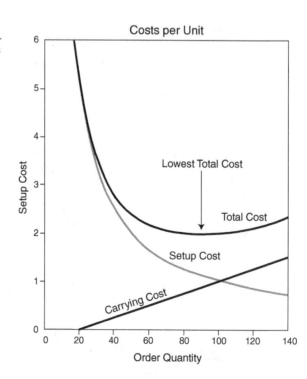

[2] T. Gorham, "Dynamic Order Quantities," *Production & Inventory Management.* 9(1), 1968.

of LUC will be larger than the carrying-cost element. The LTC technique, which seeks to equalize these elements, therefore is biased toward larger order quantities.

## Part-Period Balancing (PPB)

This technique employs the same logic as LTC, and its computation of order quantities is identical except for an adjustment routine called *look-ahead/look-back*.[3] This feature is intended to prevent stock covering peak requirements from being carried for long periods of time and to avoid orders being keyed to (i.e., starting coverage with) periods with low requirements. The adjustments are made only when the condition exists that the look-ahead/look-back corrects. In many cases, therefore, PPB and LTC will yield identical results. This would be the case with the demand data used in the preceding examples, and in order to demonstrate look-ahead/look-back, it is necessary to use different series of net requirements. The look-ahead adjustment would be operative with the following net requirements schedule:

| Period: | 1 | 2 | 3 | 4 | 5 | 6 | 7 | 8 | 9 |
|---|---|---|---|---|---|---|---|---|---|
| Net requirement: | 20 | 40 | 30 | 10 | 40 | 30 | 35 | 20 | 40 |
| Planned orders: |  | X |  | X | → |  |  |  |  |

With an EPP of 100, the first lot of 90 would cover periods 1 through 3, and the next lot would be keyed to period 4. But before this is firmed up, a look ahead to period 5 is made. The 40 units in period 5 would have to be carried in inventory for one period, which would cost 40 part-periods. If the 10 units in period 4 were added to the first lot, they would be carried for three periods at a cost of 30 part-periods. It appears that it would be more economical to key the second lot to period 5. The complete planned-order schedule, adjusted for look-ahead, is shown in Figure 8-11.

The look-ahead test is repeated for successive pairs of period demands until it fails. In our example, the second test (for periods 5 and 6) fails in that it would be more costly to carry 40 for four periods than 30 for one period. If this were not so, the lot would be keyed to period 6. To prevent the look-ahead feature from trying to overcome a steep upward trend in the demand (this would create very large order quantities and defeat the logic of LTC), an additional test is made. The part-period cost of the last period demand

**FIGURE 8-11**

Part-period balancing with look-ahead.

| Period | | 1 | 2 | 3 | 4 | 5 | 6 | 7 | 8 | 9 | Total |
|---|---|---|---|---|---|---|---|---|---|---|---|
| Net Requirements | | 20 | 40 | 30 | 10 | 40 | 30 | 35 | 20 | 40 | 265 |
| Coverage | without look-ahead | 90 | | | 80 | | | 95 | | | 265 |
| | with look-ahead | 100 | | | | 105 | | | 60 | | 265 |

---

[3] J. J. DeMatteis, "An Economic Lot-Sizing Technique: The Part-Period Algorithms." *IBM Systems Journal.* 7(1):30–38, 1968.

covered by the prospective lot is compared with the EPP, and the look-ahead process is stepped if the cost equals or exceeds the EPP.

The look-ahead test is always made first. If it fails (i.e., if the possibility of covering an additional period is ruled out), the look-back test is made. Now what is being checked is the possibility of adding the quantity of requirements in the last period covered by the order in question to the next lot, that is, decreasing the size of the first lot. The look-back can be demonstrated on the following net requirements schedule:

| Period: | 1 | 2 | 3 | 4 | 5 | 6 | 7 | 8 | 9 |
|---|---|---|---|---|---|---|---|---|---|
| Net requirements: | 20 | 40 | 30 | 15 | 30 | 25 | 50 | 20 | 40 |
| Planned orders: | | X | ← | X | | | | | |

With an EPP of 100, the first lot of 90 would cover periods 1 through 3, and the next lot would be keyed to period 4. A look-back to period 3 indicates that it would cost 60 part-periods to carry stock covering this requirement, whereas it would cost only 15 part-periods to carry the period 4 requirement if the second lot were keyed to period 3. A tradeoff therefore is indicated: The second lot is keyed to period 3, and the first lot is reduced to 60 units. The complete planned-order schedule after look-back appears in Figure 8-12.

On first reading, it seems that the special features of PPB on the effectiveness of the LTC approach. On reflection, however, the look-ahead/look-back proposition proves dubious, and its logic is murky. Let us return to the example of look-ahead:

| Period: | 1 | 2 | 3 | 4 | 5 | 6 | 7 | 8 | 9 |
|---|---|---|---|---|---|---|---|---|---|
| Net requirements: | 20 | 40 | 30 | 10 | 40 | 30 | 35 | 20 | 40 |
| Planned orders: | | X | | X | → | | | | |

It is obviously true that it is cheaper to carry 10 units for three periods than 40 units for one period, but this is not the only consequence of the look-ahead adjustment. The 30 in period 6 also will be carried for one period less. The 35 in period 7, however, would not have incurred any carrying cost had the look-ahead not been employed, but now it will cost 70 part-periods. The look-ahead feature of PPB simply does not look far enough ahead. When the adjustment is made, it tends to change the timing (and coverage) of all subsequent planned orders in the schedule, with results that the technique is oblivious of.

In our example, the look-ahead adjustment saves a total of 130 part-periods and incurs a new cost of 100 part-periods. The last lot of 60, however, entails only 40 part-peri-

**FIGURE 8-12**

Part-period balancing with look-back.

| Period | | 1 | 2 | 3 | 4 | 5 | 6 | 7 | 8 | 9 | Total |
|---|---|---|---|---|---|---|---|---|---|---|---|
| Net Requirements | | 20 | 40 | 30 | 15 | 30 | 25 | 50 | 20 | 40 | 270 |
| Coverage | without look-ahead | 90 | | | 70 | | | 110 | | | 270 |
| | with look-ahead | 60 | | 75 | | | 95 | | | 40 | 270 |

ods owing to a lack of horizon. It eventually will have to be recomputed and increased so as to pick up an additional 60 part-periods. This will more than offset the net saving of 30 part-periods in periods 1 through 9. But the first three lots then will have covered a larger time span than had they not been adjusted. At this point, the basis has been lost for making a valid comparison of the alternative strategies.

The look-back proposition appears, if anything, even more dubious than look-ahead proposition. It suffers from the same shortcoming as look-ahead in that it fails to examine the consequences of the adjustment throughout the planning horizon. In our example, the look-back produces a net saving of 75 part-periods but adds a fourth setup that is worth 100 part-periods (Figure 8-12). If setup cost was larger than unit cost, the EPP would be higher and the lots spaced further apart. The last period demand covered by any prospective lot then almost always would entail more part-periods than the first period of the subsequent lot. Look-back then would be more consistently operative, resulting in more and smaller orders, which would subvert the logic of the LTC technique on which the look-back is grafted.

## Wagner-Whitin Algorithm

This technique embodies an optimizing procedure based on a dynamic programming model.[4] The procedure is too mathematically involved to be suitable for a detailed description here. Basically, it evaluates all possible ways of ordering to cover net requirements in each period of the planning horizon. Its objective is to arrive at the optimal ordering strategy for the entire net requirements schedule. The Wagner-Whitin algorithm is elegant in that it reaches this objective without actually having to consider, specifically, each of the strategies that are possible. The Wagner-Whitin solution to the net requirements schedule used in all preceding examples, except for PPB, is shown in Figure 8-13.

The Wagner-Whitin algorithm does minimize the combined (total) cost of setup and of carrying inventory, and it is used as a standard for measuring the relative effectiveness of the other discrete lot-sizing techniques. Its disadvantages, usually mentioned in the literature, are a high computational burden and the near impossibility of explaining it to the average MRP system user.

The first of these two arguments is somewhat exaggerated. While it is true that there are typically tens of thousands of inventory items in an MRP system for which planned

**FIGURE 8-13**

Wagner-Whitin algorithm.

| Period | 1 | 2 | 3 | 4 | 5 | 6 | 7 | 8 | 9 | Total |
|---|---|---|---|---|---|---|---|---|---|---|
| New Requirements | 35 | 10 | | 40 | | 20 | 5 | 10 | 30 | 150 |
| Planned-Order Coverage | 40 | | | 65 | | | | 40 | | 150 |

[4] H. M. Wagner and T. M. Whitin, "Dynamic Version of the Economic Lot Size Model." *Management Science.* 5(1), 1958.

orders have to be computed and that requirements for a given item tend to change and cause constant recomputations, computational time, once a record is in the computer's main memory, is not significant.

The second argument, however, is entirely valid. The complexity of the procedure inhibits understanding by the layperson and acts as an obstacle to its adoption in practice.[5] An inherent weakness of the Wagner-Whitin algorithm lies in its assumption that requirements beyond the planning horizon are zero. The technique is designed for a stationary horizon. It would work well, for instance, in the case of custom-designed parts in a limited number of situations, such as a one-time contract for a quantity of special machinery with a firm, staggered delivery schedule. In most cases, though, the planning horizon is not stationary, the life of the typical inventory item is quite long, and additional requirements are constantly being brought within the planning horizon by the passage of time.

Whenever a new requirement appears at the far end of the planning horizon, the Wagner-Whitin ordering strategy (which, by definition, pertains to the entire planning horizon) may have to be revised. At least one lot at the far end of the series is subject to recomputation even if the specific requirements it covers remain unchanged. The validity of a given planned-order quantity computed under this approach may prove ephemeral, lasting no longer than one planning period. This, of course, is true with some of the other algorithms as well.

In practice, the Wagner-Whitin optimal strategy proves to be wrong if it has to be changed subsequently. From an MRP point of view, instability in the planned-order schedule is undesirable. To the extent that Wagner-Whitin is more sensitive than other lot-sizing techniques to additions of requirements caused by the extension of the horizon—owing to its optimal strategy objective—it loses its practical appeal.

## LOT-SIZE ADJUSTMENTS

The planned-order quantity determined by any of the lot-sizing techniques is subject to certain adjustments dictated by practical considerations. Among these are the following:

- Minimums and maximums
- Scrap allowances
- Multiples
- Raw materials cutting factors

Any of the lot-sizing algorithms discussed previously can be constrained by the imposition of minimums and/or maximums on the quantity of the item to be ordered. One type of minimum has already been mentioned; that is, the computed quantity, if lower than the net requirements in the period to which the order is keyed, will be increased to at least equal the net requirements. Minimums and maximums may be stat-

---

[5] G. W. Plossl and O. W. Wight, *MRP by Computer*. Washington, DC: American Production and Inventory Control Society, 1971, p. 9.

ed in absolute numbers, for example, "no less than 50 and no more than 400 units," pertaining to individual inventory items. Alternatively, the limits on quantity may be stated in terms of period coverage, for example, "no less than 4 weeks' and no more than 12 weeks' requirements" or "not to exceed one year's supply." Minimums and maximums on lot sizes are frequently imposed by management in view of the fact that the lot-sizing algorithm is blind to a number of practical operating considerations.

A *scrap allowance*, or *shrinkage factor*, is a quantity added to the computed lot size that is intended to compensate for anticipated scrap or loss in process and to ensure that the required quantity of "good" pieces is received. This is important only in instances of discrete lot sizing because the order quantity covers net requirements in an integral number of periods (no remnants). The scrap allowance normally will vary from item to item according to past incidence of scrap.

The scrap allowance may be stated either in terms of pieces or as a percentage of the order quantity. Where the latter approach is used, a fixed percentage generally is undesirable if the planned-order quantities vary significantly from lot to lot. In a machine shop environment, scrap tends to be a function of the number of different setups that are required to complete the part rather than the quantity being run. In view of this fact, a *declining percentage* formula can be used, such as the following:

$$Q = L + a\sqrt{L}$$

where  $Q$ = order quantity
   $L$ = lot size computed by the algorithm
   $a$ = multiplier reflecting scrap incidence

For example, the lot-sizing technique being used might yield a quantity of 400. This would be adjusted for scrap by adding the square root of 400, that is, 20, for a final order quantity of 420. In this case, the multiplier was assumed to be 1—the value usually used unless the responsible inventory planner specifies a different one. The multiplier, acting to reflect incidence of past scrap of the inventory item in question, may be set to vary from 0 (no scrap allowance) to a decimal fraction of 1 to a multiple of 1. With a multiplier of 1, the scrap allowances for various lot quantities would be as shown in Figure 8-14. In an MRP system, the proper way to handle scrap allowances in the time-phased inventory record is simply to add them to (include them in) the planned-order quantities.

**FIGURE 8-14**

Declining-percentage scrap allowance.

| Computed lot size | Scrap allowance | Order quantity | Percentage allowed |
|---|---|---|---|
| 1 | 1 | 2 | 100 |
| 4 | 2 | 6 | 50 |
| 9 | 3 | 12 | 33 |
| 16 | 4 | 20 | 25 |
| 25 | 5 | 30 | 20 |
| 100 | 10 | 110 | 10 |
| 400 | 20 | 420 | 5 |
| 10,000 | 100 | 10,100 | 1 |

When the planned orders are eventually released, the full quantity, including the scrap allowance, should be shown as on order. This quantity then would be reduced as and if scrap transactions are posted to the record. The practice of including scrap-allowance quantities in the item's gross requirements (in order to display the projected on-hand quantities as they are expected to be after scrap) is unsound because it distorts the fundamental relationship of parent planned-order quantity to the component item's gross requirements quantity. Furthermore, whether scrap actually will occur is uncertain. Until it does occur, the MRP system should project the item's status as though it will not occur.

Another constraint that can be imposed on the lot-sizing algorithm is the requirement that a given item be ordered in multiples of some number. This may be dictated by considerations of process (e.g., so many pieces constitute a Blanchard grinder load or so many bars of steel are fed to a bar lathe simultaneously), packaging (e.g., 12 pieces in a carton), and so on. The lot size yielded by the lot-sizing algorithm is, in these cases, increased to the nearest multiple specified.

Raw material cutting factors represent another adjustment to lot size that it is desirable to make in certain instances. The lot-sizing algorithm, unaware of the form in which raw material (from which the item in question will be made) comes, may generate a quantity that would create problems in cutting this material. For example, if a certain size of sheet metal is cut into nine pieces in the manufacture of a given inventory item, a lot size of 30 will result in either an odd size of raw material being left over or, more likely, the shear operator cutting the four sheets into 36 pieces. This then will become the actual quantity on order, as opposed to the 30 on order shown by the system.

In cases where more than one type of adjustment is to be made to the order quantity for a given item, the several adjustments are made consecutively in a logical sequence. For example, if the lot-sizing technique yields a quantity of, say, 173, which is equivalent to five periods' requirements, and the item is subject to a three-period ceiling, scrap allowance, and cutting constraint (20 pieces per unit of raw material), the "raw" quantity of 173 would be adjusted as follows:

| | |
|---|---|
| Raw order quantity: | 173 |
| Reduce to three periods' requirements: | 121 |
| Add scrap allowance: | 11 |
| Total: | 132 |
| Increase to nearest multiple of 20: | 8 |
| Adjusted order quantity: | 140 |

Scrap allowances, multiples, and cutting factors tend to create an inventory excess. This excess, however, is subsequently applied by the MRP system against later gross requirements. At any point in time, there exists a slight inventory excess, but it does not accumulate.

# EVALUATING LOT-SIZING TECHNIQUES

Every one of the lot-sizing techniques reviewed earlier is imperfect—each suffers from some deficiency, as has been illustrated. In evaluating the relative effectiveness of these techniques, the difficulty lies in the fact that the performance of the algorithms varies depending on the net requirements data used and on the ratio of setup and unit costs. Furthermore, some of the techniques assume gradual, steady-rate inventory depletion, whereas others assume discrete depletion, which affects the way inventory carrying cost would have to be computed for purposes of comparison. Ignoring this distinction and basing all inventory carrying costs on discrete depletion at the beginning of each period, the performance of the economics-oriented lot-sizing algorithms for which the same data set was used in the preceding examples compares as follows (details in Figure 8-15):

|                        | Total cost |
|------------------------|-----------|
| Wagner-Whitin          | $395      |
| Least unit cost        | 420       |
| Least total cost       | 445       |
| Period order quantity  | 455       |
| Economic order quantity| 506       |

**FIGURE 8-15**

Comparison of lot-sizing algorithm performance.

| Algorithm | Number of setups | Setup cost, $ | Part-periods | Carrying cost, $ | Total cost, $ |
|-----------|-----------------|---------------|--------------|------------------|---------------|
| W-W       | 3               | 300           | 95           | 95               | 395           |
| LUC       | 3               | 300           | 120          | 120              | 420           |
| LTC       | 2               | 200           | 245          | 245              | 445           |
| POQ       | 3               | 300           | 155          | 155              | 455           |
| EOQ       | 3               | 300           | 206          | 206              | 506           |

These figures are meaningful only in relation to the net requirements schedule, the setup cost ($100), and the unit cost ($50) used in the examples. A change in these data will produce a different sequence. For example, if setup were $300, the POQ would outperform LTC and match LUC in effectiveness. If the requirements data are changed, the example can be rigged so as to produce practically any results desired, including the EOQ equal in performance to Wagner-Whitin.[6] The factors that affect the relative effectiveness of the individual lot-sizing techniques are the following:

1. The variability of demand
2. The length of the planning horizon
3. The size of the planning period
4. The ratio of setup and unit costs

---

[6] W. L. Berry, "Lot Sizing Procedures for Requirements Planning Systems. A Framework for Analysis." *Production & Inventory Management.* 13(2), 1972.

The variability of demand consists of nonuniformity (varying magnitude of period demand) and discontinuity (gaps of no period demand). The length of the planning horizon, that is, demand visibility, obviously affects the comparative performance of the various algorithms. Shorter planning periods (e.g., weeks instead of months) result in smaller requirements per period, enabling the lot-sizing technique to get closer to the best balance between setup and carrying costs. The setup/unit-cost ratio directly affects the frequency of ordering and thus the lot size.

There does not appear to be one "best" lot-sizing algorithm that could be selected for a given manufacturing environment for a class of items and in most cases even for a single specific item. For purposes of MRP, the lot-for-lot approach should be used wherever feasible, and in cases of significant setup cost (typical in the fabrication of component parts), LUC, LTC, PPB, or even POQ should provide satisfactory results. When it comes to selecting a lot-sizing technique (or techniques) to be incorporated in an MRP system, neither detailed studies nor exhaustive debates are warranted—in practice, one discrete lot-sizing algorithm is about as good as another.

Apart from the inherent weaknesses and difficulty of meaningful comparison between algorithms, the one fact of life that renders, and always will render, any lot-sizing technique vulnerable is the possibility that future requirements will change. After the planned order is released, the order quantity may prove to be wrong in light of a change in the magnitude and/or timing of net requirements.

When this happens, it does not matter how elaborately and with what precision the lot size had been computed. All the discrete lot-sizing algorithms are based on the implicit assumption of certainty of demand. This is the true Achilles' heel of lot sizing because, in most cases, the pattern of future demand is never certain. A more realistic assumption would be that the requirements schedule against which the lot size is being computed will change.

In comparing the relative effectiveness of one discrete lot-sizing algorithm with that of another, it is possible to determine which of the two is better vis-à-vis a given schedule of net requirements. When the period of time spanned by this schedule has passed into history, however, it might develop that the algorithm originally judged less effective would, in fact, have had the better performance in light of how the requirements actually turned out.

In the final analysis, it does not matter how elaborately and with what precision lot sizes can be computed. All techniques are vulnerable to changing demand, and unfortunately, this is found in most MRP environments. The best way to ensure maximum economy in ordering materials is to develop and maintain stable plans for procurement and fabrication. The spurious precision of lot-sizing technique is invalidated by what actually happens as opposed to what had been planned to happen. The relative actual effectiveness of a lot-sizing algorithm can be determined only in retrospect.

# BIBLIOGRAPHY

Berry, W. L. "Lot Sizing Procedures for Requirements Planning Systems: A Framework for Analysis." *Production & Inventory Management* 13(2), 1972.

DeMatteis, J. J. "An Economic Lot Sizing Technique: The Part-Period Algorithm." *IBM Systems Journal* 7(1):30–38, 1968.

Diegel, A. "Seven Alternatives to Dynamic Programming for Dynamic Lots." Paper presented at the 39th National Meeting of the Operations Research Society of America, May 1971.

Gleason, J. M. "A Computational Variation of the Wagner-Whitin Algorithm: An Alternative to the EOQ." *Production & Inventory Management* 12(1), 1971.

Gorenstein, S. "Some Remarks on EOQ vs. Wagner-Whitin." *Production & Inventory Management* 11(2), 1970.

Gorham, T. "Determining Economic Purchase Quantities for Parts with Price Breaks." *Production & Inventory Management* 11(1), 1970.

———. "Dynamic Order Quantities." *Production & Inventory Management* 9(1), 1968.

Kaimann, R. A. "A Fallacy of EOQ'ing." *Production & Inventory Management* 9(1), 1969.

Kaimann, R. A. "EOQ vs. Dynamic Programming: Which One to Use for Inventory Ordering." *Production & Inventory Management* 10(4), 1969.

———. "Revisiting a Fallacy of EOQ'ing." *Production & Inventory Management* 9(4), 1968.

Lippman, S. A. "Optimal Inventory Policy with Multiple Setup Costs." *Management Science* 16(1), 1969.

Peterson, R., and E. A. Silver. *Decision Systems for Inventory Management and Production Planning.* Hoboken, NJ: Wiley, in press.

Plossl, G. W., and O. W. Wight. *MRP by Computer.* Washington, DC: American Production and Inventory Control Society, 1971, pp. 6–9 and 30–31.

*Purchasing* (general Information Manual Form No. GH20-1149-1). New York: International Business Machines Corp. 1973, pp. 140–164.

Rinehard, J. R. "Economic Purchase Quantity Calculations." *Management Accounting*, September 1970.

Silver, E. A., and H. C. Meal. "A Heuristic for Selecting Lot Size Quantities for the Case of a Deterministic Time-Varying Demand Rate and Discrete Opportunities for Replenishment." *Production & Inventory Management* 14(2), 1973.

———. "A Simple Modification of the EOQ for the Case of Varying Demand Rate." *Production & Inventory Management* 10(4), 1969.

*System/360 Requirements Planning* (Application Description Manual Form No.GH20-0487-3). New York: International Business Machines Corp., 1970, pp. 24–32.

Thomopoulos, N. T., and M. Lehman. "Effects of Inventory Obsolescence and Uneven Demand on the EOQ Formula." *Production & Inventory Management* 12(4), 1971.

Tuite, M. F., and W. A. Anderson. "A Comparison of Lot Size Algorithms under Fluctuating Demand Conditions." *Production & Inventory Management* 9(4), 1968.

Wagner, H. M., and T. M. Whitin. "Dynamic Version of the Economic Lot Size Model." *Management Science* 5(1), 1958.

Woolsey, R. E. D., E. A. Silver, and H. S. Swanson. "Effect of Forecast Errors on an Inventory Model with Variations in the Demand Rate." *Production & Inventory Management* 14(2), 1973.

## CHAPTER 9

# System Records and Files

*After all, the engineers create the bill so that, by definition, somebody other than the designer can make the product. The bill of material is, therefore, really made for others in the first place. And it would seem to follow that it should be structured for the user's, not the designer's, convenience.*
—GEORGE W. PLOSSL, IN *MRP AND BILL OF MATERIAL STRUCTURE*,
FILM PRODUCED BY INTERNATIONAL BUSINESS MACHINES CORP., 1972

**A** material requirements planning (MRP) system can be thought of as a set of logically linked item inventory records coupled with a program (or programs) that maintains these records up to date. The design of the inventory record, as well as the way the data it contains are being manipulated to produce valid system outputs, is crucial to both the effectiveness of the system and an understanding of the subject of MRP.

## THE TIME-PHASED RECORD

Under the MRP approach, a separate time-phased inventory record is established and maintained for every inventory item. Each record consists of three portions, or segments, as follows:

1. Item master data (record header)
2. Inventory status data (the body of the record)
3. Subsidiary data

The inventory status segment is either reconstructed periodically or kept up to date continuously depending on which of the two basic alternatives of implementing an MRP system—schedule regeneration or net change (discussed in Chapter 7)—had been chosen. At this point in the discussion, however, we need not be concerned with the distinction. Thus far we have been dealing only with the status data (the most important segment of the record), and there are a few more aspects of this segment to be reviewed before describing the header and subsidiary segments.

## Time-Phased Record Format

The most compact format of recording and displaying time-phased inventory status data is the one introduced in Chapter 7 and used in several previous examples. It consists of four rows of time buckets representing the following:

1. Gross requirements
2. Scheduled receipts (open orders)
3. On hand (current and projected by period)
4. Planned-order releases

This format accommodates all the information that is essential for the proper manipulation of status data and for operation of the MRP system. The four rows of buckets define inventory status in summary form, and the format contains implicit information that can be inferred from the data that are displayed directly. This is the standard format favored by many MRP system users, and it is the format normally used for purposes of communication and instruction.

At the option of a given user, however, the format can be expanded in such a way as to provide more detail and/or to state more of the information explicitly. Figure 9-1 shows both the compact and expanded formats based on the same status data. The example illustrates the possibilities of expansion rather than the format of an actual record. In practice, the expansion is usually limited to adding separate net requirements and/or planned-order receipt buckets.

## Optional Fields

In addition to what is shown in Figure 9-1, the status segment of the inventory-item record may include a field labeled "Allocated on hand." Allocation was discussed in Chapter 5. The quantity allocated indicates the quantity of the item earmarked for a parent order (or orders) that has been released but for which the material requisition has not yet been filled.

The allocated parts "belong" to the respective parent orders, and they are still physically on hand in the stockroom only because of the time gap between order release (by the inventory planner) and the filling (by the stockroom) of the supporting requisition for component materials. Where a single allocation field in each inventory record is being maintained, it is understood that parent orders normally will be released during the period for which the release is planned and not before. Also, it is assumed that all the component items are required at the time the parent order is released. Both these conditions

**FIGURE 9-1**

Time-phased
record: compact
and expanded
formats.

**A  Compact Format**

| Lead Time: 3 | | Period | | | |
|---|---|---|---|---|---|
| | | 1 | 2 | 3 | 4 |
| Gross Requirements | | 10 | 15 | 75 | 17 |
| Schedule Receipts | | 8 | | 25 | |
| On Hand | 72 | 70 | 55 | 5 | −12 |
| Planned-Order Releases | 20 | | | | |

**B  Expanded Format**

| | | | Period | | | |
|---|---|---|---|---|---|---|
| | Lead Time: 3 | | 1 | 2 | 3 | 4 |
| Gross Requirements | From Parent Items | | 10 | 10 | 15 | 12 |
| | Service Part Orders | | | 5 | | 5 |
| | Interplant | | | | 60 | |
| | Total | | 10 | 15 | 75 | 17 |
| Schedule Receipts | Supply Source A | | | | 25 | |
| | Supply Source B | | 8 | | | |
| | Total | | 8 | | 25 | |
| Planned-Order Receipts | | | | | | 20 |
| Net Requirements | | | | | | 12 |
| On Hand | Stockroom #1 | 50 | 40 | 30 | | 8 |
| | Stockroom #2 | 22 | 30 | 25 | 5 | |
| | Total | 72 | 70 | 55 | 5 | 8 |
| Planned-Order Releases | | | | | | |

will prove to be true in the typical manufacturing environment, and the single allocation field is standard.

In exceptional cases where either orders must often be released prematurely for some reason or the requirement for the different components of an assembly is staggered over a long assembly lead time, the allocated quantities would be time-phased; that is, a separate row of time buckets would be maintained. An example of this is shown in Figure 9-2. Note, in this example, that the use of a single allocation field would distort the component item's projected on-hand schedule (Figure 9-2B), which would then show 15 in

## FIGURE 9-2

Time phasing of allocated quantities.

**A**

**Parent**

|  | Period | | | |
|---|---|---|---|---|
|  | 1 | 2 | 3 | 4 |
| Planned-Order Releases | 10 |  | 15 |  |

**Original Status of Component Item**

**Component**

|  |  | | | | |
|---|---|---|---|---|---|
| Gross Requirements |  | 30 | 12 | 25 | 20 |
| Scheduled Receipts |  |  |  |  |  |
| On Hand | 60 | 30 | 18 | –7 | –27 |
| Planned-Order Releases |  | 27 |  |  |  |

**B**

**Parent**

| Planned-Order Releases | 0 |  | 0 |  |
|---|---|---|---|---|

+10   +15   –15
      –10

**Component**

**Status Following Parent Order Releases**

|  |  |  | | | |
|---|---|---|---|---|---|
| Gross Requirements |  | 20 | 12 | 10 | 20 |
| Allocated |  | 10 |  | 15 |  |
| Scheduled Receipts |  |  |  |  |  |
| On Hand | 60 | 30 | 18 | –7 | –27 |
| Planned-Order Releases |  | 27 |  |  |  |

period 1 and 3 in period 2. This is accomplished with the lead-time offset available in the bill of material (BOM) function.

Another field that optionally can be maintained in the time-phased inventory record is the quantity past due. One time bucket immediately preceding the first period is provided for this purpose in each of the schedules (rows). In the on-hand schedule, this field shows the current quantity on hand because the notion of past due does not apply to these data. Past-due quantities, if any, are recorded in the respective fields, as shown in Figure 9-3. In the computation of net requirements, the past-due quantities must be added to those in the first period (the assumption here is that where performance fell behind schedule, it would catch up in the first, or current, period) so as not to distort the projected on-hand/net requirements schedule.

This is shown in Figure 9-3A. An alternative treatment is to include the past-due quantities in the first-period buckets, as shown in Figure 9-3B. In either case, the net requirements computation must produce identical results.

## FIGURE 9-3

Treatment of quantities past due.

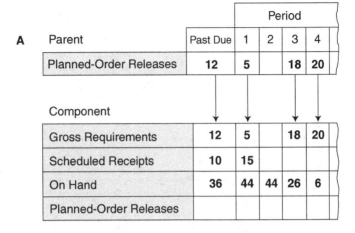

| A  Parent | Past Due | 1 | 2 | 3 | 4 |
|---|---|---|---|---|---|
| Planned-Order Releases | 12 | 5 | | 18 | 20 |

| Component | | | | | |
|---|---|---|---|---|---|
| Gross Requirements | 12 | 5 | | 18 | 20 |
| Scheduled Receipts | 10 | 15 | | | |
| On Hand | 36 | 44 | 44 | 26 | 6 |
| Planned-Order Releases | | | | | |

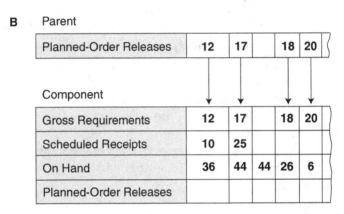

| B  Parent | | | | | |
|---|---|---|---|---|---|
| Planned-Order Releases | 12 | 17 | | 18 | 20 |

| Component | | | | | |
|---|---|---|---|---|---|
| Gross Requirements | 12 | 17 | | 18 | 20 |
| Scheduled Receipts | 10 | 25 | | | |
| On Hand | 36 | 44 | 44 | 26 | 6 |
| Planned-Order Releases | | | | | |

The respective quantities are recorded as past due either as a result of the lead time exceeding the available time in the case of planned-order releases or because of lack of planned performance in any of the schedules. If the master production schedule (MPS) that is being input to the MRP system for processing contains past-due buckets, it also will cause quantities on component levels to be timed as past due. The past-due timing in the time-phased inventory record should be avoided except where it may aid in following up and expediting behind-schedule performance on open-order completions and planned-order releases. Recording a planned order as past due when there is insufficient time left for the full lead-time offset, as shown in Figure 9-4, makes little sense because there is no delinquent performance involved that should be expedited, and the order-release action cannot take place in the past but only in the present, that is, in the current period. This situation is mitigated with establishment of the necessary strategic buffers to protect manufacturing from demand and supply variability. See Chapters 4 and 22 for how these buffers are established.

It is possible to time phase past-due quantities and to display them in the inventory record, as shown in Figure 9-5, but there is little benefit to such practice. A past-due

**FIGURE 9-4**

Past-due order release caused by insufficient lead time.

| Lead Time: 5 | Past Due | Period 1 | 2 | 3 | 4 |
|---|---|---|---|---|---|
| Net Requirements | | | | | **25** |
| Planned-Order Releases | **25** | | | | |

**FIGURE 9-5**

Time-phased quantities past due.

| | Weeks Past Due 4+ | 3 | 2 | 1 | Week 34 | 35 | 36 | 37 |
|---|---|---|---|---|---|---|---|---|
| Scheduled Receipts | **6** | | | **50** | | | **60** | |

order (assuming that its due date is valid) is past due and needs to be expedited for earliest possible delivery—irrespective of how much past due it is. Its original due date can be carried in the subsidiary segment of the inventory record. The status segment of this record is being unnecessarily complicated by the inclusion of several past-due fields.

In MRP systems of the net-change variety, the function of the past-due field is taken over by the control-balance bucket, in which both delinquent and premature performance is being recorded in terms of, respectively, positive and negative values. The last special field in the status segment of the inventory record that merits brief mention is the total bucket, which appears at the end of each time-bucket row. Inclusion of this field is optional. It would be used for purposes of reconciliation or a validity check of the status data that the computer would be programmed to carry out whenever the net requirements are recomputed.

For example:

| | | |
|---|---|---|
| Current on hand: | | 115 |
| Total scheduled receipts: | | 100 |
| | | 215 |
| Allocated on hand: | 35 | |
| Total gross requirements: | 380 | –415 |
| Total net requirements: | | –200 |
| Total planned orders: | | 225 |
| Planned coverage excess: | | 25 |

This type of reconciliation will detect irregularities in the status data or conditions calling for scrutiny and possible action. For example, if the projected on-hand quantity (equivalent to net requirements) at the end of the planning horizon is a positive value ("negative" net requirements), it is indicative of excessive coverage, probably caused by a reduction in gross requirements. If there are any open orders outstanding, they should be reduced or canceled, if possible.

## The Complete Logical Record

In addition to the status-data segment, the item master-data (header) and subsidiary-data segments make up the item inventory record. All these data together are termed the *logical record* (data that are logically related) as opposed to the physical record or records stored in possibly different formats and different locations of computer storage. The data that constitute a logical record are not necessarily stored together physically. Some of them may not be stored at all but are re-created in the computer's main memory for purposes of computation and/or display. This is a matter of programming and the design of database software that the system user, generally speaking, need not be concerned about.

# UPDATING INVENTORY RECORDS

The inventory status data are maintained up to date by means of processing (posting) inventory transactions against the item inventory record. An *inventory transaction* is defined as a notice of an event that changes the inventory status. External inventory transactions are reported to the system, whereas internal transactions are generated by the system itself in the course of requirements planning. Reports of certain events that do not affect inventory status but are posted to the subsidiary data segment of the record are called *pseudotransactions*.

## Transactions and Other Entries

The status data on which an MRP system depends are maintained up to date by means of processing transactions against item inventory records. Inventory transactions are not, however, the only entries processed by the system that affect these records. The several types of entries that the system processes in order to update inventory records may be categorized as follows:

1. Inventory transactions
2. User-controlled exceptions to regular processing logic
3. Pseudotransactions
4. Final assembly schedule entries
5. Error-correction entries
6. File-maintenance entries

Inventory transactions act to modify the status of inventory items; that is, status is changed following the processing of any inventory transaction. A given transaction may cause subsidiary records to be processed in addition to processing the status-data segment of the inventory record. A transaction may change the status in such a way as to also require the updating of component-item status (in net change implementations of MRP systems), thus affecting multiple inventory records. A transaction may report a normal or planned event, such as a stock receipt, or an unexpected event, such as a stock return.

While both may have the same physical effect, the regular processing logic must be modified (as will be shown later) to register unexpected events.

User-controlled exceptions to regular processing logic represent another category of entries processed against inventory records. Such entries serve as a means of intercession by the inventory planner. In certain situations, human judgment is required to evaluate and solve a problem, and the planner must be able to override the system's regular logic. In this category belong several types of commands that the MRP system can be programmed to obey. One example is a hold command to prevent a (mature) planned order from being issued, perhaps because of a contemplated substitution in raw material. Another example is a scrap-tag command that tells the system not to call for release of a new order if its quantity is smaller than the scrap allowance of an existing open order. A firm planned-order command, which freezes a planned order in place, is another example. The use of this command was discussed in Chapter 8.

Pseudotransactions are entries to the subsidiary-data segment of the item inventory record. Pseudotransactions do not affect inventory status. Examples are a purchase-requisition issue (status will be affected only on the release of the purchase order) and a change in an open-order detail. Another example is the recording of a subcontractor's work authorization. These transactions do not affect inventory, but they do have a significant financial impact. Final-assembly-schedule entries apply to highest-level items only in those cases where the end products themselves (because of their complexity) do not appear in the MPS. When the final assembly schedule, which is stated in terms of product models, is put together, the high-level components on which it will draw may be allocated in the respective inventory records. In another type of manufacturing business, a customer order may be processed this way on receipt. In an assembly-line environment, a day's or week's final production may be broken down into high-level components consumed, summarized by component, and processed against the respective component inventory records in lieu of stock-disbursement transactions, which are not otherwise reported for highest-level items.

Error-correction entries are not genuine transactions because they do not affect real status. In some MRP systems, special transaction codes are used to distinguish error-correction entries from genuine transactions that have the same effect. For example, the inventory planner releases an order for item A but erroneously reports it as item B. In the record of B, there now appears an open order. The error is corrected by processing an entry that reverses the previous transaction rather than by an order-cancellation transaction. The effect would be the same, but the distinction is made for purposes of record.

File-maintenance entries affect the item master-data segment (header) of the item inventory record. Such entries update the record for changes in the attributes of the item, for example, standard cost, classification, item description, and so on, or for changes in planning factors such as lead time or scrap allowance. File-maintenance entries do not affect inventory status, or rather, their processing does not trigger the replanning process in standard implementations of MRP.

## Transaction Effects

The designer of an inventory control system must decide how many different types of transaction are to be recognized, how they are to be coded, and how they are to be processed by the system. The choices are virtually unlimited, and dozens of transaction types may be recognized in a given system. The range and treatment of both transactions and pseudotransactions will be reviewed later in this section. While there is no limit to the number of different transaction types that may be used, there is a limited number of effects that these transactions can have on inventory status. Thus a number of different transaction types will affect inventory status the same way.

For example, a stock receipt of an overrun, a customer return, and an inventory adjustment up (the result of a physical count) will increase the quantity on hand and reduce net requirements. The different effects that various transactions can have on a time-phased inventory record are as follows:

### External transactions affecting one record:

1. Change quantity of gross requirements.
   - *Secondary effect:* Recompute projected on hand; recompute planned-order releases.
2. Change quantity of scheduled receipt.
   - *Secondary effect:* Recompute projected on hand; recompute planned-order releases.
3. Reduce scheduled receipt and increase quantity on hand.
4. Change quantity on hand.
   - Recompute projected on hand; recompute planned-order releases.
5. Reduce quantity on hand and reduce gross requirements.
6. Reduce quantity on hand and reduce quantity allocated.

### External transactions affecting multiple records:

7. Change quantity of planned-order release (parent record) and change quantity of gross requirements (component records).
   - *Secondary effect:* Recompute projected on-hand and planned-order release in component records.
8. Reduce quantity of planned-order release and increase scheduled receipts (parent records); reduce gross requirements and increase quantity allocated (component records).
   - *Secondary effect:* Recompute projected on-hand in parent record.
9. Increase quantity of planned-order release and reduce scheduled receipts (parent record); increase gross requirements and reduce quantity allocated (component records).

### *Internal transactions affecting multiple records:*

**10.** Change quantity of planned-order release (parent record) and change quantity of gross requirements (component records).
  - *Secondary effect:* Recompute projected on-hand and planned-order releases in component records.

Any given inventory transaction has one (and only one) of the 10 possible effects listed. The comments that follow refer to these effects by their number.

Effect number 1 (change quantity of gross requirements) is a result of either increasing or reducing the contents of a gross requirements bucket or multiple buckets. Note that a change in the timing of a gross requirement is effected by reducing the quantity in the original bucket and increasing the quantity in the new bucket. Addition of a new requirement in a new bucket is tantamount to increasing that bucket's contents from the original zero to the new quantity. Effect number 1 results from transactions reporting demand for the item (including an increase, reduction, or cancellation of this demand) originating from external sources. Orders or forecasts for service parts, interplant items, and so on are examples.

Effect number 2 (change quantity of scheduled receipt) results from increasing, reducing, canceling, or rescheduling an open order. Rescheduling, as in the case of gross requirements, means reducing the contents of one bucket and increasing the contents of another one. Transactions that will have this effect are, for example, a purchase-order increase, a scrap report, and a change in the order due date.

Effect number 3 (reduce scheduled receipt and increase quantity on hand) is caused by a stock receipt, partial or full, of an order. Note that this does not apply to an unplanned receipt for which no order had been placed previously nor to the quantity of an overrun or overdelivery. Unless delivery is premature, neither the projected on hand nor the planned-order release schedules need be recomputed.

Effect number 4 (change quantity on hand) is the result of transactions that increase or reduce the quantity on hand without affecting any open orders. Stock returns, overdeliveries, inventory adjustments up or down, and unplanned disbursements belong in this category. The unanticipated change in the quantity on hand causes a recomputation of the projected on-hand schedule and, consequently, the planned-order release schedule.

Effect number 5 (reduce quantity on hand and reduce gross requirements) results from a disbursement or shipment of an external (i.e., service part, intersystem plant, etc.) order. There are no secondary effects on the other status data in the record.

Effect number 6 (reduce quantity on hand and reduce quantity allocated) results from a planned (anticipated) disbursement of a component item against a parent order. As the material requisition or picking list, previously released to the stockroom, is filled, the transaction reporting it reduces the quantities on hand and allocated.

Effect number 7 (change quantity of planned-order release in the parent record and change quantity of gross requirements in the component record) is a result of an intervention by the inventory planner, who, as discussed in Chapter 8, solves certain problems by changing the quantity or timing of a planned order and "freezing" this change so that the MRP system will not try to recompute or reposition this particular planned order the next time the net requirements change. The transaction reporting this intervention to the system is the previously discussed firm planned order. The change in the planned-order schedule affects the gross requirements of component items and causes their status to be recomputed. The firm planned order is one of several types of inventory transactions that affect more than one record in cases of manufactured items. Purchased items have no components (in the system), and transactions reported against them never affect other inventory records.

Effect number 8 (reduce quantity of planned-order release and increase scheduled receipts in parent records; reduce gross requirements and increase quantity allocated in component records) is caused by a release of a planned order, which the respective transaction converts to an open order (scheduled receipt) in the inventory record. This transaction also affects component records, whose gross requirements are reduced and the allocated quantity increased.

Effect number 9 (increase quantity of planned-order release and reduce scheduled receipts in the parent record; increase gross requirements and reduce quantity allocated in component records) nullifies a previous order-release transaction. This happens when the inventory planner for some reason decides to rescind the release of an order. Once a shop order has started into the manufacturing process, this, of course, is no longer possible except in very unique circumstances.

Effect number 10 (change quantity of planned-order release in the parent record and change quantity of gross requirements in the component records) is caused by the only internal transaction that exists, that is, a change in a parent planned-order schedule being reflected in the gross requirements of component items. This effect is identical to effect number 7 except that here the "transaction" is generated by the system internally in the course of requirements planning (explosion).

As pointed out earlier, different transaction codes may be used for several entries that are identical logically, that is, that have the same effect on inventory status. The reason for creating a transaction set larger than the minimum that is essential lies in the desirability of being able to log transaction history (audit trail) by recording and measuring their sources, reasons, and so on, as well as being able to trigger different treatments of these various transactions in the subsidiary-data segment of the inventory record.

## Reporting Receipts and Disbursements

As mentioned in Chapter 5, an MRP system is based on the assumption that each item under its control passes into and out of stock and that reports of receipts and disbursements, that is, transactions, will be generated. In many manufacturing operations, however, it is not practical to route each inventory item through a stockroom. In fact, this can be the source of significant waste that can be eliminated. In these cases, the reporting, which is mandatory under an MRP system, may be based on events other than physical arrival in stock and departure there from. Under the following options in the treatment of receipts and disbursements, the posting of transactions is

1. Initiated on report from the stockroom
2. Initiated on report from the receiving dock
3. Triggered by shop floor events
4. Anticipated from other transactions

Reporting from the stockroom is the normal practice. Receipts of purchased items alternatively may be reported from the receiving dock, but if the stockroom is to be bypassed entirely, such transactions signal both a receipt and a disbursement. The posting of receipts and disbursements to inventory records can be triggered by certain designated events on the shop floor. Completion of the last (or other designated) operation on a shop order may be considered as a receipt or a simultaneous receipt and disbursement. Completion (receipt) of a parent order may be treated as a disbursement of component items. A production report (mentioned earlier in this section in connection with assembly lines) may be broken down and translated into component-item disbursements. A disbursement also can be anticipated from the posting of a related transaction. For example, the release of a parent planned order may be treated as tantamount to component-item disbursement.

## THE DATABASE

In a computer-based system such as MRP, files constitute the foundation on which the superstructure of the application is built. As with any foundation supporting a structure, it codetermines the soundness and utility of that structure. The effective operation and efficiency of an MRP system is, to a considerable degree, a function of system file quality. This quality, in turn, is reflected in the relative accuracy, "up-to-dateness," and accessibility of file-record information.

The importance of file organization and file management to the success of a computer-based system is great, particularly so because of a universal tendency on the part of management to underestimate both the importance and the requirements of this part of the system. Lack of file integrity is one important reason why some MRP systems installed in industry have failed to live up to expectations. Emphasis therefore must be put on system files and their organization, maintenance, and accessibility. Computer software manufacturers have invested heavily in the development of file management programs (database software), tools that help greatly in coping with the problem of maintaining file integrity.

### Problems of File Maintenance

A computer-based system such as MRP will not work satisfactorily with poor files, but in the average manufacturing company, the files in question, at the time the system is being implemented, normally are in a rather poor condition. This seems invariably true, particularly with manually maintained files of records related to product structure (i.e., BOMs), to inventory status, and to the manufacturing process proper (i.e., routings and operation sheets). This is a result of the fact that the rate of change affecting these records typically is not matched by a corresponding capacity of the responsible departments to incorporate the changes in the files fully and properly.

When the functions of inventory management and production control that use the information contained in these files are to be automated, the respective files normally must be overhauled, restructured, recoded, and updated. This is usually recognized, and file cleanup is carried out as a subproject of the system-implementation effort. But fixing something is one thing and keeping it fixed is another. Files tend to deteriorate following the conclusion of the fix-up effort. The reason for this is the extreme difficulty of manually maintaining any file that contains other than static information.

The task of file maintenance is not only difficult but requires such a large effort that it becomes, in practice, virtually impossible to keep a voluminous file complete and updated under manual methods with the meager resources normally allocated for this function. A classic example is the typical BOM or a manufacturing routing file, both of which consist of several tens of thousands of records encompassing active, inactive, semi-obsolete, and obsolete parts. These files are constantly affected by so many changes that true file maintenance can become a nightmare. This is so because many types of changes literally explode throughout the file. A single such change may affect hundreds, and sometimes thousands, of individual records.

The key to this problem is the staffing and budget provided for file maintenance, which, in most cases, is simply insufficient. The planners of a new MRP system usually recognize that pertinent files may have to be reorganized and updated, that in some cases they must be enlarged and the complexity of their structure must be increased (see Chapter 11), and that these files will have to be maintained rigorously in the future if the new system is to function properly. Such demands often run into strong opposition by the heads of departments responsible for maintaining these files because they foresee the increase in file-maintenance costs that has not been budgeted.

Furthermore, these department heads are sensitive to the cost performance of their functions, and this is why they may be reluctant to request additional funds for the maintenance of files to new, higher standards, particularly when they themselves are not the primary beneficiaries of increased file data integrity. Even when forced to augment their capacity for file maintenance, they may tend to bleed it at times of various departmental crises and particularly during cost-cutting drives. File maintenance may be economized on without the consequences becoming immediately apparent—but it eventually may prove very costly in terms of impaired system effectiveness.

Management traditionally has been reluctant to face up to the problem of file maintenance, yet even the old manual systems were built around the implicit assumption of file data integrity, and violations of this assumption impaired the effectiveness of such systems. Because these systems were entirely people-based, however, they got by without rigorous file maintenance thanks to the ability of human beings to improvise and to make up for deficiencies of procedure and file information. Because a machine lacks this ability, however, rigorous file maintenance is imperative in computer-based systems.

## INPUT-DATA INTEGRITY

Because an MRP system entails processing of data on a massive scale, it is virtually impossible to implement such a system without a computer. A computer, however, functions with full success (unlike a human being) only in a "perfect" environment, which would include error-free, complete, and timely data. When data lack integrity, any computer system tends to fail. In a system such as MRP, the computer is programmed to make many decisions (e.g., order size, order release, etc.), and thus because day-to-day tasks and functions of the business are being performed automatically, system failure may have far-reaching consequences.

Low-quality input data contribute heavily to such failures and particularly plague newly developed systems once these go into operation. The quality of input data varies with their source, and the incidence of errors is always the highest in the data being generated in factory operations. For purposes of MRP, data are being contributed by planners, stockroom employees, expediters, dispatchers, inspectors, truckers, and foremen, all in a position to introduce errors into the system.

Input-data errors cannot be prevented entirely, but it is important that their impact on the functioning of the system be minimized. It is feasible to incorporate a variety of external and internal system checks as part of the overall MRP system design, and a qualified programmer can incorporate many auditing, self-checking, and self-correcting features into a program. The "war" against input errors should be conducted on three separate fronts, that is:

1. Erection of a barrier to keep errors from entering the system
2. Programmed capability to detect internally most of the errors that got through the barrier
3. A procedure for washing out of the system the residue of undetected errors

The barrier, or filter, against input errors may consist of a number of procedures and techniques. Some form of input audit, testing the formal correctness (Does such a part number exist? Is this a legitimate transaction code? Are any of the data missing?) is always desirable. The barrier against the entry of erroneous input is a programmed capability of the computer system to detect and reject incorrect transactions at the point of entry, that is, immediately following the input step and before processing begins. Beyond a formal check at the point of entry, so-called diagnostic routines can be programmed that will conduct other tests prior to the actual processing of input data. For example, the part number, transaction code, and so on may be correct, but a diagnostic test against open-order records indicates that no order has been issued for this item. Diagnostic tests conducted against files other than those to be updated or against special tables set up for this purpose cost something in terms of extra processing time, but for computer applications in areas of high input-data error rates, by all means they should be programmed. A great variety of this type of check is possible, and when carried out in a computer, it is the swiftest and most efficient way of catching errors.

Internal detection during the actual processing of errors that got past the barrier discussed earlier is also an important system capability that usually can be programmed. It is distinguished from diagnostic tests mainly by the fact that the checks are made against the file being updated. An example might be a stock-disbursement transaction that passes both input audit and diagnostic tests but is substantively incorrect, that is, reports a withdrawal quantity that exceeds the quantity previously on hand. An entirely different kind of test, called a *test of reasonableness*, sometimes also can be employed. For example, if the use of a given inventory item averages 100 per period, a gross requirement of 1,000 or 5,000 for period $X$ is almost certainly invalid. A human can spot and question such absurdities immediately. A computer can be programmed to do the same. As far as a test of reasonableness is concerned, the computer program, by applying this test, always can flag the results that are suspect. The computer can tell the recipient of its output not to use the information without verifying it first.

Washing out residues of errors that escaped detection by other means is a must if the MRP system is to be kept from gradually (perhaps very gradually) deteriorating. It always should be assumed that at least some small proportion of errors in input data will penetrate the system despite all barriers and checks. These errors may be forever undetectable as such, but procedures should be devised to detect their effect on system files, and it is this effect that must be removed.

This is accomplished by means of various reconciliation, purging, and closeout procedures, which are analogous to writing off, periodically, miscellaneous small unpaid balances in an accounts-receivable file. Examples of this type of procedure are the reconciliation of planned versus actual requirements for an item at the MPS level and the closing out of ancient shop or purchase orders that still show some small quantity due.

It is a safe assumption that most MRP systems installed in industry contain some (small, we hope) percentage of error at all times. This can be tolerated as long as such system-resident errors do not accumulate. Even an accumulation at a minute rate eventually will smother an MRP system. As far as the washing out of error residues is concerned, it is less important how soon after the fact the effect of an error is removed but all-important that it be removed at some scheduled interval. If a certain level of residual error is inherent to the system, it should be kept constant.

Equally important as the technical aspects of preserving data integrity—or perhaps more so—is the training, discipline, and attitude of people. Those who contribute input data to the system and those who use the system in the performance of their job must be educated to the fact that a computer's outputs cannot be better than its inputs. If an MRP system is to be successful, management has to accept responsibility for convincing everyone who interacts with the system that he or she has an important new role in "feeding" the computer and thus helping to keep the system effective.

# BIBLIOGRAPHY

*Data Base Organization and Maintenance Processor* (General Information Manual Form No. GH20-0771). New York: International Business Machines Corp., 1971.

*Introduction to Data Management* (student Text Form No. SC20-8096). New York: International Business Machines Corp., 1970.

Orlicky, Joseph. *The Successful Computer System.* New York: McGraw-Hill, 1969, pp. 151–168.

*System Data Base, Communications Oriented Production Information and Control System (COPICS)*, Vol. 8 (Form No. G320-1981). New York: International Business Machines Corp., 1972.

# PART 3

# Managing with the MRP System

## CHAPTER 10

# A New Way of Looking at Things

**B**efore the advent of the computer, production and inventory control methods and systems were relatively ineffective. However, capital was available and relatively cheap, costs were not a serious problem, and customers were more tolerant of poor deliveries. Planning and control methods were crude and constrained by primitive data-processing tools (clerks and paper files) that were incapable of handling massive amounts of data. Theory and principles were lacking, and the only rigorous techniques known were machine loading (based on work standards, ca. 1900), economic order quantities (1915), and statistical safety stocks (1934). The ability to correlate multiple plans was nonexistent.

The availability of computers in the 1950s represented a lifting of the previous information-processing constraint, which presaged the impending obsolescence of older methods and techniques of production and inventory management. Business computers and software became available in the early 1960s and were applied by many companies to their planning activities. Forecasting, economic order quantity (EOQ), and safety-stock calculations were speeded up and made more frequently. Unfortunately, the gains were minimal at a fairly high cost. These methods and systems had been devised in light of the information-processing tools available at the time, and they suffered from a lack of ability to correlate and handle data on the massive scale required. This constraint of the tools, which affects the efficacy of methods and systems, also governs the way people look at things, perceive problems, and formulate solutions to those problems at a given point in time. The constraint of the tools is reflected in the thought and literature of an era.

The introduction of computers into production and inventory control work represented a sudden increase—by orders of magnitude—in the power of available tools. In the late 1950s, the constraint of the tools was lifted and a new era began. The new tools were applied to solving old problems, and eventually, solutions were devised for even those which in the past had been the most baffling and stubborn. Today, there exist solutions to problems that not only could not have been solved in the past but that no one at the time could conceive how to solve.

The essence of the production and inventory control problem in the past was not so much a lack of ability to plan as to replan—to respond to change. Today, there exists the capability to update for change easily, quickly, and correctly, thanks to the computer and the techniques of time-phased material requirements planning (MRP). This capability of timely replanning in response to change now must be examined in all its implications. The time has come to rethink certain traditional concepts, axioms, and theorems. Many of these are no longer relevant or valid because they fail to take into account the recent great enhancement in the ability to update for change. Traditional views that now must be revised pertain to the following topics:

- Manufacturing lead times
- Safety stock
- Queue analysis and queue control
- Work-in-process
- Forecasting of independent demand

These now appear in a new light, and the discussion in this chapter will attempt to describe this environment given currently available technology.

## PLANNED VERSUS ACTUAL MANUFACTURING LEAD TIME

In the classic problem environment of a job shop or general machine shop, the queue-time element of lead time may account for 90 percent or more of the total time elapsed. It is by compressing queue time that the overall lead time can be reduced. In the case of an individual shop order, it is important to distinguish between:

- Planned lead time
- Actual lead time

*Planned lead time* is the value supplied to the MRP system, and it is this lead time that the system uses for planning order releases. The original due date of an order reflects the planned lead time. *Actual lead time* reflects a revised due date that coincides with the date of actual need, if the latter has changed since the time of order release. The difference between the two lead times, planned and actual, can be major.

For example, at a plant producing a line of machinery, where all fabricated parts, large or small, traditionally had been allowed a 12-week lead time, the largest component items that required the most machining operations were a group of large steel shafts. The several product models required one such shaft each, different shafts being used in the assembly of different models. One day an assembly-line foreman noticed that the available supply of shafts used on the model being built at that time was low. A quick count indicated that the supply would run out by about 10 o'clock the next morning. An expediter was sent to look for a replenishment order in the machine shop but could find none. He called the responsible inventory planner, who, on consulting his records, informed the expediter that for some reason or other there were no outstanding orders for the shaft in question. Everyone understood that if more shafts could not be produced by the next morning, the assembly line would have to be shut down. Some 100 men would have to be sent home at full pay according to the union contract, not to mention the loss of production. The reaction to this threat was swift and decisive. While the shop-order paperwork was being prepared, a truck was dispatched to the nearest steel warehouse to pick up raw material for the shafts. An emergency order was launched promptly in the shop, and an expediter was assigned to stay with the order and see that it moved from operation to operation without delay. The shafts were produced overnight. They were rather expensive shafts because the way had been cleared for them in the shop by tearing down existing setups on all the machine tools required for performing the work, but the assembly line was kept going. At that time, the shaft that had a planned lead time of 12 weeks was made in one day. At another time, the same shaft had been in process for six weeks when management changed the master production schedule (MPS) and moved the respective product lot six months back. The shaft's actual lead time then turned out to be 30 instead of 12 weeks. Still later, another product lot was rescheduled into the next fiscal year, when a new model design would be effective and the particular shaft would no longer be used. The shafts in process at the time of the schedule change were never finished (they were scrapped eventually and written off), and thus their actual lead time proved to be incalculable—infinite.

The planned lead time was 12 weeks, but actual lead time varied from one day to infinity. What determined actual lead time was, of course, priority. When the shaft was being manufactured overnight, it had the highest priority in the house. The following definitions may be formulated by generalizing from this example:

- *Individual planned lead time* represents an estimate of the time that will elapse between start and completion of an order. This lead time is used in the planning process, and it determines order release.
- *Individual actual lead time* is a function of order priority. However, there can be steps in the process that cannot be compressed simply due to priority. The book, *The Mythical Man Month*, made famous the observation that nine women working together no matter how motivated cannot produce a baby in one month.

With MRP, individual actual lead times are determined by the order due dates established and revised by the system. The need for revision is detected as early as it arises, which mitigates extremes in the lead time, such as those related in the preceding story. While it is true that individual actual lead time is a function of priority, it must be remembered that priority is relative. The lead time of only a small number of orders (those with highest priority) can be significantly compressed at any one time in view of limited capacity. Individual actual lead time (for a specific order) therefore should be distinguished from average actual lead time.

The average actual lead time of successive orders for a given item should, with adequate capacity planning, approximate planned lead time. The average actual lead time of all orders that are in process simultaneously is a function of capacity and the level of work-in-process. Care must be taken when determining this lead time for purposes of projecting work-in-process levels or of expressing the relationship between capacity, work-in-process, and lead time algebraically because average actual lead time, when simply measured historically, will tend to be distorted. It will be inflated to the extent that it includes orders with priorities that have dropped significantly subsequent to release, that is, orders for items that had requirements deferred into the far future or for which requirements disappeared entirely. More will be covered on this later.

The point of this discussion is that the old concept of a "good" or "accurate" lead time, that is to say, an accurate planned lead time, must be discarded. Planned lead times need not and should not necessarily equal actual lead times. Actual lead time is flexible.

## SAFETY STOCK IN A NEW LIGHT

The venerable concept of safety stock needs to be rethought in light of the fact that actual lead time is flexible and that modern MRP and time-phased order-point inventory management systems have the ability to realign open-order due dates with shifting dates of need. The traditional approach to inventory control assumes that lead time is fixed and known and that only demand is variable. But what happens to this approach when it turns out that actual lead time is, in fact, flexible when it can be made to expand and contract with demand? What is to become of safety stock, and how is it to be calculated now?[1] When there is a system capable of replanning priorities by revising open-order due dates, there certainly is no longer any need for safety stock to cover the period of planned replenishment lead time.

This would seem to suggest that the traditional techniques of determining safety stock are obsolete and that new techniques and a new approach are needed. Safety stock, where it applies, should be susceptible to reduction, across the board, without an adverse effect on service.[2] Safety stock on the item level is not, of course, normally planned by an MRP system and is not intended to be. It is, however, planned under a time-phased order point. In either case, the MRP logic common to both systems tends to defeat the purpose of safety stock by preventing it from ever actually being used—if the system can help it. This will be demonstrated in an example illustrated by Figures 10-1, 10-2, and 10-3.

The item in this example has a planned lead time of four periods, demand during lead time is forecast as 40 units, and safety stock is 20. The order point therefore is 60. Figure 10-1 shows, in graphic form, the inventory projection and the position of the replenishment order at the time order point is reached. Depending on whether the item is under statistical order point or time-phased order point (equivalent to MRP), the pro-

[1] O. W. Wight, *Oliver Wight Newsletter* 12, May 1972.

[2] A. O. Putnam, E. R. Barlow, and G. N. Stilian, *Unified Operations Management*. New York: McGraw-Hill, 1963, p. 183.

**FIGURE 10-1**

Implications of safety stock.

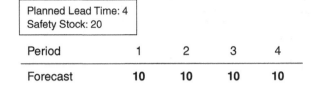

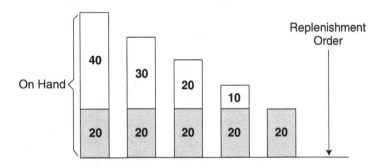

**FIGURE 10-2**

Safety stock: demand exceeds forecast in first period.

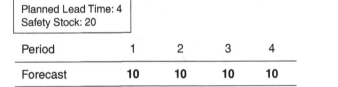

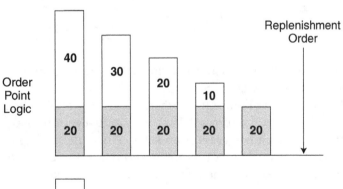

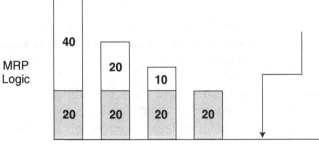

**FIGURE 10-3**

Safety stock:
demand exceeds
forecast in
second period.

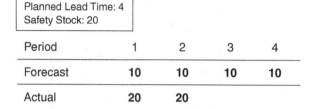

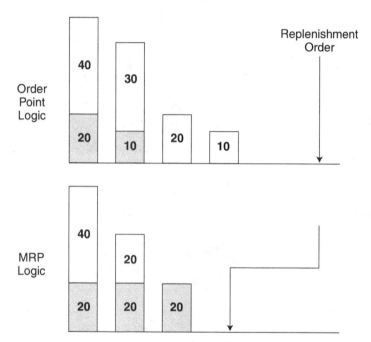

jection of safety stock and the position of the replenishment order will vary if actual demand exceeds forecast.

This is shown in Figure 10-2, where actual demand in the first period has turned out to be 20 rather than the forecast 10. Under statistical order point, the excess demand is thought of as having been met from safety stock; the timing of the replenishment order continues unaffected. Under time-phased order point, the system has reacted by moving the due date of the order one period forward, which keeps the safety stock at the original 20.

Figure 10-3 shows what happens if actual demand also exceeds forecast in the second period. Under statistical order point, safety stock is considered used up, and the order due date remains firm. Under time-phased order point, the replenishment order is rescheduled again, and safety stock remains at 20. It can be seen that if demand in periods 3 and 4 is as forecast, inventory under statistical order point will be depleted but not under time-phased order point. The latter technique strives to preserve safety stock intact. Provided that the replenishment order actually can be completed as rescheduled, safety stock proves to be "dead" inventory that could be reduced drastically, if not eliminated.

## A FRESH LOOK AT QUEUES

Queue analysis and queue control appear in a new light once the new ability to maintain valid work priorities is taken into account. Figure 10-4 shows the familiar tank (sometimes a funnel) that has been used frequently to illustrate the phenomenon of a queue and, by extension, work-in-process. On closer scrutiny, this tank example is found to be rather badly oversimplified and quite inaccurate. Consider the assumptions in this analogy: The jobs in the tank are homogeneous and interchangeable, first-in/first-out, and the total queue determines average actual lead time. This analogy holds true for water in a tank or cans in a vending machine, but not for units of work in a factory, which are stratified by priority.

Figure 10-5 shows an up-to-date version of the tank example in which there are priority strata, including sludge at the bottom. The total queue is composed of "live," "dormant," and "dead" elements. Only the live portion of the queue is meaningful, and only this portion determines average actual lead time (see next section). The "liquid" is pumped always from the top of the surface, and sludge is drained (scrap and write-off) through a separate hole in the bottom. This sludge can harden into a rocklike substance and can take significant effort to remove—like removing winestone from the bottom of a red wine fermentation tank.

Let us now consider a queue in front of a work center. Figure 10-6 illustrates the two faces of such a queue. The traditional view is from the left-hand side, and the six jobs sup-

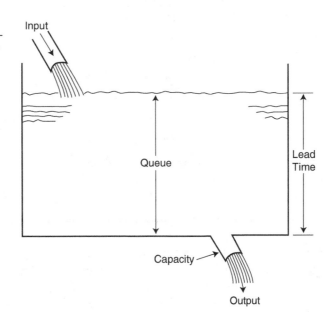

**FIGURE 10-4**

Analogy of queue and water in a tank.

Input

Queue

Lead Time

Capacity

Output

**FIGURE 10-5**

Queue with
priority strata.

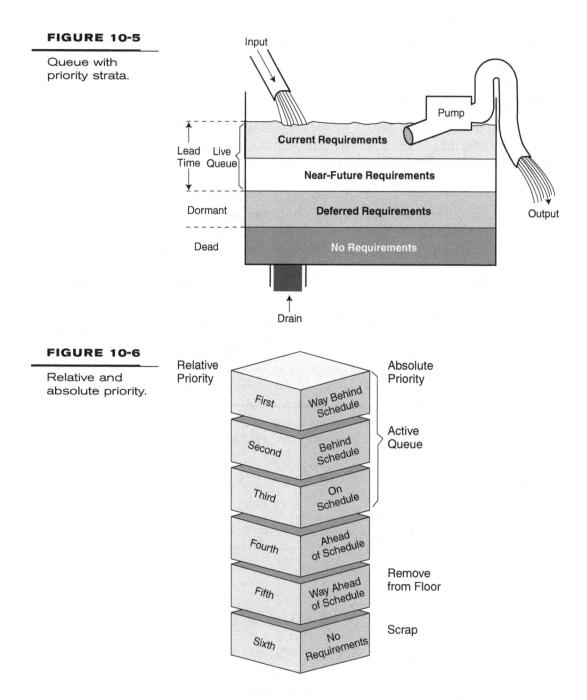

**FIGURE 10-6**

Relative and
absolute priority.

posedly represent current load. The assumption is that every one of the jobs is eligible, in priority sequence, to be worked on in the current period. In reality, this is not necessarily so. For purposes of valid queue analysis, it is not sufficient to stratify the queue by merely relative priority—absolute priority also must be taken into account. Relative priority is determined by the ranking of a group of jobs. Absolute priority is given by the relation of a job to its date of need. This is reflected in the right-hand view of the blocks in Figure 10-6, which shows that the active queue consists, in this case, of only one-half the jobs.

The conventional notion of queue control, represented by Figure 10-7, must be reconsidered once priorities are taken into account. The traditional theoretical approach to this problem is to measure a queue at a work center over a period of time (e.g., minimum 60 and maximum 100 standard hours) and to remove its fixed portion (60 standard hours) through overtime, subcontracting, and so on. The controlled queue then consists of the variable portion that fluctuates between zero and its upper limit (0 to 40 standard hours). This is the minimum queue required to prevent running out of work.

In reality, it would be foolhardy to assume that standard hours of work adequately describe a queue. The units of work are not necessarily homogeneous and interchangeable, as has been demonstrated in previous examples. If the fixed portion of the queue were to be worked off, it certainly would be the jobs with the highest relative priority; that is, the queue would be reduced from the top rather than from the bottom, which is shown in Figure 10-8. When looked at this way, this entire approach to the problem proves nonsensical because what is left at the work center are the dormant and dead portions of the queue.

Important to note is that this queue analysis has relevance only at the bottleneck or constraint in the plant. Having the nonbottlenecks run out of work has zero impact on the overall throughput of the plant. Running the bottleneck out of work never can be recovered, and the throughput is lost for the entire plant.

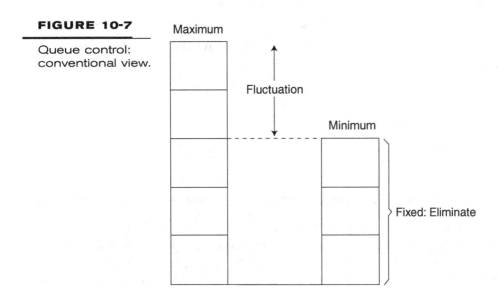

**FIGURE 10-7**

Queue control: conventional view.

Maximum

Fluctuation

Minimum

Fixed: Eliminate

**FIGURE 10-8**                    Maximum

Reducing the
queue to
unneeded work.

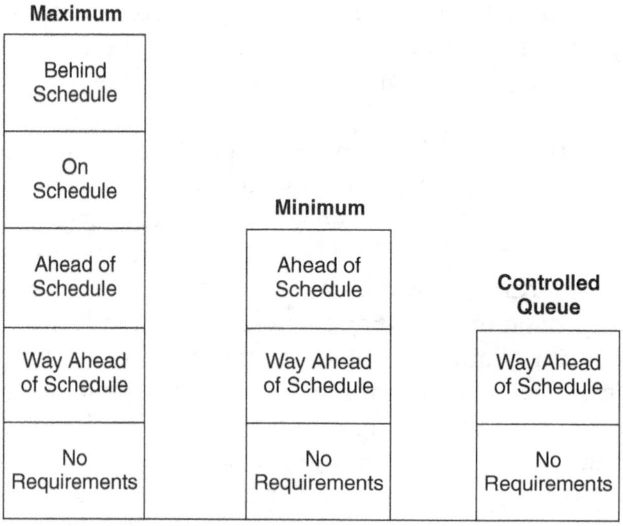

| | | |
|---|---|---|
| Behind Schedule | | |
| On Schedule | **Minimum** | |
| Ahead of Schedule | Ahead of Schedule | **Controlled Queue** |
| Way Ahead of Schedule | Way Ahead of Schedule | Way Ahead of Schedule |
| No Requirements | No Requirements | No Requirements |

## WORK-IN-PROCESS REVISITED

Work-in-process, in its relation to lead time, traditionally has been viewed as conforming to the following theorem:

$$L = W/R$$

where  $L$ = lead time (in days, weeks, or months)
$\quad\quad$ $W$ = work-in-process inventory (in units, hours, or dollars)
$\quad\quad$ $R$ = rate of output (per period of $L$, in units of $W$)

$$L = 1{,}200 \text{ units}/200 \text{ units per week} = 6 \text{ weeks}$$

The average lead time arrived at in this way is not always meaningful. When work-in-process is stratified by priority, the formula will be seen to be in need of modification. Average actual lead time is a function of the live portion of work-in-process and of the rate of production. For example:

| | |
|---|---|
| Active queues: | 800 |
| Deferred requirements: | 300 |
| No requirements: | 100 |
| Total | 1,200 |

$$L = 800 \text{ in (active) process}/200 \text{ units per week} = 4 \text{ weeks}$$

Work-in-process and lead time codetermine one another. This is a kind of chicken-and-egg relationship, and the difficulty with the preceding equation is that it assumes that work-in-process (or, in another version, lead time) is given. In reality, of course, work-in-process is variable, and its level is a function of the relationship between input to and output from the shop. It is obvious that when input exceeds output, work-in-process (and lead time, with the exception noted earlier) goes up, and vice versa.

It follows, therefore, that in order to reduce work-in-process and lead times, output must exceed input temporarily. In order to control work-in-process so as to keep it from exceeding a given level, input must be held to the existing rate of output. In an MRP environment, input to the shop is represented by the work (and cost) contents of the shop orders being released (actually, being recommended for release) by the system. The flow of these orders, however, must not be held to the rate of current production.

The flow of orders being generated by an MRP system actually cannot be controlled or regulated directly by anyone. It will not be possible to release an order prematurely if the system has planned the component materials to become available only at the time of need. To hold back on orders that the system is trying to release may create an increasing backlog of orders past due for release. All such orders later will have to be released with less than planned lead time, which is likely to cause difficulties in the shop and missed order due dates.

With MRP, unimpeded order release is required. The system is releasing orders with correct priorities, in correct sequence, and at what is, by definition, the right time. Priority, that is, need, rather than the work-in-process level or any other consideration should govern the releasing of orders. When the priority strata of work in process are taken into consideration (work with deferred or nonexistent requirements is, at any given time, "immovable" regardless of current input and output), the level of work-in-process assumes secondary importance. What are important are priorities, order completions, shipments, and customer delivery service.

When there is a significant change in requirements as a result of revising the MPS, the MRP system may pump a large number of new orders into the shop on top of an already high work-in-process. As long as the MPS is valid in terms of both marketing needs and capacity, let work-in-process be what it will, and let the MRP system do its job without interference at the order-release point.

When input to the factory exceeds output, it is an indication of the fact that the MPS is overstated in terms of capacity and should be changed unless capacity can be increased promptly. An updated view of the relationship between input and output is presented in Figure 10-9. The desired balance between the rates of input and output should exist not at the point of the MRP system's output but at the point of its input. What should be regulated in relation to the factory's output is the amount of future work implicit in the MPS. This measure of capacity has relevance only at the bottleneck for the plant. Having the other resources temporarily overloaded or underloaded—provided that it does not affect the schedule at the bottleneck—is not cause for concern.

## Living with Bad Forecasts

Forecasting of independent demand, a classic problem of inventory management, appears in a decidedly different light in view of the new capability to update for change.

**FIGURE 10-9**

Input/output and
the MRP system.

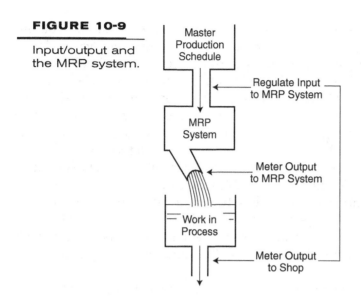

A forecast is required when an advance commitment (to procure or to manufacture) has to be made. The less flexibility there is in subsequently modifying the original plan, the more important the dependability of the forecast is. When it is easy to modify the plan, however, and to keep modifying it correctly and continuously based on actual developments, the quality of the forecast loses in importance.

MRP and time-phased order point have the ability to replan and to keep replanning quickly, accurately, and automatically. Time-phased order point in particular replans with equal ease whether the replanning is due to a change in the forecast or a disparity between forecast and actual demand. The self-adjusting capability of this technique makes the relative forecast accuracy almost unimportant. Time-phased order point depends on a forecast of demand for being able to function, but it does not depend on this forecast's accuracy for its effectiveness.

This is a highly significant and fortunate development because forecast accuracy, regardless of the technique used, remains generally unsatisfactory. Over the past decades, there has been improvement in forecasting as far as sophistication of technique is concerned. Improvement in forecasting effectiveness, on the other hand, has been rather modest. Working with poor forecasts is still the order of the day and likely will continue to be. If so, refinements in forecasting technique are a lot less important than the development of planning methods that enhance the ability to live with poor forecasts.

Time-phased order point provides an excellent example of an inventory control technique that works almost independent of the quality of the forecast. The next example will illustrate this point. Figure 10-10A, an inventory record of an item under time-phased order point, shows a forecast of 30 in every period and a quantity currently on hand of 140. Safety stock is 15, and it is projected to be reached in period 7. The next replenishment order is scheduled for release in period 4.

**FIGURE 10-10**

Time-phased order point: actual demand is less than forecast.

| Lead Time: 3 Safety Stock: 15 | | Period | | | | | | | |
|---|---|---|---|---|---|---|---|---|---|
| | | 1 | 2 | 3 | 4 | 5 | 6 | 7 | |
| Gross Requirements | | 30 | 30 | 30 | 30 | 30 | 30 | 30 | A |
| Schedule Receipts | | | | | | 80 | | | |
| On Hand | 140 | 110 | 80 | 50 | 20 | 70 | 40 | 10 | |
| Planned-Order Releases | | | | | 100 | | | | |

Actual Demand
In First Period: 0

| | | | | | | | | |
|---|---|---|---|---|---|---|---|---|
| Gross Requirements | | | 30 | 30 | 30 | 30 | 30 | 30 | B |
| Schedule Receipts | | | | | | →80 | | |
| On Hand | | 140 | 110 | 80 | 50 | 20 | 70 | 40 |
| Planned-Order Releases | | | | | →100 | | | |

Suppose that actual demand in the first period turns out to be zero rather than the forecast 30, as shown in Figure 10-10B, which shows the status of the item at the beginning of period 2. On-hand inventory at that time will be 140 instead of the previously projected 110. This will affect net requirements and coverage. Note that both the open and the planned orders have been moved back one period because the date of need has receded. If there is no demand in period 2, these orders again will be moved back; if demand equals forecast, the orders will stay as presently scheduled, but if demand exceeds forecast (in our case, by 6 or more units), they will be moved forward.

Suppose that actual demand for the same item turns out to be 90 in period 1 instead of the forecast 30, as shown in Figure 10-11B. This, of course, will have changed the quantity on hand at the end of period 1 to 50 as opposed to the previously projected 110. This, in turn, changes both the open-order and planned-order schedules. Since the date of need has advanced, all orders have been moved forward, and the first planned order will be released in the current period, which would not have been the case if actual demand had equaled forecast.

These examples show that no matter how large the forecast error may prove to be, the time-phased order-point technique makes an automatic adjustment and goes on from there. The self-adjustment characteristic of this technique, as it applies to open orders, is particularly significant because the gross and net requirements for independent-demand items are expected to keep changing owing to forecast error. In the first of the two examples, it is important to note that should there never be any more actual demand, the open order will never be finished (no more cost incurred), and no planned order will ever be released.

**FIGURE 10-11**

Time-phased order point: actual demand exceeds forecast.

| Lead Time: 3 Safety Stock: 15 | | Period | | | | | | | |
|---|---|---|---|---|---|---|---|---|---|
| | | 1 | 2 | 3 | 4 | 5 | 6 | 7 | |
| Gross Requirements | | 30 | 30 | 30 | 30 | 30 | 30 | 30 | A |
| Schedule Receipts | | | | | 80 | | | | |
| On Hand | 140 | 110 | 80 | 50 | 20 | 70 | 40 | 10 | |
| Planned-Order Releases | | | | 100 | | | | | |

Actual Demand
In First Period: 90

| Gross Requirements | | 30 | 30 | 30 | 30 | 30 | 30 | B |
|---|---|---|---|---|---|---|---|---|
| Schedule Receipts | | | 80◄——— | | | | | |
| On Hand | 50 | 20 | 70 | 40 | 10 | -20 | -50 | |
| Planned-Order Releases | | 100◄—— | | | 100◄——— | | | |

Time-phased order point is able to work with a "bad" forecast. An MRP system is able to work with "inaccurate" item lead times. This is something entirely new. Older techniques were wedded to demand and lead-time values with a basic assumption of their validity and accuracy. Their effectiveness suffered primarily because these values have always been and continue to be inherently volatile. The new techniques use forecast demand and planned lead time merely as points of departure. These data serve as the rawest of raw materials for the construction of a preliminary plan, which then is modified and modified again in the face of reality. These techniques depend more on what is happening than on what was planned to happen—they are truly adaptive.

## TOTAL PLANNING HIERARCHY

MRP programs, made practical by computers, were implemented in many firms beginning in the 1970s. Interest was greatly stimulated by an MRP crusade conducted by the American Production and Inventory Control Society (APICS) nationwide. The powers of the technique were stressed, but too little emphasis was given to the supporting activities of master scheduling, structuring bills of material, getting accurate data, shop floor control, and capacity planning and control.

Little attention was given to integrating the MPS and MRP into the total planning hierarchy that embraces:

1. *Strategic planning.* Long-range, broad-based, focusing on types of businesses, markets, and future directions.

2. *Business planning.* Long-range, focusing on product-family marketing and services.

3. *Production planning.* Midrange, focusing on facilities, technologies, and all types of resources.

4. *Master schedule planning.* Short-range, defining specific end items and driving detailed plans for resources.

In many companies, interfaces between these were weak at best and often missing. Strategic plans, developed by top management, had little influence on production and master schedule planning. Marketing/sales business plans were largely ignored by production people. Their production plans, often used only for budgeting, were not linked to the MPS. Pre-computer manual systems were very difficult to link, and habits developed then were carried on long after powerful computers and software were available. Pre-MRP planning methods not only were fragmented and crude, but they also completely lacked an ability to replan—to respond to change.

More attention was given to integrated planning and control systems in the late 1970s and to the task of educating people in the body of knowledge, language, principles, and techniques of manufacturing planning and control. APICS began to certify practitioners through a set of examinations covering these areas.

With the tools of planning and control understood and available, the focus in the 1980s was on solving the problems of execution. Long believed inevitable and unsolvable, the causes of upsets and interference with smooth, fast flow were attacked successfully by many companies. Unfortunately, too many others had already succumbed to competitors, most of whom were in foreign countries.

Computer-based MRP programs made possible six revolutionary advances:

1. Masses of data could be stored and manipulated cheaply at blinding speeds.

2. Plans could be integrated over products and processes.

3. Complex product structures were easily loaded, stored, and retrieved for multiple uses.

4. User options were available for classification, lot-sizing, safety-stock, and other techniques.

5. Myriad of data were available for many uses.

6. Frequent, rigorous replanning was possible easily, quickly, and inexpensively.

Early applications of MRP focused on items 1, 3, 4, and 6; items 2 and 5 in the list were neglected, although these had great potential benefits and met many important needs. MRP eliminated the greatest excuse for not executing the plan—that it was not valid. Overenthusiasm for MRP created the unfortunate impression that it was a system that could improve customer service, reduce inventories, and cut manufacturing costs simultaneously, along with performing other miracles. The truth, of course, is that MRP is not a system; it is a technique that can help people to do their jobs by:

1. Recommending when orders should be released
2. Maintaining validity of order priorities
3. Providing data for planning capacity and other uses

The early promise of MRP also was dimmed by emphasis on using "standard MRP systems." Self-proclaimed but underqualified system evaluators rated commercial software packages against the raters' specifications, which included many features inappropriate for some businesses. Repetitive manufacturing with no need for lot-order identity is an example. Reluctant to risk low ratings by not meeting all standard specifications, software suppliers included trivial features that added cost and complexity with little value to most users. This was exacerbated further with the advent of client-server technology and the accompanying evolution of enterprise resource planning (ERP) systems.

Implementation was viewed as "getting software running on the computer," not as using the programs to run the business more effectively, so users were poorly prepared, incomplete systems were installed, and proper foundations were not put in place. This part of this book will cover the effective use of the currently available MRP tools and techniques.

## BIBLIOGRAPHY

E. Goldratt, E. Schragenheim, and C. Ptak. *Necessary but Not Sufficient*. North River Press, 2000.

## CHAPTER 11

# Product Definition

Throughout the discussions in previous chapters it has been assumed that a master production schedule (MPS) exists to which a material requirements planning (MRP) system can be geared and that such a schedule states the overall plan of production completely and unambiguously. Implied in this assumption is a bill of material (BOM) that defines the product line not only from the customer's (and final assembly) point of view but also in a way that is suitable for purposes of procurement, fabrication, and subassembly. In other words, if an MRP system is to function properly, the product must be defined in such a way as to make it possible to express a valid MPS in terms of BOM numbers, that is, assembled-item numbers.

Unlike the order-point approach, MRP works with products and the relationships of their component items using the BOM as the basis for planning. MRP thus puts the BOM to a wholly new use, and the BOM acquires a new function. In addition to serving as part of product specifications, it becomes a framework on which the entire planning process depends. In some cases, however, the BOM maintained by the engineering department is not usable for purposes of MRP without a certain amount of modification.

As an important input to the MRP system, the BOM must be accurate and up to date if the system's outputs are to be valid. In addition, it must be unambiguous and so structured as to lend itself to MRP. The mere existence of a BOM is no guarantee that an MRP system can function properly. The BOM is essentially an engineering document, and its traditional function has been to define the product from the design point of view. With the advent of MRP, the product may have to be redefined so as to fit the needs of planning and manufacturing. Such redefinition is termed *structuring* or *restructuring* of the BOM. As companies change their manufacturing approach, the expectation is that BOMs also will be restructured. For example, it is common for companies embracing lean concepts to remove levels from their BOMs in a process called *flattening the BOM*.

The term *BOM structure* pertains to the arrangement of component-item data within the BOM file rather than to the organization of this file on a storage medium or in a storage device of a computer. BOM processor software packages, mentioned previously, edit, organize, load, maintain, and retrieve BOM records but do not structure them. These programs assume that the BOM file is already structured properly to serve the needs of MRP. This chapter attempts to clarify the subject of BOM structuring and to describe the basic techniques used to achieve good BOM structure.

When an MRP system is about to be introduced into a manufacturing company or plant, the existing BOM should be reviewed to ascertain its suitability for purposes of MRP. The following checklist will aid in spotting any structural deficiencies:

- The BOM should lend itself to the forecasting of optional product features. This capability is essential for MRP purposes.
- The BOM should permit the MPS to be stated in the fewest possible number of end items. These items will be products or major assemblies, as the case may be, but in either case, they must be stated in terms of BOM numbers.
- The BOM should lend itself to the planning of subassembly priorities. Orders for subassemblies have to be released at the right time with valid due dates, and the due dates should be kept up to date.
- The BOM should permit easy order entry; it should be possible to take a customer order that describes the product either in terms of a model number or as a configuration of optional features and translate it into the language that the MRP system understands: BOM numbers.
- The BOM should be usable for purposes of final assembly scheduling; apart from MRP, the final assembly scheduling system needs to know specifically which assemblies (assembly numbers) are required to build individual units of the end product.
- The BOM should provide the basis for product costing. Standard costs and variances are derived from the BOM.

In a given case, when these criteria are applied to the existing BOM, it will often be found that some, but not all, of these requirements can be satisfied. The BOM may have to be restructured, and this can be done without affecting the integrity of product specifications. The severity of the problem of BOM structure varies from company to company depending on the complexity of the product in question and the nature of business. The term *BOM structuring* covers a variety of types of changes made in the BOM and several different techniques for effecting such changes. The subject of BOM structure, as reviewed in this chapter, consists of the following:

- Assignment of item identities
  - Elimination of ambiguity
  - Levels of manufacture
  - Treatment of transient subassemblies
- Product model designations
- Modular BOM
  - Disentangling product option combinations
  - Segregating common from unique parts

- Pseudo-BOMs
- Interface to order entry

## ASSIGNMENT OF IDENTITIES TO INVENTORY ITEMS

If a BOM is to be used for MRP purposes, each inventory item that it covers must be uniquely identified. One part number must not identify two or more items that differ from each other, if ever so slightly. This includes raw materials and subassemblies. The assignment of subassembly identities tends to be somewhat arbitrary because a new entity is actually created every time another component is attached in the course of the assembly process. The product designer, the industrial engineer, the cost accountant, and the inventory planner might each prefer to assign them differently.

### Elimination of Ambiguity

The question is when do unique subassembly numbers have to be assigned for purposes of MRP and when do they not. In reality, it is not the design of the product but the way it is being assembled that dictates the assignment of subassembly identities. The unit of work, or task, is the key here. If a number of components are assembled at a bench and are then forwarded, as a completed subassembly, to storage or to another bench for further assembly, a subassembly number is required. Without it, the MRP system could not generate orders for these subassemblies and plan their priorities.

Some engineering departments are overly conservative in assigning new part numbers, and the classic example of this, encountered quite often, is a raw casting that has the same part number as the finish-machined casting. This may suit the engineer, but it is difficult to see how an automated inventory system such as MRP is supposed to distinguish between two types of items that must be planned and controlled separately. They have different lead times, different costs, and different dates of need.

Another requirement is that an identifying number define the contents of the item unambiguously. Thus the same subassembly number must not be used to define two or more different sets of component items. This sometimes happens when the original design of a product subsequently becomes subject to variation. Instead of creating a new BOM for the assembly affected, with its own unique identity, the original BOM is specified with instructions to substitute, remove, and add certain components. This shortcut method, called *add and delete*, represents a vulnerable procedure that is undesirable for MRP purposes, as will be discussed later.

### Levels of Manufacture

The BOM should reflect, through its level structure, the way material flows in and out of stock. The term *stock* in this connection does not necessarily mean a stockroom but rather a state of completion. Thus, when a fabricated part is finished or a subassembly is completed, it is considered to be on hand, that is, in stock, until it is withdrawn and associated with an order for a higher-level (parent) item as its component. An MRP system is constructed in a way that assumes (as discussed in Chapter 5) that each inventory item under its control goes into and out of stock at its respective level in the product structure. MRP also assumes that the BOM accurately reflects this flow. Thus the BOM is expected to specify not only the composition of a product but also the process stages in that product's manufacture. It must define product structure in terms of levels of manufacture, each of which represents the completion of a step in the buildup of the product. This is vital for MRP because it establishes, in conjunction with item lead times, the precise timing of requirements, order releases, and order priorities.

Sometimes there is reluctance toward assigning separate identities to semifinished and finished items, where the conversion to the finished stage is minor in nature. An example might be a die casting that is first machined and then receives one of three finishes, chrome, bronze, or paint, as shown in Figure 11-1.

The three finished items will have to be assigned separate identities if they are to be ordered and their order priorities planned by the MRP system. This is an example of a situation where unique item identity (of the finished casting) normally would not exist but should be an established prerequisite to MRP because otherwise such items would fall outside the scope of the system and result in loss of control.

## Treatment of Transient Subassemblies

Another example of an item identity problem that is almost the opposite of the preceding one is the transient subassembly, sometimes called a *phantom*. Assemblies of this type never see a stockroom because they are consumed immediately in the assembly of their

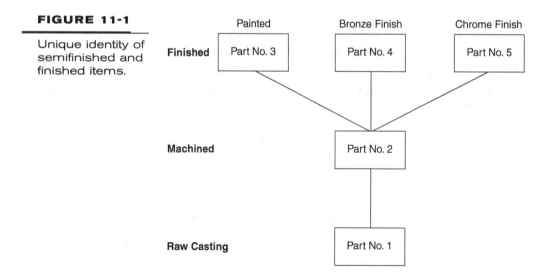

**FIGURE 11-1**

Unique identity of semifinished and finished items.

parent items. An example of this is a subassembly built on a feeder line that flows direct-ly into the main assembly line. Here the subassembly normally carries a separate identi-ty. Because it is recognized in the BOM, the MRP system would treat it in the same way as any other subassembly. This may be undesirable because if this kind of item is planned under an MRP system, its logic assumes that each component item goes into and out of stock and that all receipts and disbursements are being reported. This is the way the time-phased inventory record is designed and updated, and the question arises as to how to handle such subassemblies within an MRP system. Additional detail can be found in Chapters 15 and 18.

A transient subassembly would not have to be identified in the BOM at all if there were never an overrun, a customer return, or service part demand. Otherwise, it must be separately identified in the BOM, and its inventory status must be maintained. In MRP systems of the net-change variety, this would pose a particular problem because all trans-actions for transient subassemblies would have to be reported continuously to the system to maintain the respective inventory records up to date. This is really quite unnecessary and a waste of effort in the case of order releases, order completions (receipts), and dis-bursements in view of the ephemeral existence of transient subassemblies. Fortunately, there is no need to do this, thanks to a technique called the *phantom BOM*. While trans-actions of the type mentioned do not have to be reported and posted under this approach (this applies to assembly activities but not to stockroom receipts and disbursements), the system will pick up and use any transient subassemblies that happen to be on hand. Service-part requirements also can be entered into the record and will be handled cor-rectly by the system. Otherwise, it will, in effect, bypass the phantom item's record and go from its parent item to its components directly.

To describe the application of this technique, let us assume that assembly $A$ has a transient subassembly $B$ as one of its components and that part $C$ is a component of $B$. This item $B$, for purposes of illustration, the phantom, is envisioned as being sandwiched between $A$, its parent, and $C$, its component. To implement this technique, the transient subassembly is treated as follows:

1. Lead time is specified as zero.
2. Lot sizing is lot for lot.
3. The BOM (or the item record) carries a special code so that the system can rec-ognize that it is a phantom and apply special treatment. The special treatment referred to means departing from regular procedure, or record update logic, when processing the phantom record. The difference between the procedures can best be described through examples.

In Figure 11-2, inventory status data for items $A$ (top), $B$ (middle), and $C$ (bottom) are shown. Note that the zero-lead-time offset on the item in the middle places the planned-order release for 18 pieces in the same period as the net requirement. This, in turn, corresponds to the requirement for 18 $C$'s in the same period.

**FIGURE 11-2**

Transient
subassembly B,
its parent, and
component.

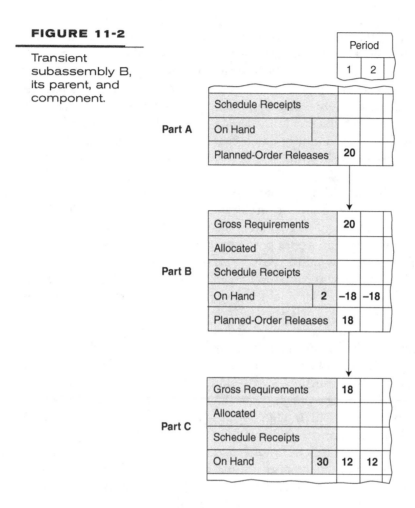

Following release of the planned order for *A*, the update procedure for item record *B* will vary depending on whether or not it is coded as a phantom. In the absence of such a code, regular logic applies. The regularly updated records of items *A* and *B* are shown in Figure 11-3.

Record *C* continues unchanged. Following release of the planned order for *B*, item record *C* is updated, as shown in Figure 11-4.

Had item *B* been coded as a phantom, all three records would have been updated in one step, as illustrated in Figure 11-5, as a result of the planned-order release of item *A*. Note that the release of planned order *A*, which normally would reduce only the corresponding gross requirement *B* (as in Figure 11-3), in this case also reduces the gross requirement for *C* as though *C* were a direct component of *A*.

Note also that the two units of *B* in stock (perhaps a return from a previous overrun) are applied to the gross requirements for A and that the allocation has been distributed between *B* and *C*. On closer examination of these examples, it can be seen that the phantom logic is nothing more than a different treatment of allocation. (Zero-lead-time and

**FIGURE 11-3**

Regular update
logic following
release of order
for item *A*.

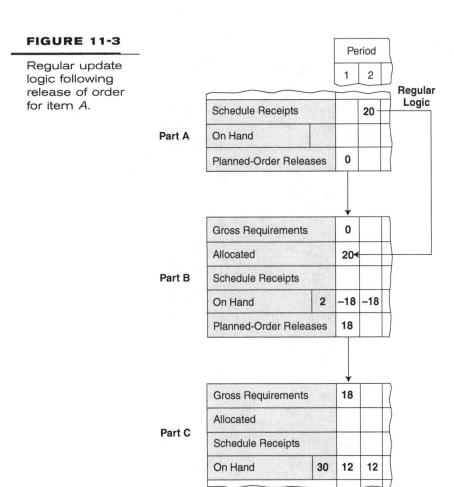

lot-for-lot ordering are assumed, but these can be specified for some regular subassemblies also.) Once this step is carried out, regular processing logic applies, causing the records to be updated and their status data aligned correctly.

The phantom BOM technique, as pointed out earlier, applies primarily to net-change MRP systems. In regenerative systems, the question of posting or not posting transactions to phantom records to cover assembly activities is not crucial because a planned-order release does not update component gross requirements data. Hence the problem of rebalancing (realigning) the planned-order and gross requirements data of the three records does not arise. Following the planned-order release of the transient subassembly's parent, the next requirements planning run will wash out both the gross requirement and the planned-order release for the transient subassembly.

The objective of not having to report phantom transactions still remains, however, and it can be achieved by again specifying lead time as zero, lot-for-lot ordering, and coding the inventory record of the transient subassembly so that notices for planned-order releases are either suppressed or flagged to be disregarded. The MRP system will function correctly.

**FIGURE 11-4**

Regular update
logic following
release of order
for item *B*.

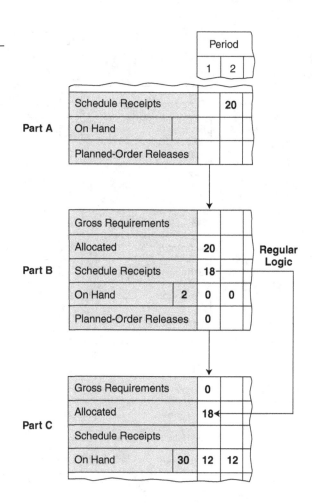

The problem then becomes one of component requisitioning (for the transient sub-assembly's parent orders), and it must be solved by modifying the procedure that generates material requisitions. When some transient subassemblies are on hand, two requisitions will have to be generated, one for the quantity of the transient subassembly on hand and one for the balance of the order for the subassembly's component items. In the Figure 11-5 example, these quantities are 2 and 18, respectively.

## PRODUCT MODEL DESIGNATIONS

A product line consists of a number of product models or product families. The marketing organization normally forecasts sales in terms of models, management thinks in terms of models, and the MPS also may be stated in terms of models. In cases of highly engineered products with many optional features, however, model identities are not fully meaningful for purposes of MRP because the model designations fail to provide a precise and complete product definition.

**FIGURE 11-5**

Simultaneous updating of item B's parent and component records.

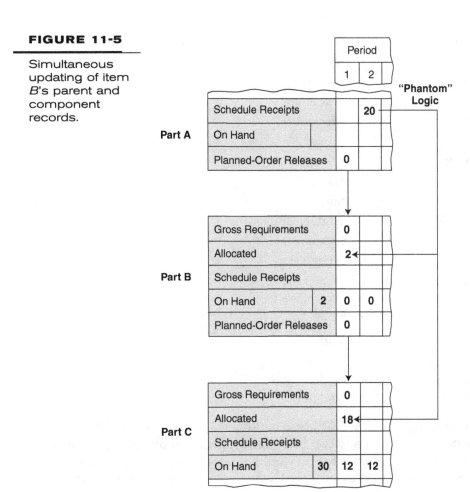

Model-code assignment is basically a matter of pointing up differences between products belonging to a product family. As a result of optional features of each line of product, in many cases there can be an almost endless variety of product buildup owing to the large number of option combinations that are possible. For example, a farm tractor in a given model category might have the following optional features:

| Function | Options |
|---|---|
| Wheel arrangement | Four-wheel construction |
| | Three-wheel construction, single front wheel |
| | Three-wheel construction, double front wheel |
| Fuel | Gasoline |
| | Diesel |
| | LP gas |

| Horsepower | 56 hp |
| | 68 hp |
| Transmission | Stick shift |
| | Automatic |
| Steering | Mechanical |
| | Power |
| Rear platform | Regular |
| | Low |
| Axles | Standard |
| | High-clearance |
| Hitch | Mechanical |
| | Hydraulic |
| Power takeoff (PTO) | With, type A |
| | With, type B |
| | Without |
| Radiator shutters | With |
| | Without |
| Operator cab | With |
| | Without |

Based on these options, it is feasible to build $3 \times 3 \times 2 \times 2 \times 2 \times 2 \times 2 \times 2 \times 3 \times 2 \times 2 = 6{,}912$ tractors without any two being identical; each represents a unique configuration of optional features. In this case, it would be possible to use any number of model designations from 1 to 6,912. For instance, two models could be recognized based on horsepower, such as:

Model A: 56 hp
Model B: 68 hp

or six models could be recognized, such as:

Model A: 56 hp, four-wheel construction
Model B: 56 hp, three-wheel construction, regular
Model C: 56 hp, three-wheel construction, special
Model D: 68 hp, four-wheel construction
Model E: 68 hp, three-wheel construction, regular
Model F: 68 hp, three-wheel construction, special

The number of models could be raised to 12 by adding, say, axle clearance, to 18 by adding the fuel option, or to 36 if both were included. In each case, the model code would tell something—but not everything—about the features of the machine. Depending on which options are considered important enough to be reflected in model designations, creating model codes is an arbitrary matter.

A multitude of model designations is an otherwise harmless thing (it looks good in the sales catalog) except that all the models tend to get into the process of forecasting and master scheduling. Separate figures then are shown for each model, resulting in lengthy and cumbersome documents that are laborious to prepare, difficult to interpret, and even more difficult to evaluate. Furthermore, since each model represents a certain combination of optional features, any forecast expressed in terms of models will tend to be grossly inaccurate, necessitating constant revisions in the forecast and changes in the MPS.

It is difficult enough to forecast demand for any single option, but to forecast, with any degree of dependability, what other options it will be combined with is virtually impossible. In order to improve the quality of forecasting and to simplify the process of master scheduling, the number of models within each product family should be reduced (at least for internal purposes) to just a few and ideally to one. It is much more difficult to forecast by model than by basic product and option. The basic product represents components, if there are any, common to all the possible product buildups and serves to indicate how many units of the product are expected to be sold and how many are scheduled to be built.

This principle pertains only to forecasting, master scheduling, and planning for the procurement, fabrication, and subassembly of components. For purposes of final assembly scheduling, specific combinations of options must be specified for each unit to be built. Reducing or abolishing product models for purposes of internal planning does not mean that such model designations necessarily would have to be eliminated from price lists and sales literature. To implement the principle of forecasting and planning by basic product and option, the BOM used by the MRP system would have to be modularized accordingly. The principles and techniques of modularization will be reviewed next.

## MODULAR BILLS OF MATERIAL

A modular BOM is arranged in terms of product modules, that is, sets of component items, each of which can be planned as a group. The process of modularizing consists of breaking down the BOMs of highest-level items (products and end items) and rearranging them into modules. There are two somewhat different objectives in modularizing a BOM, namely:

1. To disentangle combinations of optional product features
2. To segregate common from unique, or peculiar, parts

The first is required to facilitate forecasting or, in some cases, to make forecasting at all possible under the MRP approach. The second is aimed at minimizing inventory investment in components that are common to option alternatives, that is, that are used in either optional choice. Demand for product options must be forecast, and this makes it necessary to plan safety stock in which the common components may be duplicated.

The preceding two objectives and the techniques used to achieve them will be reviewed separately in the discussion that follows.

## Disentangling Option Combinations

Under the MRP approach, product variations or optional features must be forecast at the MPS level; that is, it must be possible to forecast end items rather than their individual component items. When a product has many optional features, their combinations can be astronomical, and forecasting these combinations becomes impractical. A valid MPS could not even be established and stated in terms of such BOMs. This problem is solved by means of a modular BOM. Instead of maintaining BOMs for individual end products, under this approach, the BOM is restated in terms of the building blocks, or modules, from which the final product is put together. The problem and its solution can best be demonstrated by an example. The farm tractor discussed in the preceding section had 11 optional features and a total of 25 individual choices, making it possible to build 6,912 unique product configurations.

There is no special difficulty in writing a BOM for any one of these configurations, but it is not practical to store and maintain thousands of BOMs for a single product family. Many of the 6,912 possible configurations may never be sold during the life of the product, and thus their BOMs would never be used. Furthermore, design improvements and engineering changes could add additional BOMs to the file. Consider this: The tractor, as described, has only one type of fender, but if the engineers create an option of special fenders with mudguards, the number of possible option combinations will double from 6,912 to 13,824. This means that another 6,812 BOMs would have to be added to the file. This is one reason why BOMs for end products should not be maintained in this case. But the other reason mentioned earlier is equally important; that is, with these thousands of BOMs, it would not be possible to state a valid MPS in terms of end products.

If the tractor manufacturer produces 300 of this type of tractor per month, which 300 of the 6,912 possible configurations should he or she select as a forecast for a particular month? This is simply not a practical proposition. Note that volume is part of the problem. A product family with 100 possible option combinations constitutes a problem if volume is 20 per month. If volume were 10,000 per month, the forecasting problem would not be nearly as serious.

The solution of this problem lies in forecasting each of the higher-level components (i.e., major assembly units such as engines and transmissions) separately and not attempting to forecast end products at all. This amounts to forecasting the various choices within the optional product features and translating such forecasts into the MPS.

Specifically, if 300 tractors of the type in question are to be produced in a given month, 300 so-called basic tractors (including fenders, hoods, rear wheels, etc.) would be scheduled. A BOM for this *module* would be required to match the schedule. There are two choices of transmission, however, and let us assume that past demand has averaged, say, 75 percent stick shift and 25 percent automatic. Applying these percentages to the transmission option, 225 and 75 units, respectively, could be scheduled. But the batch of

300 customer orders in any one month is unlikely to break down exactly this way, and thus some safety stock would be desirable.

As mentioned previously, the proper way to handle safety stock under the MRP approach is to plan it at the MPS level. Thus the transmissions would be deliberately overplanned (the statistical technique of determining standard deviations for a binomial population[1] may be used to establish the correct safety stock), and transmission quantities such as 275 and 100 would be put into the MPS. This would not be done in every period because unused safety stock is *rolled forward*, that is, applied in subsequent periods. The same approach would be followed for the other optional features. Each of the optional choices would have to be covered by an appropriate module of the BOM for use by the MRP system. Under this approach, the total number of BOMs would be as follows:

| | |
|---|---|
| Basic tractor | 1 |
| Wheel arrangement | 3 |
| Fuel and horsepower | 6 |
| Transmission | 2 |
| Steering | 2 |
| Rear platform | 2 |
| Axles | 2 |
| Hitch | 2 |
| Power takeoff | 3 |
| Radiator shutters | 1 |
| Operator cab | 1 |
| Total | 25 |

This total of 25 compares with 6,912 if each tractor configuration had a BOM of its own. Now, if the engineers add special fenders, it would add only two BOMs to the file (regular fenders, part of the basic tractor, would become an optional choice and would have to have a BOM of their own) instead of doubling it.

At this point in the discussion, the reader may be wondering how this type of problem is being handled in a real-life situation if the manufacturer does not have BOMs set up in modular fashion. The chances are that there would be several BOM for some of the 6,912 possible configurations covering the (arbitrarily established) models that are being recognized. These BOMs also would be used for all other configurations by adding and subtracting optional components.

The *add-and-delete technique* solves some but not all of the problems. Its principal disadvantages are vulnerability to human error, the slowing down of order entry, and awkwardness in establishing proper historical data for option forecasting purposes.[2] Under

---

[1] In this case, the standard deviation is based on the *proportion* of the population's individuals who possess one of two qualities—male or female, Democrat or Republican, stick shift or automatic transmission.

[2] D. Garwood, "Stop: Before You Use the Bill Processor," *Production & Inventory Management* 11(2), 1970.

this approach, the add-and-delete components most likely would be maintained under order points and safety stock. This is highly undesirable because it deprives the user of some important benefits (discussed in Chapter 13) of an MRP system.

## Modularization Technique

The technique of restructuring end-product BOMs into a modular format will be demonstrated next. For this purpose, the preceding tractor example will be scaled down so that the solution may be seen clearly. Let us assume that the tractor has only two optional features, the transmission and steering, each of which has two choices. The customer can choose only between stick shift and automatic transmission and between mechanical and power steering. Figure 11-6 shows the four BOMs.

The first combines stick shift and mechanical steering; the second, stick shift and power steering; and so on. In the product structure, the end-product (model) numbers, 12-4010 and so on, are considered to be on level zero. The level 1 components, A13, C41, and so on, represent assemblies, but their components are omitted from the chart to keep it simple. To restructure these BOMs into modules, they are broken down, their level 1 components are analyzed and compared, and these components then are grouped by use. For example, it can be seen that the first component item in the first BOM, A13, is common to all models, and therefore it would be assigned to the "common" group. The next item, C41, is found in stick-shift/mechanical and stick-shift/power combinations but not in the automatic transmission models. This indicates that C41 is unique to the stick-shift choice. The item that follows, L40, is used only with mechanical steering. The remaining component items are similarly examined and assigned to groups. The result is shown in Figure 11-7.

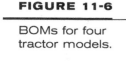

**FIGURE 11-6**

BOMs for four tractor models.

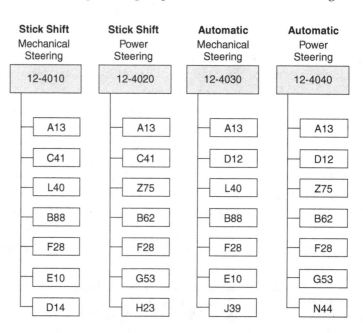

**FIGURE 11-7**

Level 1 items
assigned to
groups by
options.

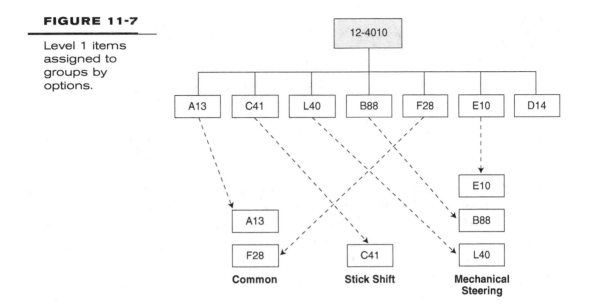

## Items Used Only with Specific Option Combinations

Note that the last level 1 component item, D14, does not fit into any of the groupings. When all four of the BOMs are broken down this way and their level 1 components are grouped by use, items D14, H23, J39, and N44 remain unassigned because each of them is used only with one or the other of the option combinations. Here the process must be carried one step further. These items are broken down themselves, as illustrated in Figure 11-8, and their (level 2) components are assigned to the groupings by use.

The final result is shown in Figure 11-9, where all the items involved in our example are grouped into the respective modules. In this example, the entire problem has been solved by applying the technique of breakdown and group assignment.

However, if items D14, H23, and so on had not been subassemblies but single parts, it would have been impossible to break them down. In cases such as this, the part that is used only with a certain combination of options should, if possible, be redesigned. This will not always be feasible, however, as illustrated by the engine in the original tractor example. Special engine-block castings are used for diesel, and the horsepower option entails different cylinder bores; the fuel and horsepower combinations cannot be disen-

**FIGURE 11-8**

A breakdown of
unassigned level
1 items.

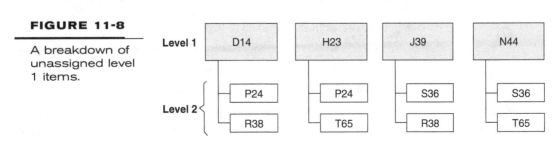

**FIGURE 11-9**

Completed
modularization.

| Common | Stick Shift | Automatic | Mechanical Steering | Power Steering |
|--------|-------------|-----------|---------------------|---------------|
| A13 | C41 | D12 | L40 | Z75 |
| F28 | P24 | S36 | B88 | B62 |
| | | | E10 | G53 |
| | | | R38 | T65 |

tangled. Either such combinations must be forecast (and form separate modules), or the items in question can be assigned to more than one grouping in the modularizing process. For example, item D14 (Figure 11-6) could be duplicated in both the stick-shift and mechanical steering modules (Figure 11-9), ensuring that it would never be under-planned. Such duplicating is particularly indicated for inexpensive items of this sort, in preference to separate forecasting or redesign.

## Options Within Options

The option-combination items just discussed represent one type of complication from the modularization point of view. Another one is options within options. The tractor used in our example can have four-wheel or three-wheel construction, and an option in the latter is a single or double front wheel. This is an option within an option, and it calls for establishing three modules, that is:

1.  Items common to the three-wheel-construction option
2.  Items unique to the single-wheel suboption
3.  Items unique to the double-wheel suboption

The proper treatment of optional product features is overplanning, that is, forecasting and safety stock. This means that a suboption will have to be even more overplanned than an option. For example, the following sets of items might be scheduled when 300 tractors are to be produced:

| | |
|---|---|
| Basic tractors (common items): | 300 |
|    Option | |
|       Four-wheel construction: | 100 |
|       Three-wheel construction: | 275 |
|    Suboption | |
|       Single front wheel: | 200 |
|       Double front wheel: | 125 |

Note that the option is overplanned by 75 sets of components, but when the suboption is taken into account, the overplanning amounts to 125 sets (100 + 200 + 125 = 425).

The suboptional parts are, by necessity, doubly overplanned. Breaking out suboptions is an alternative to treating the suboption as an option in its own right, that is:

Option

Four-wheel construction:                            100

Three-wheel construction, single wheel:             200

Three-wheel construction, double wheel:             125

Under this approach, items common to the three-wheel options will be unnecessarily overplanned. In our example, 275 sets of items common to three-wheel construction would be planned under the option/suboption approach and 325 sets under the straight option approach.

It can be seen that the lack of modularization in product design (so-called integrated design), of which option combinations that cannot be disentangled and options within options are examples, entails more investment in safety-stock inventory and makes inventory management more difficult. This is due to a high forecast error in forecasting option combinations and the need for double overplanning in the case of options within options.

## Planning BOMs

To return to the previous modularization example and to recap the steps taken up to this point, end-product numbers (codes) have been abolished, and their BOMs have been done away with as unnecessary for purposes of MRP, where the final product that formerly served as the end item in the BOM level 1 items (and in one case level 2 items) has been promoted to end-item status. This procedure established a new, modular planning BOM suitable for forecasting, master scheduling, and MRP.

The job of restructuring is not finished, however. The former level 1 items used with option combinations, D14, H23, and so on, that have been excluded from the planning BOM cannot simply be abolished. These items eventually will have to be assembled, and the production control system has to be able to place orders for these items, schedule them, and requisition their components. These BOMs therefore must be retained for the purposes just mentioned, as well as for industrial engineering and cost-accounting purposes.

## Segregating Common from Unique Parts

Earlier it was mentioned that one of the two reasons for modularization is to disentangle option combinations for purposes of forecasting and master production scheduling. The other objective of modularization, that is, segregating common from unique (optional) parts for purposes of inventory minimization, however, has not been fully met in the example we have been working with.

In modularizing the BOMs, level 1 items in the example were assigned to groups by option. However, at least some of those items were assumed to be assemblies, and they may contain common components. For example, a subassembly that is used only with the stick-shift choice may have some common parts with another subassembly used only with automatic transmissions. Requirements for such common items will be overstated because they are included in the safety stock of both options. In order to segregate such parts, the BOMs would have to be modularized further, that is, torn apart. In some cases, it might be desirable to do this, but if this technique is carried to its extreme, the planning BOM might end up containing single parts only and no subassemblies at any level. The ultimate module of the product is, of course, the individual component part.

BOM modularization may be a complex task if the product itself is highly complex, if it is engineered on the "integrated design" principle (nonmodular design), and if it entails a proliferation of optional features. Judgment must be used in deciding what should be modularized and how far. Particularly in attempting to segregate common or semi-common (an item used with diesel and gasoline choices but not with LP gas) parts, the approach should be conservative; that is, excessive modularization should be avoided.

## Effects of Modularization

Modularization affects the timing of subassembly completion. A subassembly promoted to end-item status takes the place of its former parent in the MPS, which means that it will be finished later than had its lead time been offset from the timing of the parent. Each time an item is broken down one level, its lead time somehow must be accounted for. While subassembly lead times are not significant in most cases, end items in the modular (planning) BOM should, strictly speaking, be advanced in the MPS by the amount of their lead time if they have originally been on level 2 or lower. This would complicate master scheduling, and in practice, it is not done when the lead times of the subassemblies in question are very short, as they usually are.

The crucial question in modularizing the BOM is how far downward in the product structure to go. What is really being done is to determine the correct level in the (original) BOM at which to forecast, master schedule, and plan material requirements. Whether a given subassembly should be forecast and planned or just its lower-level components is the fundamental question, and it depends on when it needs to be assembled.

There are two alternatives. One is to assemble the subassembly as a function of executing the MPS via the MRP system. This means assembling to stock or preassembling before the end product itself is scheduled to be built via the final assembly schedule following the receipt of a customer (or warehouse) order. The other alternative is to defer making the subassembly until such time as the end product itself is scheduled to be built. Making the subassembly then becomes a function of executing the final assembly schedule. The choice between these two alternatives should be dictated by the nature of the product in question, as well as by the nature of the business. Lead times and the economics of subassembly operations (Is it feasible to make the subassemblies one at a time?) will determine, in each case, whether the item should be preassembled or it can wait until final product assembly.

The MPS (further discussed in Chapter 12) is essentially a procurement, fabrication, and subassembly schedule. Its object is to furnish the component items required for final assembly of the product. Different categories of subassemblies are under the control of this schedule and the final assembly schedule. When the BOM is being modularized, a given subassembly is, in effect, being assigned to one or the other of these two schedules, that is:

1. To the MPS, by retaining it in the planning BOM
2. To the final assembly schedule

Thus the question of how far modularization should go tends to answer itself when the BOM for a particular product is analyzed and when the nature of the various subassemblies in a particular manufacturing environment is examined.

Level 0 products, those sold to customers, are rarely used in master scheduling. Instead, particularly for complex products with options, level 1 and level 2 modules are established, associated with individual options, and promoted to end-item status in the MPS. However, BOMs for level 0 products and other subassemblies excluded from planning BOMs are needed in execution; these products have to be ordered, scheduled, and assembled, and requisitions must be issued for components. Final assembly schedules may match customer or warehouse orders in quantities and delivery dates; larger quantities, however, may be assembled and shipped in smaller lots. Marketing, sales, industrial engineering, cost accounting, and others need level 0 BOMs as well.

Modular BOMs, called *M-bills*, meet these needs. M-bills are coded to distinguish them from planning BOMs that MRP uses exclusively. M-bill items can be components of end products or of other M-bill items. M-bill components may be other M-bill items or end items in the MPS, the top-level items in planning BOMs. Purchased and manufactured items procured to support execution of final assembly schedules (rather than to support the MPS) belong in M-bills, but they also may be included in kits in planning BOMs to ensure procurement in time. M-bills are not involved in planning; they are intended for use in execution. Planning defines the resources needed to support the plans; execution assigns available resources to produce what customers have ordered.

Ideally, planning should result in available resources being adequate for execution; this never happens. At best, planning is good enough to limit shortages of resources to amounts that can be acquired during execution with minimal cost and few harmful effects on customer deliveries.

To conclude the present discussion, it may be proper to reflect on the objectives of modularizing the BOM. In addition to the specific objectives brought out earlier, there is another, broader one. And that is to maintain flexibility of production with a minimum investment in component materials inventory. The goal is to be able to offer a wide choice of products and to give maximum service to customers and at the same time keep component inventories low. Modular BOMs are designed to help achieve just that.

## PSEUDO-BOMS

When the BOM is broken down in the process of modularization, various subassemblies are promoted and become end items, that is, highest-level items with no parent in the planning BOM. This tends to create a large number of end items. Since it is the end item that has to be forecast, and since the MPS has to be stated in terms of end items, the hundreds (or thousands) of new end items would prove too many to work with. Fortunately, there is a simple solution to this.

The objective always is to have the smallest possible number of items to forecast and the smallest possible number of end items shown in the MPS. To meet this objective, the technique of creating *pseudo-BOMs* is used. Going back to Figure 11-9, where the newly created end items are grouped by option, there is no obstacle to taking any such group and creating a pseudo-BOM (assigning an artificial parent) to cover it. This is illustrated in Figure 11-10, where a new series of (pseudo) BOMs has been established.

These BOMs, sometimes called *super-BOMs* or *S-bills*, are an example of nonengineering part numbers being introduced into a restructured BOM. An S-number, such as S-101 in Figure 11-10, identifies an artificial BOM for an imaginary item that will never be assembled. The sole purpose of the S-number is to facilitate planning. With the S-bills established, when the transmission option in the tractor example is being forecast, only S-102 and S-103 would be involved. These pseudo-BOM numbers then represent this optional product feature in the MPS, and the MRP system will explode the requirements from this point on, using the S-bills in the BOM file.

Another pseudo-BOM term in industrial use is the so-called kit number or K-number. This technique is used in some manufacturing companies where there are many small, loose parts on level 1 in the product structure. These are often the fasteners, nuts, bolts, and cotter pins used to assemble the major product units together. Under an MRP system, to deal with such items individually on the MPS level would not be practical.

**FIGURE 11-10**

Super-BOMs.

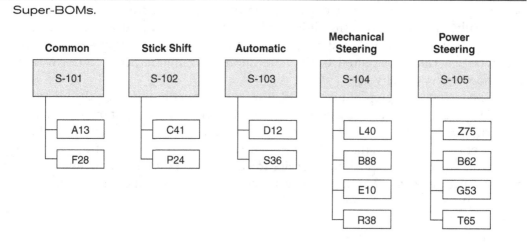

These parts therefore are put into an imaginary bag, as it were, and a part number is assigned to identify this bag or kit. A (pseudo) BOM is established for the kit, which is then treated as an assembly for purposes of master scheduling and MRP.

The principle involved here is the same as in the case of the S-bill, that is, assigning a single new identity code to individually coded items that constitute a logical group and employing the format of a BOM to relate such items to one another. K-numbers may be used to advantage within a modular BOM (e.g., to streamline material requisitioning), or they may be used even when there is no need for a modular BOM. The K-number is another nonengineering part number. These artificial identity codes have little to do with the design of the product and are not part of product specifications but are created for more convenient forecasting, planning, and master scheduling.

These newly created BOMs, along with the M-bills discussed earlier, represent a superstructure in the BOM file that, once established, then must be maintained along with the rest of this file. This is a new function that increases the cost of file maintenance.

## INTERFACE TO ORDER ENTRY

Procedures governing customer order entry and backlog management, called *order entry* for short, are outside the boundaries of the MRP system except in cases where customer orders or contracts constitute the MPS itself or where such orders are substituted for items originally planned via this schedule. Otherwise, order entry interfaces with the MRP system through the final assembly scheduling system. The latter then calls on the MRP system for components and, where implemented properly, checks the availability of these components by accessing the respective inventory records during the course of the final assembly scheduling process.

To make this feasible, however, and to be able to back up the final assembly schedule with part numbers (BOM numbers) of the highest-level component items required for each specific unit of product being scheduled, it is first necessary to translate customer or warehouse orders into manufacturing language, that is, BOM numbers.

With most products of some complexity (assembled products), customers—and sales personnel—normally specify orders in descriptive English, in terms of model numbers, or by means of a so-called generic code that serves as shorthand for an English description. An example of a generic or product-description code is shown in Table 11-1.

The generic code is convenient to use in a marketing environment, and it has the additional advantage of not being subject to variation between models or to engineering changes. Power steering remains power steering, but the respective BOMs will vary between models, and their identity coding, as well as contents, will tend to change over a period of time. For purposes of manufacturing and costing, the generic code must be translated into a specific code, that is, into BOM and part-number terms.

In a modular BOM situation, each generic code would have one or more S-number counterparts, and the generic-to-specific conversion could be effected either by means of decision tables or through inverted BOMs (another instance of pseudo-BOMs), as illustrated in Figure 11-11.

## TABLE 11-1

Product-Description Code

Example: 3AG11AP3

| Position | Code | Option |
|----------|------|--------|
| 1 | 1 | Model 450 tractor |
|   | 2 | Model 550 tractor |
|   | 3 | Model 650 tractor |
| 2 | A | 4-wheel construction |
|   | B | 3-wheel construction, regular |
|   | C | 3-wheel construction, special |
| 3 | G | Gasoline |
|   | D | Diesel |
| 4 | 1 | 56 hp |
|   | 2 | 68 hp |
|   | 3 | 76 hp |
| 5 | 1 | Stick-shift transmission |
|   | 2 | Automatic transmission |
| 6 | A | Regular axles |
|   | B | High-clearance axles |
| 7 | M | Mechanical steering |
|   | P | Power steering |
| 8 | 1 | No power takeoff |
|   | 2 | Power takeoff type A |
|   | 3 | Power takeoff type B |

## FIGURE 11-11

Inverted BOMs.

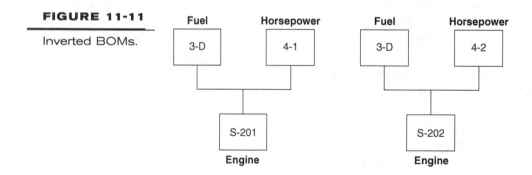

In this example, a diesel option (letter $D$ in position 3 per Table 11-1) coupled with 56 hp calls out S-bill 201, the engine, and so on. In this way, the final assembly scheduling system can be integrated, or logically linked, with the MRP system.

In conclusion, let us briefly discuss who does and who does not have to restructure BOMs as a precondition for successful MRP system operation. Where the product line consists of a limited number of items or models, modularizing the BOM or any other changes for the sake of BOM structure may be unnecessary. Master scheduling and MRP can be based on models and on their BOMs, provided that the model codes define products uniquely.

On the other hand, BOM restructuring is called for where the product line consists of a virtually unlimited number of end-product configurations owing to the complexity of design and a proliferation of optional features. The study of how BOMs should be constructed then becomes a vital part of the work of designing and implementing an MRP system.

## BIBLIOGRAPHY

Orlicky, J. A., G. W. Plossl, and O. W. Wight. "Structuring the Bill of Material for MRP." *Production & Inventory Management* 13(4) 1972.

*Structuring the BOM* (Brochure Form No. G320-1245). New York: International Business Machines Corp., 1973.

## CHAPTER 12

# Master Production Schedule

**A** master production schedule (MPS) is to a material requirements planning (MRP) system what a program is to a computer. The MPS is, technically speaking, only one of three principal inputs to an MRP system (see Chapter 6), but whereas the other two, that is, inventory status and product structure, supply reference data to the MRP process, the MPS constitutes the input that "drives" it. It is the prime input on which an MRP system depends for its real effectiveness and usefulness. MRP is the first step in the implementation of the overall manufacturing program of a plant, which is what the MPS represents. In the upstream/downstream relationship of information flow between systems, the MPS is furthest upstream, and it acts as a wellhead of the flow of manufacturing logistics planning information.

A given MPS is the determinant of future load, inventory investment, production, and delivery service. It is the cause of certain inevitable consequences in the areas just mentioned, and it may contain the seed of future problems and failures. As pointed out in Chapter 6, downstream systems are unable to compensate for deficiencies of their input. An MRP system will carry out its functions of inventory ordering, priority planning, and (indirectly) capacity requirements planning with great efficacy, provided that it is presented with a realistic, valid MPS to be processed.

## MASTER PRODUCTION SCHEDULING CONCEPTS

Does every manufacturing company or plant have an MPS? If such a schedule is defined as the overall plan of production, it would be difficult to conceive of a plant operating without one. In any manufacturing operation, the sum total of what a plant is committed to producing at any given point in time is equivalent to an MPS. What some manufacturing managers really mean when they say that they do not have an MPS is that in their case the overall plan of production is not being expressed in one formal document. For purposes of MRP, the creation and maintenance of a formal MPS is a prerequisite.

## Definitions

An MPS should not be confused with a forecast. A forecast represents an estimate of demand, whereas an MPS constitutes a plan of production. These are not necessarily the same. A distinction therefore should be maintained between the functions of developing a forecast and laying out a schedule of production despite the fact that in some cases the two may be identical in content.

An MPS is a statement of requirements for end items by date (planning period) and quantity. In Chapter 11, an *end item* was defined as the highest-level item (i.e., an item that is not a component of any parent) recognized in the bill of material (BOM) that the MRP system uses for exploding requirements. There must be a correspondence between such items in the BOM and the terms in which the MPS is stated. End items may be products, major assemblies, groups of components covered by pseudo-BOMs (see Chapter 11), or even individual parts used at the highest level in the product structure. A component item may double as an end item when it is subject to service part, interplant, or other demand from sources external to the plant. Orders and/or forecasts for all external-demand items are technically part of the MPS, although they are not normally listed in the formal document but exist in the gross requirements schedules of the respective inventory records.

Where the product line consists of complex assembled products with many optional features, it is not practical to state and maintain an MPS serving as input to the MRP system in terms of the products (product models) themselves for reasons discussed at some length in Chapter 11. In such cases, the schedule is expressed in terms of major components rather than end products. For example, a machine tool manufacturer would specify in its MPS the quantities of columns, knees, tables, beds, and other major assemblies from which individual machine tools eventually will be built to customer order.

The format of an MPS is normally a matrix listing quantities by end item by period. The meaning of these quantities in relation to the timing indicated is fixed by convention—in a given case, it may represent end-item availability, end-item production, or end-item component availability. Depending on which one it is, the form of interface between the MPS and the MRP system is affected, as discussed in Chapter 7.

The sales forecast and the MPS that management and the marketing organization use are often developed and stated in terms of weeks, months, or quarters, however. They are usually also stated in terms of product models. The MPS then must be broken down and restated in terms of days and weeks and specific end-item numbers. Thus it may exist in two versions or layers, as illustrated in Figure 12-1.

The period of time that the MPS spans is termed the *planning horizon* (discussed in Chapters 7 and 14), and it may be divided into a firm portion and a tentative portion, also indicated in Figure 12-1. The firm portion is determined by the cumulative (procurement and manufacturing) lead time. It is not necessarily firm in the sense of being "frozen," but it does represent quantities of end items committed to and started in manufacture.

For the MPS, the frequency of maintenance (i.e., updating or revision) usually is geared to the forecasting cycle, which is almost always monthly. Between these "official"

**FIGURE 12-1**

A master production schedule (MPS).

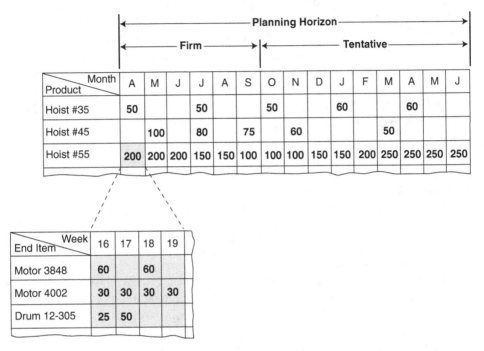

issues of an updated version of the MPS, however, there may arise a need for revision at any time, as a result of the particular mix of new customer orders and the various unplanned developments in procurement and manufacturing. It is highly desirable, therefore, for the MRP system to be able to process intervening and intermittent "unofficial" changes in the MPS on a more frequent basis than that afforded by the forecasting cycle. As discussed in Chapter 14, an MRP system loses much of its effectiveness if it is not used to replan daily. Arguments for continuous replanning of material requirements were presented in Chapter 7.

## THE FINAL ASSEMBLY SCHEDULE

In order to comprehend the essence and the true function of the MPS, a distinction must be drawn between it and the final assembly schedule. This was touched on earlier in connection with other topics, but at this point, a more thorough discussion is warranted. The distinction between these two schedules is a source of frequent confusion because in some cases the schedules, although always different in concept, may be identical in reality; that is, the final assembly schedule may serve as the MPS.

There may be no difference between these two schedules where the product line is limited or where the product itself is small and/or simple. Lawnmowers, hand tools, bicycles, vacuum cleaners, and clocks are examples of this situation, in which the shippable product is the end item. Interestingly enough, the MPS and the final assembly schedule may be identical in the case of highly complex products that are engineered and manufactured to customer order, such as turbines and weapons systems.

Between these extremes, though, there lies the broad middle ground of complex products assembled from standard components into a variety of configurations, often to specific customer order. In this category belong vehicles, machinery of all sorts, electrical equipment, and a long list of others. Here, the two schedules are distinct. The MPS is expressed in terms of high-level components (assemblies, etc.), and typically, because of the disparity between manufacturing lead time (long) and customer delivery time (short), it must be formulated and committed long before the final assembly schedule is prepared.

Whereas the typical MPS extends a number of months into the future, the final assembly schedule usually covers only days or weeks. It is stated in terms of product models or specific configurations of optional product features, often in serial-number sequence. The MPS is based on anticipated customer demand. The final assembly schedule responds to actual customer demand and is constrained by the availability of components provided by the MPS via the MRP system.

The MPS is essentially a procurement, fabrication, and subassembly schedule. Its function is to provide component availability, and it therefore may be viewed as a component-availability schedule. In this context, the term *component* means any inventory item below the end-product level. The MPS may be said to "produce" the mentioned components in support of the final assembly schedule. This is true to the extent that these components are part of the BOMs reflected in the MPS. The exceptions to this rule are items excluded from the planning BOM during the process of modularizing the BOM, as discussed in Chapter 11. One of the points made there is worth repeating: A given subassembly may be assigned either to the planning BOM (used by the MRP system) or to the M-bill (used by the final assembly scheduling system). This is tantamount to putting the item in question under the control of one or the other of these two systems. If the item is part of the M-bill, the final assembly schedule, rather than the MRP system, is "responsible" for producing it.

This rule extends to selected manufactured items and purchased items, which may be put under the control of the final assembly scheduling system. Then they will be manufactured or procured as a function of executing the final assembly schedule in correspondingly small lot quantities. Such items are characterized by:

- High unit cost
- Short lead time
- Short assembly lead time of the item's parent (if any)
- Absence of significant setup or quantity discount considerations

Examples of component items governed by the final assembly schedule, related to products referenced in previous discussions, are as follows: A horizontal milling machine of certain design has a so-called overarm that is required in the fourth week of the machine's final assembly. The overarm is a simple steel cylinder involving little machining with minor setup, but it is a massive and relatively expensive part. It is assembled into the milling machine by inserting it into the proper hole in the column and fastening it inside the column. Such an item is assigned properly to final assembly schedule control and is machined in quantities perhaps as small as one or two during the final assembly cycle for specific machines being built.

An example of a purchased part under final assembly schedule control is a tractor rear tire, a very expensive item. These tires (of which there are many varieties, makes, sizes, and tread patterns) are shipped by the vendor at very short notice in any quantity needed to meet the current requirements of the tractor assembly schedule. Quantity discounts may apply to total annual consumption of all tire models rather than to individual orders.

In both of the preceding examples, significant inventory investment is avoided or minimized and the possibility of surplus is precluded by gearing the manufacture or procurement of the items in question to the final assembly schedule.

## FUNCTIONS OF MASTER PRODUCTION SCHEDULING

An MPS serves two principal functions, namely:

1. Over the *short horizon*, to serve as the basis for the planning of material requirements, the production of components, the planning of order priorities, and the planning of short-term capacity requirements
2. Over the *long horizon*, to serve as the basis for estimating long-term demands on the company's resources such as productive capacity (i.e., square footage, machine tools, personnel), warehousing capacity, engineering staff, and cash

These two functions relate to the "firm"' and "tentative" portions of the MPS mentioned earlier. A well-implemented MRP system encompasses the entire planning horizon; that is, both the firm and tentative portions of the MPS are reflected in the time-phased inventory records. While only the firm portion of the planning horizon is, strictly speaking, required for purposes of order release and order-priority planning, the system maintains data on tentative (but formally planned per the master schedule) requirements and planned orders to provide visibility into the future on an item-by-item basis. These data can be put to a variety of uses, including lot sizing, projections of capacity requirements and inventory investment, serving to guide the negotiation of blanket-order contracts with vendors, determining inventory obsolescence and indicated write-offs, and others.

The MPS should strive to maintain a balance between the scheduled load (input) and available productive capacity (output) over the short horizon, and it forms the basis for establishing planned capacity over the long horizon. This represents long-term estimates of resources required to execute the MPS. Some of these resources, such as plant and new machinery, may take a year or more to acquire, and this is why an MPS should extend beyond the total cumulative production lead time. The long-horizon function of resource requirements planning will be reviewed in the next section.

# MPS DEVELOPMENT

The specific method of developing an MPS tends to vary from company to company. The general procedure, however, consists of a number of logical steps (described below) that can serve as the basic blueprint on which modifications are made depending on the nature of a particular manufacturing business.

## Preparing an MPS

An MPS represents, in effect, the future load on production resources. The load arises from requirements placed on the plant that reflect the demand for the product being manufactured. The method of establishing these requirements varies depending on the industry. In the manufacture of products to stock, future requirements generally are derived from past demand. In the manufacture to order, the backlog of customer orders may represent total production requirements. In custom assembly of standard components, a mixture of forecasting and customer orders generates requirements. The organization of the distribution network and field inventory policy also directly affects production requirements (see discussion in Chapter 4). In most manufacturing companies, the requirements placed on a given plant derive from several sources. Identification of these sources and of the demand they generate constitutes the first step in developing an MPS. These sources are as follows:

- Customer orders
- Dealer orders
- Finished-goods warehouse requirements
- Service-part requirements
- Forecasts
- Safety stock
- Orders for stock (stabilization inventory)
- Interplant orders

*Customer orders* may constitute the MPS in the case of custom-engineered products, in contract manufacturing for the government, in industry-supplier situations, or in any case where the order backlog extends beyond the cumulative production lead time. In other cases, customer orders are filled by the plant but create requirements via the final assembly scheduling system on final assembly facilities only. Requirements on the rest of the factory are conveyed by the MPS, which anticipates component-item demand.

*Dealer and warehouse requirements* for products constitute another source of demand that, for purposes of master production scheduling, sometimes may be treated the same way as customer orders. In most cases, however, the difference lies in the practice of dealers and distribution warehouses of indicating their requirements (quotas) in advance of orders actually being issued. These advance commitments normally are stated in terms of product models without specific choices of optional features. Then these have to be forecast, for master scheduling purposes, by the plant. In the case of simple products without optional features, planned order schedules of a time-phased order-point system employed by the warehouse represent demand on the plant.

*Service-part requirements* by either customers or a service warehouse normally bypass the MPS development process. They are entered in the form of either forecasts or orders directly into the respective inventory records. An exception might be the case of large, expensive service-part assemblies that would be master scheduled along with regular products. Where a service warehouse uses time-phased order point, requirements are best conveyed via the planned-order schedules of the warehouse system.

*Forecasts* in some cases may constitute a source of requirements placed directly on the plant. In many manufacturing businesses that either ship directly to customers from a factory warehouse or assemble to order, a sales forecast is the sole source of production requirements reflected in the MPS. In many other cases, however, forecasting also generates requirements that are being conveyed by the MPS. This pertains to product variations or to optional product features, which usually are forecast by the plant even though the MPS is based on commitments for product units by dealers or field warehouses, as mentioned earlier. In these cases, the exact configurations of optional features are supplied just prior to shipment.

*Safety stock*, as mentioned earlier, should be planned on the MPS level rather than on the component level. Safety-stock requirements therefore must be viewed as a separate source of demand on the plant. Safety stock, in terms of end items, is incorporated into the quantities stated in the MPS.

*Orders for stock* may be the principal source of production requirements in cases where the product is being stockpiled in anticipation of future need. In businesses subject to highly seasonal demand, products and/or components normally are produced to stock during the off-season in order to be able to meet the peak demand with a level load on productive capacity throughout the year. The resulting inventory is known as *stabilization stock*.

*Interplant orders* normally are confined to component items rather than products, which may include anything from single-component parts to assembled end items appearing in the MPS. The treatment of these requirements parallels that of service parts. In cases where the "customer" plant uses an MRP system, this type of demand is conveyed more effectively via the planned-order schedules for the interplant items, as was discussed in Chapter 7.

Demands from all the sources just reviewed, when consolidated, represent the so-called schedule of factory requirements. The creation of this schedule constitutes the second step in the development of the MPS. The latter is derived from the former but is not necessarily identical to it for the following reasons:

- A part of the demand expressed in the schedule of factory requirements may be met from plant inventory.
- Product lot-sizing considerations, important from the manufacturing point of view, are obviously not reflected in the schedule of factory requirements. The demand is shown by quantity and date without regard for production economics. In the process of developing the MPS, product lot sizes are established that may deviate in both quantity and timing from the requirements of the various sources of demand. Additional lot sizing may take place subsequently at the component-item level.

The load represented by the schedule of factory requirements either may exceed productive capacity or be below the capacity to which the plant is committed. This load may fluctuate excessively. The schedule of factory requirements may be stated in terms of product models that will have to be translated into end-item BOM numbers. The schedule of factory requirements may not specify optional product features the demand for which must be forecast before being incorporated into the MPS.

The schedule of factory requirements serves as the basis for final preparation of the MPS, which constitutes the third and final step in MPS development. Thus a specific manufacturing program is created that will then be processed by the MRP system to plan all subsequent component procurement, fabrication, and subassembly activity. In transforming the schedule of factory requirements into an MPS, the predominant consideration is that of capacity availability. The process and techniques used to achieve a balance between load and capacity over the long horizon are described next.

## Resource Requirements Planning

An MPS must be considered in relation to the load it places on available or planned resources, including capacity, space, and working capital. If available resources are not adequate to meet the requirements represented by a given MPS, they must be increased, or the schedule should be reduced. Unless solid planning of resource requirements takes place before the planning of production, there is a likelihood of failure in delivery service, a logjam in work-in-process, a disruption in the production control system, and increased manufacturing costs.

The resource requirements planning concept entails a long-range planning function intended to keep in balance the ability to meet demand and a reasonably level load on the company's resources. The technique of resource requirements planning consists of five steps as follows:

1. Defining the resources to be considered
2. Computing a load profile for each product that indicates what load is imposed on what resources by a single unit of the product
3. Extending these profiles by the quantities called for by a proposed MPS and thus determining the total load, or resource requirement, on each of the resources in question

4. Simulating the effect of alternative MPSs
5. Selecting a realistic schedule that makes the best use of (existing or planned) resources

Defining the resources to be considered is a management function. Resources range from engineering personnel to cash to capital equipment and plant square footage. In the discussion that follows, only one such resource, *productive capacity*, will be referenced. For purposes of resource requirements planning, productive capacity may be subdivided into individual capacity resources or groups. For example, the entire machine shop may be defined as a resource, and the impact of a given MPS then is measured in terms of total load on the shop. Or the shop may be defined as two or more resources by function, such as heavy casting machining, miscellaneous machining, and sheet-metal operations. Individual departments or groups of departments may constitute measured resources. A still finer breakdown would identify work centers or even individual machines. A single critical machine legitimately may be identified as a resource under this approach.

Resource requirements planning, however, is intended for relatively large group-ings because its purpose is not to determine the exact load on an individual resource but rather to evaluate the overall impact of a given MPS. Resource requirements planning is conducted on a macro level using rough approximations of load, and a precise fit is not sought. The important thing is to be able to develop the alternative loads quickly so that several different MPSs may be tried out.

Computing load profiles for individual products is based on the simple proposition that each (quantity of) product in the MPS generates measurable load and that the same procedures that are used to arrive at a machine load report can be used to compute a product load profile. A given load profile consists, for instance, of the standard hours of fabrication required, by period, to produce one unit of product measured against what-ever fabrication resource is selected.

The load profile for a given product may be thought of as a load report for a simu-lated product lot of one computed on a relative time scale. In Figure 12-2, the end of peri-

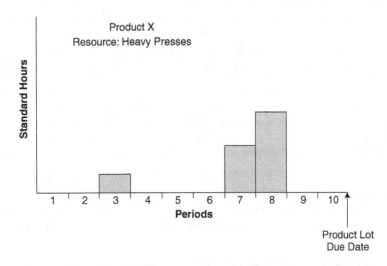

**FIGURE 12-2**

A product load profile for a single resource.

Product X
Resource: Heavy Presses

Standard Hours

1   2   3   4   5   6   7   8   9   10

Periods

Product Lot
Due Date

od 10 corresponds to whatever period the product would appear in the MPS, and the overall fabrication load it generates is distributed over eight periods (the product's fabrication lead time) preceding the completion date. The figure shows a load profile for the same product related to heavy presses, a critical resource or bottleneck.

A load profile is computed by using the MRP system as well as the operations scheduling and loading systems as simulators. The quantity of one each of the end items making up the product (using a typical combination of optional features), arbitrarily assigned to some future period, is processed by the MRP system using blank inventory records or a special program routine suppressing the netting function. The gross requirement of one is exploded through all levels of the product structure, bypassing any lot-sizing computations. The resulting output is a planned-order receipt schedule (all minimum quantities) for all items at whatever level that would be used in the production of one each of the end items in question. These planned-order schedules then serve as input to the regular scheduling and loading systems, and using whatever scheduling rules and loading conventions are in effect, a special load report is generated. This load report, when summarized, represents the product load profile, which then is stored for future use.

In developing the load profile, the treatment of setup time will vary depending on whether setup standards exist, that is, whether setup is considered direct labor or overhead. Where the routings contain setup standards, setup hours are part of the load profile created by the method just discussed. The setup load, however, is stored separately from run-time load because of the different treatment each will receive when the total load for a product lot is calculated. If setup standards are not maintained, empirical setup-hour data can be apportioned to the respective run times, or else the latter can be increased by some percentage to account for setup.

The final load profiles of all products are stored so that they can be used repeatedly in resource requirements planning without a need for the detailed computation. The development of product load profiles is a one-time job. Unless the product in question is redesigned drastically, its load profile will serve throughout product life because engineering changes normally would have only a trivial effect on the load involved.

Extending load profiles by the quantities called for by a given (version of the) MPS and summarizing them by period are simple matters. They are accomplished very quickly with a computer that has access to the file in which the profiles are stored. The result is a report (printed or conveyed through a visual-display device) showing the effect of the MPS over the entire planning horizon on the various resources for which profiles are maintained. These are called *resource requirement profiles*. They provide a fair indication of the loads that can be expected. The loads may be segregated by individual product lots to show which of these are causing potential capacity problems. This is portrayed graphically in Figure 12-3.

Note that load generated by service-part and interplant requirements is added to that derived from product lots. This could be an empirically verified percentage of the load, it could be forecast, or it might be computed through separate load profiles if the service-part and/or interplant items are large and their demand significant.

**FIGURE 12-3**

A resource requirements profile.

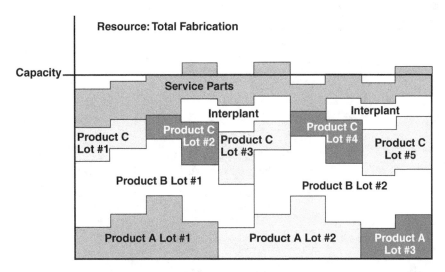

Simulating the effect of alternative MPSs is part of the selection/decision process. If the load generated by a proposed MPS is unsatisfactory (because of significant overload or underload in one or more periods), the schedule is changed usually on a trial-and-error basis, and the procedure is repeated. Note that in a business with a line of simple products made to and shipped from stock, planned-order schedules of a time-phased order-point system may be used to represent the schedule of factory requirements. The latter would be converted into an MPS through use of the firm planned-order technique, which also would be used to modify the schedule. This is diagrammed in Figure 12-4 at the end of the next section.

In the absence of such a procedure, the MRP system, of course, will process any MPS, and load will be calculated subsequently. The load report (assuming it is based on both open and planned orders) then serves as a resource requirements profile, but if it proves unsatisfactory, the MPS would have to be changed and reprocessed—an unwieldy and costly procedure. Thanks to the availability of load profiles, a large number of potential MPSs can be tried for fit in a very short time.

Selecting a feasible MPS is the final step in this process. This ensures that the schedule roughly fits capacity constraints. Further capacity adjustments will be made subsequently in the course of short-range capacity requirements planning (see Chapter 13), when overtime, work transfer, subcontracting, and so on will compensate for the fluctuation of load from period to period. In the typical manufacturing business, the MPS decided on by management corresponds to some specific rate, or level, of production (e.g., 60 machines a month, 80 vehicles a day) to which all activities then are geared.

The purpose of resource requirements planning is twofold, as indicated earlier. Over the short horizon, it is to keep load within the bounds of available capacity. Over the long

**FIGURE 12-4**

Relating
production
problems to the
MPS.

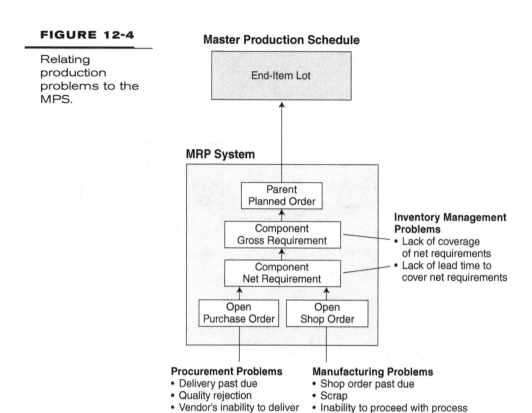

horizon, it is to help decide what additional capacity, if any, will have to be added and when.

## CLOSING THE LOOP

In managing the MPS and in using it to manage inventories and production, the following basic "law" always should be observed:

> The MPS should be a statement of what can and will be produced rather than what management wishes had been produced in the past and/or would like to be able to produce in the immediate future.

This law, which stipulates that the MPS must be realistic, is still honored mostly in the breach. This is a result of long tradition because in the past—before computers and before MRP—the realism of an MPS was not easily ascertained or measured. The schedule simply set a goal that everyone in the plant was supposed to scramble to reach. The pressure that the MPS exerted to "get the product out the door" by keeping the manufacturing organization under pressure and off balance was considered by management to be beneficial, and it still often is. This approach to managing production employs what an old machine shop saying calls "brute force and ignorance." The MPS then acts as a brain that can transmit action commands to muscle-equipped members but lacks feedback. It drives the organism blindly because it is insensitive to obstacles and injury. This is, of course, extremely inefficient and costly. Today it is also unnecessary.

## The Plan and Reality

The relationship between the MPS (the master plan) and the many elements of its execution is clearly visible and in precise form, thanks to the modern MRP system (the existence of standard-variety scheduling, loading, and work-assignment subsystems is assumed). Such a system converts the master plan into a detailed plan of execution and helps to monitor the execution proper. The linkage between plan, execution, and progress of execution now can be maintained, and the connection can be seen at all times. This means that it has become both desirable and feasible to close the loop, something that has never been practical in the past. The situation in the real world of procurement and manufacturing can and should be fed back to the master plan so that it may be adjusted to better reflect reality.

In a manufacturing environment, most difficulties and problems are caused either by obstacles encountered in carrying out procurement and manufacturing tasks or by the MPS itself. For the overall manufacturing logistics system to function properly, the MPS must be realistic in three ways. What can be produced (as opposed to what it would be nice to produce) is a function of the availability of

- Material
- Time
- Productive capacity

Each one of these is equally important. A lack of critical material or lead time or capacity precludes production, and if the MPS insists on such production, it will incapacitate the MRP system in its priority planning function, leading to a collapse of the shop priority system. The manufacturing organization then reverts to form: staging, stock-outs, assembly shortages, hot lists, expediting, confusion, and an increase in manufacturing cost. The informal system takes over because the formal system, of which the MPS is a critical part, is not doing its job.

In a manufacturing plant, probably the most commonly encountered problem is difficulty in or inability of meeting the monthly plan of shipments (shipping budget) caused by an inability to complete final assembly owing to a shortage of components. This problem is highly visible, but it is not of primary nature. Rather, it is a symptom of a variety of specific problems in earlier stages of the production process. These may be classified as follows:

- Problems in inventory planning
- Problems in procurement
- Problems in manufacturing

Inventory planning problems are represented by either lack of coverage of net requirements or lack of lead time to cover net requirements. Procurement problems consist of past-due deliveries, rejections of vendor shipments on the basis of quality, and a vendor's inability (usually temporary) to produce and deliver. Manufacturing problems take the form of past-due shop orders; scrap; inability (usually temporary) to proceed with manufacture owing to a lack of tooling, machines, or other facilities; and overloads. Every one of these types of problem affects the integrity of shop priorities (defined in Chapter 13), which is most important for the efficient and smooth operation of a plant.

As pointed out previously, the objective in managing inventories and production via the MPS is to establish and maintain a realistic relationship between plan and execution. Whenever a disparity develops between what the MPS calls for and the likelihood of being able to do it, an effort at reconciliation should be undertaken. The first step always should be to determine what, if any, extraordinary action can be taken to solve the problem at the execution level so that the MPS may remain intact. This is the usual course of action when overtime, subcontracting, expediting, and so on are resorted to and every effort is made to meet the schedule, and it is a completely proper effort as long as there is a reasonable probability that the schedule actually will be met.

A different case entirely is the situation where it develops that some part of the MPS cannot and in fact will not be met. Here, the schedule must be changed promptly if it is to remain realistic. At this point, the question is exactly what in the MPS to change and how. The answer can be ascertained accurately through the MRP system. The pegged requirements capability (discussed in Chapter 14) allows any of the specific problems just enumerated to be traced and related to the MPS.

Some problems may be solved below the MPS level by revising planned-order data in parent-item inventory records, as will be shown in Chapter 14. In other cases, it will be necessary to use pegging to step through all the higher levels to pinpoint the end-item lot (the quantity in a specific MPS bucket) that has to be changed to restore harmony between the schedule and reality.

## Restoring the Schedule to Valid Status

The effect of any of the problems mentioned earlier, whether caused by the MPS itself or by unforeseen developments in the production process, is reflected in and can be traced through the time-phased inventory records. In an MRP environment, these records provide the information that triggers all procurement and manufacturing activity, and obstacles encountered in the course of this activity can be related back to the respective records, as can inventory management problems. From the record in question, pegged requirements provide the trace to the MPS. This was illustrated in Figure 12-4.

Problems that are caused by the MPS itself (as contrasted with problems caused by poor performance in meeting this schedule) are the result of overstating the schedule. The latter may be overstated in its totality (exceeding overall capacity in every period), it may be overstated in certain periods only, or it may be overstated in terms of specific capacities at a specific time (e.g., boring bars in weeks 45 to 51).

Figure 12-5 shows an MPS that is obviously overstated in the first period, when 6,000 units are scheduled, compared with an average of 3,000 in subsequent periods. An indicator of capacity is past output, which also averages 3,000 per period. The phenomenon of overstated current periods is quite common in manufacturing companies. Sometimes the first period carries a load equal to several periods' capacity!

Another way of overstating the MPS is to carry a behind-schedule backlog, as illustrated in Figure 12-6. Here, the backlog amounts to 4,500 units, which is equivalent to 150 percent of capacity per period; in addition, a full load is assigned to the first period. Presumably the plant is supposed to get back on schedule—by producing 250 percent of its capacity in the current period! This example illustrates a situation that is even more common than the preceding one—the custom of carrying behind-schedule buckets in the MPS is prevalent. In reality, of course, nothing can be produced yesterday, only today and

**FIGURE 12-5**

An MPS overstated in the first period.

| End Item | 6-Months Average Production per Period | Period 1 | Period 2 | Period 3 | Period 4 |
|----------|----------------------------------------|----------|----------|----------|----------|
| A | 100 | 180 | 100 | 90 | 100 |
| B | 200 | 480 | 200 | 160 | 180 |
| | | | | | |
| | | | | | |
| Total | 3000 | 6000 | 3000 | 3100 | 2900 |

**FIGURE 12-6**

An MPS overstated in the backlog.

| End Item | 6-Months Average Production per Period | Behind Schedule | Period 1 | Period 2 | Period 3 |
|----------|----------------------------------------|-----------------|----------|----------|----------|
| A | 100 | 150 | 100 | 100 | 90 |
| B | 200 | 180 | 240 | 200 | 160 |
| | | | | | |
| | | | | | |
| Total | 3000 | 4500 | 2900 | 3000 | 3100 |

tomorrow. The MPS should reflect this reality. The behind-schedule column is best abolished unless care is taken that the total of behind schedule and the first period does not exceed capacity for one period.

Both examples represent a gross overstatement of the MPS, which is disastrous to the shop priority system. In such cases, most orders are no doubt past due, and most jobs in process are behind schedule and marked "Rush." Expedite lists are long, and because of this, there is a special expedite list within the expedite list. Work-in-process inventory is excessive. Manufacturing costs are high. Although the company has the ability to plan priorities (an MRP system is being assumed), the formal priority system has collapsed, if anyone ever took it seriously in the first place. When everything has high priority, nothing has high priority.

Disparities between the MPS and the realities of production will arise even when the schedule is not overstated. This is caused by a miscellany of unplanned events that tend to take place in the typical manufacturing operation. Delays in the progress of work owing to the condition of tools and machinery are not uncommon. Neither is a temporary lack of adequate specific capacity. Neither is scrap, nor lack of material. There may be quality problems. Vendors fail to deliver. Interplant shipments get lost in transit. No system can prevent such obstacles from developing, but they can and should be adjusted and compensated for.

With the aid of an MRP system, the remedy is straightforward, Whenever one of the mentioned difficulties occurs, and when it becomes clear that some task (usually an open order) will not, in fact, be completed as planned, the item in question is traced to the MPS (assuming that the problem cannot be solved via pegging and firm planned order at an intermediate parent level), the schedule is revised and subsequently reexploded to establish up-to-date requirements and priorities. Note that it does not suffice to reschedule the order in question because of dependent priorities. The example of scrap in Chapter 13 (Figure 13-1) illustrates this problem and its solution.

Failure to realign dependent priorities is the most common reason why shop personnel consider a formal priority system unreliable and may decide to work around it. If, for any reason, a given component item definitely will not be available at the time of need, the real relative priority of its cocomponent is, in fact, lower than it would be otherwise. The priority of these orders depends on the availability of the item in question, and if the formal priority system disregards this, it loses credibility in the eyes of the shop people. They always find out. If these people cannot work according to the formal priorities and be satisfied that they are working on the right jobs at the right time, the priority system must be considered in a state of collapse. On-time delivery to the customer will suffer even when the company has sufficient capacity. Customers then will go elsewhere.

Keeping the MPS in harmony with the realities of production is a classic problem of manufacturing management. With older conventional methods, it is difficult or impossible to identify the specific end-item lot (or several lots that use a common component) that is linked to some minor disaster on the shop floor or on the receiving dock. With MRP, all the tools are there.

The closed-loop approach made possible by the capabilities of an MRP system applies equally to MPS development (reviewed in the discussion of resource requirements planning earlier in this chapter) and to MPS implementation. This is illustrated in Figure 12-7, a logic diagram of the procedures described in this chapter thus far.

**FIGURE 12-7**

MPS development and implementation.

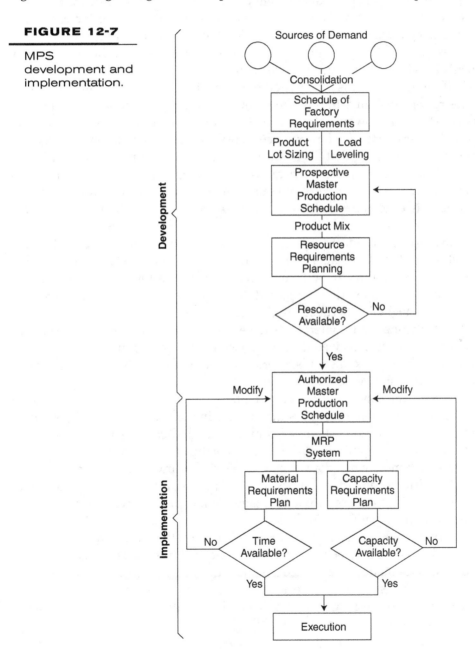

## Changing the Schedule for Marketing Reasons

The changes in the MPS discussed thus far have been related to problems of production. Changes also will be made, however, for reasons of marketing. It is common for a marketing manager to request and for management to authorize changes in the MPS so as to accommodate a customer or to make a sale. Such changes usually call for increasing the quantity or advancing the timing of an end-item lot. From the point of view of the company, these are desirable changes, but if they are made arbitrarily, the MPS again may become unrealistic, with all the adverse consequences discussed earlier.

Each schedule change of this type, whether merely intended or put into effect, represents the desire for flexibility, that is, freedom in changing previous decisions. The flexibility is constrained, however, by the realities of commitment. Another way of expressing this is to say that the consequences (cost) of a previous decision constitute the practical limits of changing that decision. The limits of flexibility contract with passage of time, making it less and less practical to effect changes as the end item nears its scheduled completion date. The reality of commitment acts as a funnel with ever-narrowing walls that, as times goes on, leave less and less room for deviation from the original plan.

If lead time is, say, four months, there is a world of difference in impact and cost of changing something in the MPS that is four months away from completion as opposed to something that is three months away from completion. In the former case, the consequences of the previous decision (having placed the original end-item lot into the respective MPS bucket) are negligible because tangible commitments have not yet been made. In the latter case, only one month later, costs have already been incurred in the processing of requisitions and orders, as well as in purchasing and manufacturing activities. In addition, the action already taken entails a certain amount of committed-for investment for materials that it may not be possible to cancel.

This is related to the concept of firm and tentative portions of the MPS mentioned earlier in this chapter. End-item quantities that appear in the firm segment of the schedule represent products in various stages of commitment and corresponding degrees of amenability to change. The tentative portion of the schedule represents merely a plan for the execution of which money has not yet been expended or invested as far as materials are concerned. The firm portion of an MPS is of the same length at any point in time; that is to say, it moves along the time scale with passage of time, progressively covering what previously had been the tentative area.

In order to avoid making marketing-motivated changes that would render the MPS unrealistic, the MRP system may be used to make a so-called trial fit. This means that the contemplated change is made in the schedule and exploded in a simulation mode. An MRP system has the inherent capability to act as a simulator, in that it will accept any version of an MPS and, by processing it, will indicate the specific consequences—material availability, order action, and lead-time availability.

Special programming (or a modification of the regular program) normally would be used for this purpose, but an MRP system always can be used (less efficiently) as a simulator without any modification. If the consequences of a trial fit prove unattractive, the system is restored to its previous condition simply by reversing and reprocessing the trial entries.

In some manufacturing businesses where it is feasible to match incoming customer orders with the MPS or to incorporate them into it, trial fitting may be a regular procedure. A trial-fit report then indicates which orders may be accepted with the customer's delivery-date request and which orders should be renegotiated for later delivery by a specific number of periods. The MRP system determines this based on the availability of component materials and lead time.

## The Master Scheduler

An MPS, as mentioned earlier, may have two layers, the lower, more detailed of which serves as input to the MRP system. With the advent of such a system, the position of master scheduler in the production and inventory control department assumes special importance.

The master scheduler (and his or her staff, if any) is responsible for creation and maintenance of the lower-layer MPS. He or she converts product models into specific end-item BOM numbers, divides monthly into weekly quantities, and forecasts product options not specified in the (upper-layer) MPS or not forecast by marketing. He or she keeps track of the use of safety stock provided at the MPS level, accounts for differences between quantities of end items produced via the MPS and those consumed by the final assembly schedule, and generally keeps the MPS up to date and valid.

An important function of the master scheduler is to evaluate problems of priority integrity, as described earlier, that are brought to his or her attention by inventory planners who have traced a problem to the MPS by means of pegged requirements. He or she determines which end-item lot should be changed and how, and he or she initiates a recommendation to management that such a change be authorized.

The position of master scheduler may be a new one, necessitated by the introduction of MRP. It constitutes an organizational link vital to closing the loop in the logistics planning system. Other organizational aspects of operation under an MRP system will be reviewed in the section that follows.

# MANAGEMENT AND ORGANIZATIONAL ASPECTS

The MPS (or schedules, if there is more than one plant) documents the overall manufacturing program of a company. The development and administration of such a program, viewed as a broad function, should be the joint responsibility of all four of the basic functional divisions of a manufacturing enterprise, that is, marketing, manufacturing, finance, and engineering. The first three of these are involved on a continuous basis, whereas engineering enters the picture occasionally, when redesign or the introduction of new products affects the manufacturing program.

## Assignment of Responsibilities

The general responsibilities, relative to the manufacturing program, of the three divisions that are continuously involved may be specified as follows:

### Marketing:
- Responsibility for forecast of customer demand, which is basically the answer to the question of what can be sold and when
- In some instances, responsibility for finished-goods inventory in terms of units, model mix, and storage location (This responsibility is sometimes retained by manufacturing or given to a special organization in charge of distribution.)

### Finance:
- Responsibility for financing and control of finished-goods inventory in terms of total investment, credit, and receivables
- Responsibility for financing of the manufacturing program

### Manufacturing:
- Responsibility for development of MPSs within the constraints established by the preceding
- Responsibility for performance to MPSs

While finance is concerned with the broader aspects of its function and deals with the problem in terms of dollars within the framework of fiscal periods, marketing and manufacturing are involved more closely in that they must deal in terms of specific units of product and cope with the day-to-day problems of producing and selling. This means that once the broad plans for sales volume, supporting production, and overall financing have been made, responsibility for their administration and execution rests with marketing (with the exceptions noted earlier) and manufacturing. If the operation is to be successful, it becomes quite important to arrive at the proper modus operandi between these two divisions and to define and clarify their respective specific responsibilities in the following three areas:

1. Forecasting versus scheduling
2. Inventories of component materials versus inventories of finished product
3. Component materials for optional product features

Forecasting of demand is clearly a responsibility of marketing, whereas the scheduling of production is (or should be) up to manufacturing. A forecast and an MPS are two different things, but in practice, they are sometimes confused in that in some types of manufacturing business the raw forecast is allowed to act as an MPS in disregard of production considerations. In other types of manufacturing businesses, marketing goals rather than a forecast of demand are reflected in the makeup of the MPS. The preceding remarks also pertain to (the statements of) these goals.

The authority for specifying and changing the contents of the MPS is sometimes improperly assigned or, perhaps even more typically, remains unassigned. In such cases, marketing tends to influence and change existing MPSs directly, possibly creating a number of undesirable consequences in production. The principle of separation of forecasting from scheduling production means that the only thing ever changed by marketing should be the forecast or some other expression of marketing requirements. Such a change need not necessarily always result in a schedule change.

Inventories represent another area of responsibility that in many businesses can be divided between marketing (or a distribution organization) and manufacturing. In such cases, manufacturing exercises control over and is held responsible for plant inventories of raw materials, work-in-process, and finished components carried to support current MPSs. Marketing, on the other hand, assumes responsibility for both field and plant inventories of finished product.

The responsibility for component materials of optional product features often can also be divided between marketing and manufacturing. In dividing this responsibility, management tries to apply the rule that whoever is in a better position to determine the quantities of materials to be ordered for a given optional feature should assume responsibility for it.

To determine how this responsibility is to be assigned, the options for each product are ranked according to their relative weight, that is, the percentage of the total cost of the product. Options representing a significant portion of total product cost are then forecast by marketing; the balance, by manufacturing. Marketing is in the best position to estimate the trend in future demand for major options, whereas manufacturing typically has better historical statistics on the use of a host of minor options.

The logistics system that a manufacturing company needs to regulate the flow of materials through the entire cycle from vendor to finished-goods inventory to customer must act to coordinate activities of several functional divisions of the company. The MPS, which "drives" the entire system, serves as a basis for resolving the inevitable conflicts between the functional divisions and represents a contract between them.

This is why the various steps involved in the development and finalization of an MPS, reviewed earlier in this chapter, are in most cases carried out by a master scheduling committee or a hierarchy of committees composed of representatives from the interested marketing, manufacturing, and finance organizations. The creation of an MPS is too important and critical a function to be entrusted to any one functional division of the company.

## Management and the MPS

On occasion, it has been suggested that the MPS, that is, its preparation and maintenance, could be automated and brought under complete computer control. This is envisioned as an extension of the process of automating systems and procedures in the area of manufacturing logistics. Where statistical forecasting of demand applies, so the reasoning goes,

the automated forecasting procedures could be integrated into a program of MPS creation, including preparation of the schedule of factory requirements, netting, product lot
sizing, and so on. The logic of the procedures can be clearly defined, and all the required
data are available.

This notion must be repudiated. All the required data are, as a matter of fact, not
available. Information on a multitude of extraneous factors, current company policy, and
seasoned managerial judgment—all of them bearing on the contents of an MPS—cannot
be captured by a computer system. This is why management should be involved in the
creation and maintenance of the MPS every step of the way.

The MPS represents the overall plan of production to which all subsequent detailed
planning is geared. Inventory management action, procurement action, and manufacturing action—all these are directly or indirectly dictated by the contents of the MPS.
Developing and maintaining the best possible MPS is the premise on which depends the
success of the manufacturing logistics system. This, it would seem, will always be too
important to entrust to a computer program.

The MPS represents the main point of management entry into the overall system. It
is through this schedule that management provides (or can provide) direction, initiates
changes in production, exercises control over inventory investment, and regulates manufacturing and procurement activities. It was pointed out earlier that given properly
implemented and properly used systems for planning and execution, the MPS is virtually the sole determinant of what will happen in areas of capacity, production, and customer delivery service. Inevitable consequences flow from an MPS because it, in effect,
contains within itself the scenarios that will be acted out later. Management has the
opportunity and responsibility to manage all this through the MPS.

Coupled with a modern MRP system, the MPS constitutes a new tool for the solution of many problems that traditionally have had to go unattended to in a manufacturing operation. To take advantage of this tool, it is important to understand the relationship between factors of production, especially of open orders, and the MPS and the desirability of maintaining a correspondence between this schedule and the realities of the
manufacturing floor. The key to this is willingness on the part of management to change
the MPS.

This calls for a departure from the traditional view of the MPS as representing a goal
not subject to change and one that, if somewhat overambitious, acts to spur the factory to
greater efforts. In the modern view, an MPS should represent a feasible goal subject to
continuous review and adjustment. The MPS no longer must be considered a sacrosanct
document; on the contrary, it should be treated as a flexible, living plan, adaptive to actual developments. Even in the presence of an MRP system, inventory, priority, and capacity planning will be invalidated in the face of an inflexible MPS.

This is a new situation, brought about as a consequence of applying the principles
and techniques of time-phased MRP. It calls for a new way of looking at things in a manufacturing business environment. Management has been given a new, powerful tool, and
it should step up to its responsibility for using it well. Management is responsible for

keeping the MPS valid, realistic, and up to date. Changes, additions, and adjustments in this schedule should be managed because of their effect on inventory investment, manufacturing costs, and delivery service to customers.

## BIBLIOGRAPHY

Everdell, Romeyn. "Master Scheduling: Its New Importance in the Management of Materials." *Modern Materials Handling*, October 1972.

"MPS Planning," in *Communications Oriented Production Information and Control System (COPICS)*, Vol. 3 (Form No. G320-1976). New York: International Business Machines Corp., 1972, Chapter 4.

## CHAPTER 13

# More Than an Inventory Control System

The early material requirements planning (MRP) systems were conceived and used as replacements for their predecessor inventory control systems that were relatively primitive and/or ineffective. In use of the new systems, the emphasis almost exclusively was on order-release action. As the systems were further developed and refined, and as the users gained experience in using them, it became apparent that an MRP system yields information that can be of value for several purposes other than just inventory control. Moreover, users discovered that with some minor additional programming, the system could provide outputs in a number of functional categories and thus can serve as a planning system in areas well beyond the boundaries of traditional inventory control.

An MRP system that is properly designed, implemented, and used actually functions on three separate levels:

1. It plans and controls inventories.
2. It plans open-order priorities.
3. It provides input to the capacity requirements planning system.

These are the three principal functions and principal uses of an MRP system. Optionally, the system also can serve certain other functions briefly described below. The three principal functions of the system will later be reviewed in more depth in separate sections of this chapter.

## USE OF SYSTEM OUTPUTS

An MRP system can provide a great number of outputs in a variety of formats at the user's option. It is not practical to list and describe all the specific outputs and formats generated by MRP systems found in industry because outputs represent an aspect of the system that lends itself to tailoring, individualization, and infinite modification. An MRP system's files in general and the inventory-status records in particular contain a wealth of information that provides an opportunity for extracting or further processing the data for a whole spectrum of possible outputs.

In the discussion that follows, outputs of an MRP system, which take the form of reports, individual messages (notices), and displays on monitors, will be reviewed by functional category rather than individually. Six such categories may be recognized:

1. Outputs for inventory order action
2. Outputs for replanning order priorities
3. Outputs to help safeguard priority integrity
4. Outputs for purposes of capacity requirements planning
5. Outputs aiding in performance control
6. Outputs reporting errors, incongruities, and out-of-bounds situations within the system

*Outputs for inventory order action* are based primarily on planned orders becoming mature for release. The MRP system detects such orders by examining the contents of planned-order release buckets in the time-phased inventory records. Other types of inventory order actions are increases, reductions, and cancellations of order quantities. These types of outputs are self-explanatory, and the category should be the one most easily understood in light of the contents of several preceding chapters.

*Outputs for replanning order priorities* serve to alert the inventory planner to cases of divergence between open-order due dates and dates of actual need, as indicated by the timing of net requirements. Examples of data on which outputs in this category would be based will be presented later in this chapter. In generating these outputs, the MRP system has the capability to indicate precisely by how many periods (or days) each item affected should be rescheduled and in what direction. Under its standard implementation, the system does not change open-order due dates automatically (although it can easily be programmed to do so) but depends on the inventory planner to take rescheduling action.

*Outputs to help safeguard priority integrity*, that is, to keep order priorities not only valid but also honest, relate problems of item inventory status to the master production schedule (MPS). The concept of priority integrity will be discussed further in this chapter. To keep priorities honest, the MPS must reflect the realities of production; that is, it must not contain end-item requirements that it will be impossible to meet for lack of capacity, material, or lead time. Some companies use reports in this category to provide guidance in accepting customer orders for guaranteed delivery. Such reports are generated by *trial fit* (see Chapter 12) of the order into the MPS and by letting the MRP system determine component-material and lead-time availability. If the order does not fit, the report indicates a best delivery date alternative.

*Outputs for purposes of capacity requirements planning* are based on quantities and due dates of both open and planned shop orders, which serve as input to the capacity requirements planning (loading) system. This function will be discussed further in this chapter.

The MRP system makes it possible for the load report to be complete, valid, and extending far enough into the future to allow capacity-adjustment action to be taken in time. To keep the load projection up to date and valid, it must be recomputed repeatedly as the order schedules in the MRP system change.

*Outputs aiding in performance control* are by-product outputs of an MRP system that enable management to monitor the performance of inventory planners, buyers, the shop, and vendors, as well as financial or cost performance. A net-change MRP system, through the control-balance fields it maintains in the item inventory records, has an outstanding ability to generate performance control reports by listing deviations from plan. Special reports on item inactivity, inventory investment projections, and purchase-commitment reports also belong in this category of outputs. When the inventory record contains standard cost, the quantities on hand projected by period (supplemented by planned-order receipts) are simply costed out and summarized by item group to obtain a highly accurate forecast of the inventory investment level. The same is true for open purchase orders—provided they are recorded by valid due date—which can be converted into a purchase-commitment report. The product-structure file, with its explosion and implosion chaining (see Chapter 9), serves as a basis of product costing. The entire database, usually also including the routing file, permits management to obtain profit and loss statements, if desired, by individual customer order, by customer, by market, by product, and by product family in addition to other critical business analytics.

*Outputs reporting errors, incongruities, and out-of-bounds situations* are called *exception reports* and would cover the following:

- Date of gross requirement input outside the planning horizon
- Planned-order offset into a past period but placed into current period
- Due date of open order outside planning horizon
- Allocated on-hand quantity exceeding current quantity on hand
- Past-due gross requirement included in the current period

In addition to exception reports, individual exception messages can be generated at the time of inventory transaction entry listing reasons for transaction rejections. Such messages would include the following:

- Part number does not exist.
- Transaction code does not exist.
- Part number is incorrect (rejection based on self-checking digit).
- Actual receipt exceeds quantity of scheduled receipt by $X$ percent (test of reasonableness).
- Quantity of scrap in stock exceeds (previous) quantity on hand.
- Quantity of disbursement exceeds (previous) quantity on hand.
- Order being released exceeds quantity of planned-order release.

These and similar exception messages are generated as a result of employing diagnostic routines and other system checks (discussed in Chapter 9).

# AN INVENTORY PLANNING AND CONTROL SYSTEM

This function of an MRP system has been described and discussed in some detail in the preceding chapters. We have seen how an MRP system answers the fundamental questions of:

- What to order
- How much to order
- When to order
- When to schedule delivery

An MRP system also can furnish several types of additional inventory management information, including, as mentioned earlier, a forecast (more precisely, a projection) of future inventory investment and clues to an indicated write-off of obsolete and/or inactive items.

Assuming proper system implementation and file data integrity, an MRP system's outputs are always correct and valid relative to the MPS that the system translates into material requirements, and the system signals for correct inventory action at all times. The timeliness of the system's inventory control outputs is a function of replanning frequency (discussed in Chapter 7), which is controlled by the system user.

An MRP system is self-adjusting in that it constantly replans and reallocates existing inventories to changing requirements via the netting process. Manufacturing inventories therefore are minimized relative to the management-imposed MPS, lot-sizing policy, and safety stock and the constraining factor of manufacturing lead times.

# A PRIORITY PLANNING SYSTEM

The key to priority planning and priority control of work in the factory is valid open-order due dates. The order due date establishes the relative priority of the order in question, which must contend for limited productive capacity with other orders in the shop. Each shop order entails a number of operations that must be performed to complete the order. A distinction therefore must be drawn between:

- Order priority
- Operation priority

Shop scheduling, loading, dispatching, and job-assignment techniques are based on operation priorities. These priorities, to be valid, must be derived from valid order priorities, that is, valid order due dates. An MRP system has the capability to establish valid order priorities at the time of order release and to maintain them up-to-date and valid by revising a due date that has been invalidated subsequently. This capability is inherent in any MRP system, and it exists whether or not the user takes advantage of it.

## Validity and Integrity of Priorities

An MRP system keeps reevaluating all open-order due dates (for purchase and shop orders both) automatically as a routine step in its netting process. The system "knows" when an open order is not aligned properly with net requirements and, if programmed to do so, can "tell" the user about it. This will be discussed in more detail and illustrated in examples in Chapter 14 as part of the section dealing with the role of the inventory planner.

Traditional inventory control systems, as mentioned previously, acted as "push" systems or order-launching systems (order the right item at the right time) that had to be supplemented by "pull," or expediting, systems (get the right item completed at the time of actual need). An MRP system functions as a push system and pull system rolled into one.

What the MRP system does, in concept, is to attempt to make two dates coincide, namely:

- The due date
- The date of need

The *due date* is defined as the date currently associated with the order. It is the date someone put on the order, and it represents what he or she planned, or expected, to be the date of order completion. The *date of need* signifies the time that the order is actually needed. These two dates are not necessarily the same. They may coincide at one time, but they tend to grow apart. An MRP system makes these dates coincide at the time of order release, and it monitors them afterward whenever a change in status causes a recomputation of net requirements. The MRP system detects any divergence of the due date and the date of need and, by signaling the inventory planner, causes them to be brought back together by rescheduling the order.

Note that when the dates diverge, the date of need may move in either direction—forward or backward in time. The MRP system accordingly can either expedite the order or "de-expedite" it, that is, have it rescheduled to an earlier or later date. It is obviously important to schedule some orders out when other orders must be completed earlier than scheduled originally.

An MRP system is able to keep priorities valid, but priority validity is mechanical (i.e., coincidence of due date with indicated date of need), and it is not the same as priority integrity. The validity of all data generated by an MRP system is, as pointed out earlier, relative to the contents of the MPS. Thus, if this schedule does not reflect what actually must and can be produced, the order priorities derived from it by the MRP system will be valid mechanically and at the same time untruthful or unrealistic.

The credibility of a priority planning system depends on both priority validity and priority integrity. This credibility is extremely important because the system, to function successfully, requires cooperation and trust by factory personnel. When the formal priority scheme lacks integrity, shop people soon discover this and revert to the traditional expediting/shortage-list approach. This is tantamount to collapse of the priority system.

## Dependent Priority

The integrity of priorities derives from the MPS, and there are basically two ways in which this integrity may be compromised. The MPS may contain end items that are either not actually needed at the time indicated or cannot be produced as scheduled for lack of material or components (lack of capacity results in lack of components). The effect of this becomes evident when the concept of dependent priority is considered. This concept pertains to the real priority (What is really needed when?) as opposed to the formal priority (What priority did the system assign to the order?), which may not necessarily be the same.

The concept of priority dependence recognizes that the real priority of an order depends on the availability, or lack of availability, of some other inventory item(s) at the time of order completion. For example, the supply of product $A$, which is shipped from stock, is forecast to run out in week $X$. The due date of an order for component item $B$ is week $X - 1$, allowing time to assemble a quantity of product $A$ and replenish its stock in week $X$. If sales lag and the product is still in ample supply as the component order nears completion, the real priority of the order is lower than its due date indicates.

This can be thought of as *vertical priority* dependence because the real priority in this case is a function of availability of an item (or items) on a higher level in the product structure. This concept was introduced in 1964 by means of a new technique for the dynamic updating of operation priorities called *critical ratio*.[1] This technique (as originally formulated) does not rely on due dates and establishes relative priorities by computing the value of the critical ratio for the next operation to be performed on every open shop order as follows:

Ratio $A$ = quantity on hand/order point          For example, $50/100 = 0.50$

Ratio $B$ = lead time for balance of work/total lead time          For example, $14/25 = 0.56$

Critical ratio = ratio $A$/ratio $B$ = $0.50/0.56 = 0.89$

Ratio $A$, the disparity between the order-point quantity and the quantity of stock on hand, is used as a measure of need. Ratio $B$, the disparity between the amount of time that was scheduled to be available for operations not yet completed and the total planned lead time, is used as a measure of response to that need. Ratio $A$ represents the percentage of stock depletion; ratio $B$, the percentage of work completion.

A critical ratio value of 1.00 signifies that work on the order has kept pace with the rate of stock depletion—the order is "on schedule." A value lower than 1.00 indicates that the pace of work completion lags behind the rate of demand (the order is "behind schedule"), and a value higher than 1.00 indicates the opposite ("the order is ahead of schedule"). The lower the value of the critical ratio, the higher is the priority of the job.

---

[1] A. O. Putnam, "How to Prevent Stockouts," *American Machinist* 108(4), 1964.

The critical ratio approach is ingenious because the technique establishes relative priorities accurately, and with frequent recomputation, the priorities of work on all open shop orders could be kept up to date and valid if there were gradual stock depletion at a steady rate, which assumption is implicit in the technique. In the case of dependent-demand items, however, depletion tends to be "lumpy," which renders ratio A meaningless.

The critical ratio technique just described is geared strictly to an order point, and it therefore fails to the extent that order point fails in an environment of discontinuous, nonuniform item demand. Nevertheless, critical ratio has made a contribution to the state of the art of inventory and production management. This contribution consists not of the practicality of application but of highlighting the fact of vertically dependent priorities.

In the case of assembled products, there may or may not be a vertical priority dependence, but there is always a horizontal dependence. A component is not really needed when a co-component is not available for the assembly of their common parent item. This principle can best be illustrated in an example (Figure 13-1).

The orders for the three manufactured components have different lead times but an identical due date, which coincides with the scheduled assembly of parent item X. If the order for component item A, for instance, is scrapped at a date too late for on-schedule recovery, parent item X, in fact, will not be assembled on the date planned. The real priority of orders B and C therefore has dropped because they will not be needed on the original order due date.

The MRP system is oblivious of this fact, of course, because the indicated requirements for items B and C have not changed. They have not changed because the parent planned order (for item X) has not been rescheduled, and the MPS has not changed. The system therefore will continue the original due dates, which, while technically correct, are in fact false. It is the responsibility of the system user to reestablish priority integrity.

This can be done by rescheduling the parent planned order (using the so-called firm planned-order techniques described in Chapter 14) and by letting the MRP system replan requirements and dates of need for the component orders in question. If the parent planned order cannot be rescheduled (by compressing its planned lead time) without also rescheduling its parent's planned order(s) and other related planned orders on high-

**FIGURE 13-1**

Horizontal priority dependence.

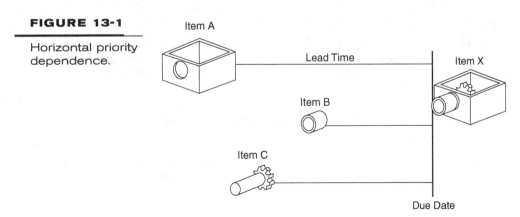

er levels, the MPS may have to be changed. This can and should be done so as to reflect the reality of assembly $X$ (and consequently, higher-level items of which it may be a component) having to be completed at some later date.

Once the MPS is again in harmony with the realities of production, the MRP system will explode it, recompute the respective requirements, and call for a rescheduling of orders $B$ and $C$ based on their new dates of need. Note that the MRP system in this case will automatically identify all open orders affected by the scrapping of the order for item $A$, will update their priorities, and will determine exactly by how many periods each individual order should be rescheduled. This is an outstanding capability of an MRP system, the full implications of which may not be evident immediately. The example in Figure 13-1 is extremely simple. In a real case, hundreds of orders (among thousands outstanding in the shop) may be involved. While the inventory planner realizes that the scrapping of the order will delay assembly of the parent item, he or she does not necessarily know

- Which parent item
- How many component orders are affected
- The identity of affected orders
- Exactly how each of these orders should be rescheduled

Unless the planner has a tool—an MRP system—to establish these facts quickly and accurately, he or she will not even try to solve the problem. The complexity of the problem is compounded by the fact that the scrapped item may have multiple parents, some of which may not figure in the current production plan at all; by the fact that several orders (in different stages of completion) for the same item may be open; and by the fact that lot sizing plus common usage obscure what rescheduling action, if any, should be taken for a given order.

This is a formidable problem that seems to defy solution, particularly when one considers that such minor catastrophes as scrapping an order, equipment breakdown, inability of a vendor to deliver, and so on occur all the time. Such events, when preventing completion of a component order, have an identical effect on priority integrity. An MRP system can solve this type of problem quickly, automatically, and with complete accuracy. Without an MRP system, it is impractical even to attempt to solve the problem, and there is no hope of maintaining the integrity of order priorities. Order priorities and operation priorities constitute a classic, chronic problem in the manufacturing industry. This problem has proven intractable, and it has defied solution by even the most sophisticated techniques until the advent of the computer and time-phased MRP.

## Severity of the Priority Problem

To recap the subject of dependent priorities and to bring it into focus, a classification of priority problems by type of manufacturing business, as suggested by Oliver Wight,[2] will

---

[2] O. W. Wight. *Oliver Wight Newsletter No. 10*, September 1971.

be reviewed briefly. For purposes of this review, manufacturing companies are divided into four categories as follows:

1. Companies producing a one-piece product made to order
2. Companies producing a one-piece product made to stock
3. Companies producing an assembled product made to order
4. Companies producing an assembled product made to stock

In the first type of manufacturing business (e.g., a foundry, a crankshaft manufacturer, etc.), the order-priority problem is the simplest. The customer places an order and requests a delivery date, and when this date is confirmed, it represents the priority of the order. Unless the customer changes the order due date, the priority remains fixed. Relative priorities of all open orders derive from their respective due dates.

In the second type of manufacturing business (e.g., a nuts and bolts manufacturer), the order-priority problem is a little more complex. To keep these priorities honest, open-order due dates would have to be related to the availability of stock for each of the items in question. In this environment, priorities are vertically dependent on the customer demand that causes stock depletion.

In the third type of manufacturing business (e.g., a special machinery manufacturer), the priority problem is quite complex because the orders for components of the assembled product have horizontally dependent priorities. The availability of all the components is prerequisite to completion of each parent subassembly and of the end product.

The priority problem is most severe in the fourth type of manufacturing business (e.g., a manufacturer of a line of power saws with a single factory warehouse) because here the order priorities are both vertically and horizontally dependent. The real priority of a shop order is a function of the available supply of the parent product as well as of the availability of co-components required for the assembly of a parent item.

The preceding classification scheme is somewhat oversimplified because many types of manufacturing businesses do not fit neatly into a single one of the four categories. Nevertheless, it is useful for purposes of analysis and exposition. From the examples, it can be seen that the more severe the priority problem, the more benefits an MRP system will be able to provide.

## Priority Control

An MRP system functions (or can function) as a priority planning system par excellence. Where it so functions, however, it must be supplemented by a priority control system in the factory. The MRP system keeps order priorities up to date by planning and replanning order due dates, but it obviously cannot cause those due dates to actually be met. The priority control system provides the procedural machinery to enforce adherence to plan, and it takes the form of what is variously known as the *dispatching system, job-assignment system, shop floor control system,* and so on.

The instrument of priority control is a dispatch list or departmental schedule, typically generated daily or weekly (the latter is of lower effectiveness), that may be in the
form of a printed report, cards, or individual work-sequence messages or conveyed
through a visual-display device. In any of these cases, the essence of the dispatch list is a
ranking, by relative priority, of jobs (material that an operation is to be performed on)
physically present in a manufacturing department and sometimes of jobs about to arrive
in the department. A sample dispatch list is shown in Figure 13-2.

The sequence of work established by the dispatch list is based on operation priorities, which, in turn, are derived from order priorities. As order due dates are being
revised via the MRP system, operation priorities must be reestablished accordingly. They
may be expressed by means of operation start dates (as in the Figure 13-2 example) or
operation finish dates, and if so, the affected orders first must be rescheduled by the operations scheduling system. Alternatively, the priorities can be expressed by means of one
of a variety of priority ratios that are geared to order due dates.

A priority control system (i.e., dispatching or job sequencing) and its related instru-
ments exist in every manufacturing plant in some form. Without input from an MRP system, however, a priority control system cannot function very effectively because the
information tends to be out of date, invalid, and untrue. Shop personnel do not and cannot rely on this information, and so the system must be supplemented by expediting of
shortage lists. The shortage lists, incomplete as they usually are, then reflect the true priorities.

## FIGURE 13-2

Daily dispatch list.

| Dispatch List | | | | | | |
|---|---|---|---|---|---|---|
| Department: **No. 12 Turret lathes** | | | | | Date: **March 15** | |
| Jobs in department | | | | | | |
| Order # | Part # | Quantity | Operation # | Start Date | Std. Hours | Remarks |
| 5987 | B-3344 | 50 | 30 | 3/7 | 3.2 | *Tooling in repair* |
| 5968 | B-4567 | 100 | 25 | 3/12 | 6.6 | *Eng. hold* |
| 5988 | F-8978 | 30 | 42 | 3/14 | 4.8 | |
| 5696 | 12-1133 | 300 | 20 | 3/15 | 14.5 | |
| 5866 | A-4675 | 60 | 20 | 3/15 | 7.0 | |
| 5996 | A-9845 | 200 | 30 | 3/16 | 6.2 | |
| 5876 | F-4089 | 25 | 40 | 4/10 | 5.4 | |
| Jobs scheduled to arrive this date | | | | | | |
| 6078 | A-3855 | 160 | 30 | 3/14 | 6.5 | *In dept. 15* |
| 6001 | D-8000 | 300 | 65 | 3/16 | 9.8 | *In dept. 08* |

In the preceding discussion of priorities, the emphasis was on manufactured items and on shop orders because they involve operation priorities, which is not the case with purchase orders. Everything that has been said about priorities of shop orders, however, applies equally to purchase orders. Purchased component items have dependent priorities, the same as manufactured items, and they should, of course, be planned by the MRP system.

# DETERMINING CAPACITY REQUIREMENTS

In Chapter 6, the point was made that an MRP system is capacity-insensitive and properly so because its function is to determine what materials and components will be needed and when in order to execute a given MPS. There can be only one correct answer to this, and it cannot therefore vary depending on what capacity does or does not exist. The MRP system can be thought of as assuming that capacity considerations have entered into the makeup of the MPS, that is, that the MPS being submitted to it for processing is realistic vis-à-vis available or planned capacity.

## Capacity Requirements Planning

The term used in connection with long-range planning of capacity at the MPS level is *resource requirements planning*, and this function was discussed in Chapter 12. Capacity requirements planning is the function of determining what capacities will be required by work center by period in the short-to-medium range to meet current production goals. The output of the MRP system indicates what component items will have to be produced and when, and this output therefore can be converted into the capacities required to produce those items.

Such a conversion results in a machine load, or work load, projection that is then compared with available departmental and work-center capacities to help answer the day-to-day operating questions, such as:

- Should we work overtime?
- Should we transfer work from one department to another?
- Should we transfer people from one department to another?
- Should we subcontract some work?
- Should we start a new shift?
- Should we hire more people?

The tool that has been used traditionally to provide information on which answers to the preceding questions could be based is the so-called load report. This report is generated by the scheduling and loading system, which schedules individual operations of orders being released, converts the scheduling into hours of work load, and accumulates them by work center by period. The traditional load report reflects only the backlog of open orders, and the typical load pattern (for a work center, a department, or a plant) looks like that shown in Figure 13-3.

**FIGURE 13-3**

Typical load
pattern.

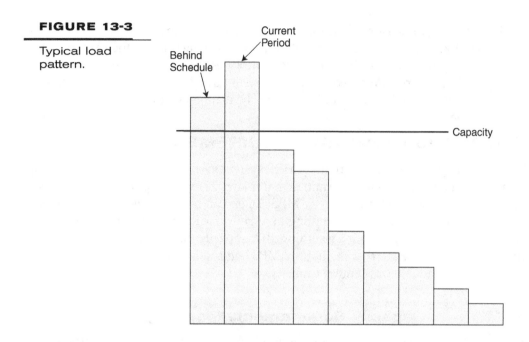

This pattern still can be found in the load reports of many manufacturing companies and plants. The manager evaluating the information understands that shop orders released in the current (first) period will add their load to the second period and those beyond, orders released in the next period will add their load to the third period and those beyond, and so on. He or she can only estimate, or guess, what the total load in any of the future periods will turn out to be. But this would seem to be less important than the question of when the behind-schedule load will be worked off. It assuredly will not happen in the current period, which is already overloaded. The manager knows, from experience, that next period's load report likely will indicate an overload in the second period, which then will be current. He or she also knows that a relatively heavy behind-schedule load appears to be a permanent condition, according to the load report. This work center may be the bottleneck or constraint for the entire plant if this is indeed the permanent condition.

The manager who tries to work with this type of load report may be baffled by the curious fact that while the load report always has indicated a highly unsatisfactory capacity situation relative to current and behind-schedule work load, shipments of the product have been more or less on schedule. Accordingly, he or she views the load data with healthy skepticism and is loath to act on the information provided by the load report. The load pattern illustrated in Figure 13-3 constitutes virtual proof that the load report exhibiting it is

1. *Incomplete* because it fails to include load that will be generated by planned orders

2. *Invalid* because priorities are not being kept up to date

When planned orders do not enter into the load report, the indicated load is bound to decline following the current period and to trail off at a point that roughly corresponds to the span of the average item lead time. This type of load projection is incomplete in a way that offers very little "visibility" into the future beyond the current period. This is such a serious shortcoming that it all but defeats the purpose of projecting the work load. Because capacity-related corrective action, such as hiring or subcontracting, entails a lead time of its own, the very information that would be most desirable, that is, a valid load picture several periods in the future, is missing.

The big "bulge" in behind-schedule and current-period load is a sure indication that priorities are not being kept valid. A good portion of the load classified as behind schedule is likely not really behind schedule if requirements have been changing. The order due dates and operation dates simply have not been revised to reflect this. The same will be true for at least some of the work that constitutes the overload in the current period.

## Usefulness of a Load Projection

A good, usable load projection has the following three attributes:

1. It is complete.
2. It is based on valid priorities.
3. It provides visibility into the future.

Under any inventory control system other than MRP, the load report tends to fail on all three counts, if it is being generated at all. Its usefulness is limited in practice to comparing successive load reports for the purpose of trend detection. Capacity-adjustment action practically always lags behind the actual load development. Owing to the load report not being trustworthy, the plant usually must get into actual trouble before management takes corrective action.

An MRP system has the potential for helping to solve the capacity requirements planning problem. An MRP system generates planned orders that can be converted into load and added to the load created by open orders. Whatever method is used to convert released orders into load can be used for planned orders as well. This satisfies the requirements of completeness and visibility because the entire planned-order schedule (spanning the full planning horizon) may be input to the scheduling and loading system.

The requirement of validity can be satisfied only if the MRP system is being used as a priority-planning system. When open-order due dates are being revised to stay valid, the entire load projection can be based on valid priorities because the MRP system maintains the timing of planned orders continually up to date.

The MRP system does not itself plan capacity requirements, but it provides input to a capacity requirements planning system, without which the latter cannot possibly function effectively. The load projection or capacity requirements report that is based on the outputs of an MRP system exhibits the kind of pattern illustrated in Figure 13-4. Depending on operations-scheduling practice, behind-schedule load may or may not dis-

**FIGURE 13-4**

Load pattern
based on both
open and
planned orders.

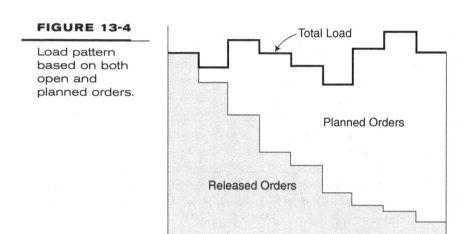

appear, but at any rate, the former "bulges" are redistributed over a number of future periods because some open orders have their due dates rescheduled to later dates.

The load projection normally is not perfectly level because actual load tends to fluctuate from period to period, but it is roughly level (or trending) compared with the trail-off pattern of the traditional load report. Production rates, by department, can be set with reasonably high confidence that the load in the foreseeable future will average what the capacity requirements report indicates. Short-term capacity adjustments must be made to compensate for load fluctuation from period to period, but there is ample notice provided by the information in the capacity requirements report.

To sum up this discussion of the multiple functions or users of an MRP system, it can be stated that such a system, in conjunction with the master scheduling function (discussed in Chapter 12), acts as a central planning system in the area of manufacturing logistics. Other systems, such as purchasing, scheduling, capacity requirements, dispatching, and so on, are designed to execute the outputs of the MRP system. For their effectiveness, these systems depend on the completeness, validity, accuracy, and timeliness of their inputs. In this and preceding chapters, the ability for a time-phased MRP system to have the capability to generate such outputs has been demonstrated.

In the area of manufacturing logistics, the inventory system is all important. The MRP approach guarantees that the inventory system will, in fact, be able to meet all the demands that management can reasonably place on it. In companies that are developing or overhauling computer-based systems for production and inventory control applications across a supply chain, an MRP system should be the first goal.

## CHAPTER 14

# System Effectiveness: A Function of Design and Use

The design, or architecture, of material requirements planning (MRP) systems has by now been standardized, and it is embodied in the application software that computer manufacturers offer to their customers. The MRP program packages are popular in the manufacturing industry, and in fact, most of the existing MRP systems use standard software; only a small minority of these systems has been designed and programmed by users.

This is not to say that a large number of installed MRP systems are identical—there probably are not two exactly alike. This stems from the fact that the user of standard software has considerable freedom in how he or she constructs his or her particular system by configuring the modules (functions) making up the package, what decisions he or she makes about certain parameters of use, and whether he or she uses so-called program exits to supply his or her own programming of procedures not provided in the package. The effectiveness of the system that results depends, in part, on the decisions the user made at the time of system construction. However, no matter how well the system may have been designed technically, its true effectiveness also depends on how well it is being used. Both these considerations will be addressed in this chapter.

## CRITICAL SYSTEM DESIGN FEATURES

The three principal functions that an MRP system can provide—at the user's option—were reviewed in Chapter 13 and for purposes of design can be summarized in the following checklist of objectives:

1. Inventory:
   - Order the right part.
   - Order in the right quantity.
   - Order at the right time.

**2.** Priorities:
   - Order with the right due date.
   - Keep the due date valid.
**3.** Capacity:
   - Determine a complete load.
   - Determine an accurate (valid) load.
   - Allow an adequate time span for visibility of future load.

A full and proper use of the MRP system, represented by the preceding checklist, will prove difficult or impossible unless the system's design anticipates such use. The intended use of the system therefore should dictate a number of critical design decisions, in particular:

**1.** The span of the planning horizon
**2.** The size of the time bucket
**3.** The coverage of inventory by class
**4.** The frequency of replanning
**5.** The traceability of requirements
**6.** The capability to "freeze" planned orders

## Planning-Horizon Span

For purposes of inventory ordering, the planning horizon should at least equal the (longest) cumulative product lead time, as defined in Chapter 7. If the horizon is shorter than this, the MRP system will be unable to time releases of planned orders for items at the lowest level correctly, with the result that orders for such items (e.g., purchased materials and component parts) will be consistently released too late. The system, in successively offsetting for lead time in the course of the level-by-level planning process, simply runs out of available time when it reaches the items on the lowest level. This is due to a lack of information input to the system rather than a computing constraint.

This is illustrated in Figure 14-1, where the cumulative lead time is 15 periods and the planning horizon is 13 periods. The order for purchased material, developed by the system through explosion of an end item inserted into the master production schedule (MPS) at the very edge of the planning horizon, should have been released two periods ago, based on the lead-time values supplied to the system. The best the system can do, under these circumstances, is to plan the order release for the current period. The order then is two periods behind schedule before it is even released.

Because of the multilevel product structure and successive lead-time offsetting, there is a partial loss of horizon at lower levels. The effective planning horizon is successively diminished as the planning process progresses from one level to the next. The lower the level, the less visibility there is into the future. For example, in Figure 14-1, a planned-order release for the fabricated item never can be farther out than period 3. The

**FIGURE 14-1**

Planning horizon
and cumulative
lead time.

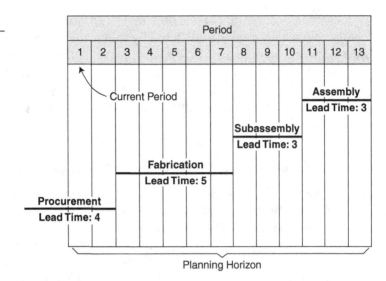

Planning Horizon

effective planning horizon for this item therefore is only three periods. Although its time-phased inventory record would show 13 time buckets for planned-order releases, the last 10 always would have zero contents.

One of the consequences of very short horizons at low component-item levels is the inability to apply a lot-sizing technique such as *least unit cost* or *least total cost* effectively simply because of a lack of sufficient net requirements data. This was discussed in Chapter 8.

Another, even more serious consequence of an inadequate planning horizon is a lack of visibility for purposes of capacity requirements planning. In the Figure 14-1 example, a complete load for fabrication operations cannot be projected beyond three periods (in period 4 and beyond, orders not at present planned by the system will be released eventually and will add to the load), which severely reduces the usefulness of the capacity requirements report or load report. It is, of course, precisely on the low (fabricated part) level that load visibility is most desirable.

A special case is the situation where the planning horizon is ample both in the MPS and in the MRP system, but management insists on specifically authorizing each "manufacturing order" or "product lot" in the MPS for release into production. This means that a quantity of an end item in the MPS cannot be processed by the MRP system unless and until management authorizes it. The lead time for this release that management recognizes tends to be arbitrary and usually on the short side. Additional delay also may be caused by the sign-off procedure.

With time-phased MRP, such a procedure for authorizing production is quite unnecessary and, in fact, undesirable. The MRP system plans the ordering of each component item on its individual merit, that is, according to its lead time and those of its parents on higher levels. The system orders the right items at the right time, not sooner and not later, which means that the production commitment goes into effect gradually. The entire product lot is not actually committed at one time (as the authorization procedure implies), and cost is being incurred gradually, as dictated by individual-item lead times. It is one of the advantages of working with an MRP system that management need not be concerned with authorizing individual product lots for production. All that is required is to maintain the best possible MPS at all times and to let the MRP system do the rest. The usefulness of this system should not be impaired by arbitrary interference and unnecessary delays in authorization.

## System Coverage of Inventory Classes

Coverage of inventory by class is another important decision that the system user must make. The ABC classification of inventory was discussed in the Introduction to this book, where it was noted that an MRP system is capable of according the same stringent treatment to any inventory item regardless of class. An MRP system user may feel, however, that C items do not warrant such elaborate treatment, and he or she may exclude them from the system. There also exist MRP systems that cover A items only, on the theory that if the most important and expensive inventory items are planned and controlled properly, the rest will take care of themselves, more or less. This is simply not so.

MRP systems that are limited in inventory-item coverage yield only a small fraction of the benefits they are capable of. Such systems cannot displace the informal system that always has been and under these circumstances will continue to be the modus operandi of the factory. For purposes of assembling the product, the lowly C item is as important as an A item. Both must be available in the right quantity at the right time. Furthermore, some A items have components classified B and C, and shortage of one of the latter will prevent the completion of the A item. As pointed out earlier in this book, no matter how much C item safety stock there may be under an order-point approach, there will be occasional shortages. As to priority planning, unless all manufactured items are covered by the MRP system, relative shop priorities cannot be established. A manufactured C item must be manufactured in the shop and must contend for productive capacity with A and B items. Unless a C item order due date is maintained up to date through the MRP system, its validity is always questionable, but it will never do to assume that A items have automatic priority over C items.

Unless the dates of actual need for both an A item and a C item are known, it is obviously impossible to tell which has priority over which. Here again, shortages and expediting will have to establish what the real priorities are.

Purchased C items, when excluded from the MRP system, do not necessarily affect the priorities of other purchase orders, and they may be considered the exception to the rule that an MRP system should cover all classes of inventory for purposes of priority planning. The C item purchase-order due dates in this case will tend to be invalid, however, causing some shortages and last-minute expediting.

Another reason why no inventory class should be excluded from system coverage lies in capacity requirements planning. All manufactured items A, B, and C must be covered by the MRP system if the capacity requirements report is to contain complete load data. If C items are controlled by order point, only open C item orders can be reflected in the load. Because some of these orders will carry invalid due dates, the scheduling of their operations will be incorrect, and this will affect the validity of the entire load projection. By excluding any manufactured items from the MRP system, the usefulness of capacity requirements planning information is impaired, if not destroyed.

## Replanning Frequency

The frequency of replanning is under complete user control, but it is of utmost importance to the effectiveness of system performance. As a general rule, the more dynamic or prone to change the environment is, the more frequently the material requirements should be replanned. In most manufacturing companies, a longer than weekly replanning cycle will prove unsatisfactory, especially if the MRP system is used for purposes of priority planning. As stated previously, the recommendation is to replan at least daily. The frequency of planner action does not need to align with the replanning frequency.

Any MRP system that is used to replan cyclically (rather than continuously) can do no more than "take a snapshot" of inventory status at the time of replanning and plan order priorities accordingly. Their validity deteriorates gradually following the replanning as the inventory status changes. If the "snapshots" are not taken frequently enough to revalidate priorities, it becomes impractical to follow the priorities established by the formal system, and the informal system must take over. As pointed out earlier, without valid order priorities, there can be no valid load projection. With insufficiently frequent replanning, the user cannot realize the potential of the MRP system. The subject of replanning frequency was reviewed in some depth in Chapter 7.

Certain special capabilities can be incorporated into the basic MRP system that will enhance its usefulness. These system features are not absolutely essential to the system's operation, and therefore they may not be included in a given MRP software package. They do, however, increase the power of the MRP system significantly as a planning tool and warrant inclusion in the system. Of the various special system features, the most important ones are pegged requirements and the firm planned order.

## Pegged Requirements

The pegging of requirements provides the capability to trace item gross requirements to their sources. The process of MRP, as described earlier, progresses from top to bottom of the product structure. The gross requirements for a component item, derived from its parent items and from external sources of demand, if any, are summarized by period. The contents of a given gross requirements bucket represent a total, the breakdown and sources of which are obscured.

Pegging requirements means saving this information, which at one point in the requirements planning process is known to the system, and recording it in a special file.

Pegged requirements may be thought of as a selective where-used file. In comparison with a regular where-used file, which lists all parents of a component item, a pegged requirements file lists only the parents that show planned orders (the source of component gross requirements) in their records. This permits the inventory planner to trace requirements upward in the product structure to determine which parents a given gross requirement came from, where their requirements came from, and so on. By following the pegs from one item record to another, the planner can trace the demand to its ultimate source, that is, a specific bucket (or buckets) in the MPS.

Requirements pegging is effected by establishing a so-called peg record for each component item in which the breakdown, or detail, of gross requirements is recorded by period and tied to its source. An example is provided in Figure 14-2. Here, the demand for item X comes from parent items A, C, and D and from an interplant or service-part order. Pegging provides a capability of specialized inquiry for the benefit of the inventory planner.

The preceding discussion of pegged requirements covers the so-called single-level peg, that is, the ability to trace the source of item demand to the immediately higher level only. With the single-level peg, a succession of peg inquiries is required to trace item demand to an end-item lot (or lots) called for by the MPS. In order to link item demand to that schedule by means of a single inquiry, the so-called full-peg capability would be required. Under the full-peg approach, each individual requirement for a component item is identified with a specific product (or end-item) lot or customer order listed in the MPS.

This principle can be extended to orders and even on-hand quantities of the component item so that it always may be known which group of parts "belongs" to which product lot. It is rarely practical to program a full-peg capability, however, because in most manufacturing environments it is intended that individual requirements for a component item stemming from multiple parents be combined, that an order cover multiple

**FIGURE 14-2**

Pegged requirements.

**Requirements Record – Item X**

| Period | 1 | 2 | 3 | 4 | 5 | 6 |
|---|---|---|---|---|---|---|
| Gross Requirements | 20 | | 35 | 10 | | 15 |

**Peg Record – Item X**

| Period | Period | Parent Record | External Order |
|---|---|---|---|
| 1 | 20 | A | |
| 3 | 15 | A | |
| 3 | 20 | C | |
| 4 | 10 | | No. 38447 |
| 6 | 15 | D | |

net requirements, and that parts on hand or in process be commingled. Lot sizing, safety stock, scrap allowances, and the level-by-level planning process itself tend to obscure (or even erase) a clean path connecting noncontiguous levels.

Full pegging is feasible and desirable in a limited number of situations, such as when the product is custom-engineered and made to order, when the different standard products have few or no common components, or when the MPS consists of special contracts. Common component usage and repetitive production tend to make full pegging impractical. With MRP, eggs are deliberately scrambled, as it were. Full pegging attempts to keep the eggs from getting scrambled in the pan—an awkward and often impossible task.

### The Firm Planned Order

This term denotes a capability by the system to accept a command to "freeze" the quantity and/or timing of a planned-order release. This is another important tool of the inventory planner by means of which he or she is able to solve certain types of problems (reviewed below).

The firm planned-order command immobilizes the order in the schedule, forcing the MRP system to "work around" it in adjusting coverage of net requirements. The firm planned order forbids the system to put another planned order into the "frozen" bucket, which in some cases may result in a given net requirement not being fully covered. This special capability therefore should be used judiciously and for a specific planned order only rather than for the whole planned-order release schedule.

## THE SYSTEM AND THE INVENTORY PLANNER

The inventory planner (also called the *inventory analyst, inventory controller*, etc.) is responsible for the planning and control of a group of specific inventory items, and in an MRP, he or she interacts continuously with the MRP system. He or she is the recipient of the system's principal outputs, and his or her first duty is to take inventory order action based on information supplied by the system. The inventory planner inquires into the system's files for the data needed for purposes of analysis, and he or she handles a miscellany of problems that arise in the course of this work. The inventory planner's specific job description varies from company to company, but in the typical case, his or her function consists essentially of the following responsibilities:

- Releasing orders for production
- Placing purchase requisitions
- Changing the quantity of orders and requisitions, including cancellation
- Changing the timing of (i.e., reschedule) open shop orders
- Requesting changes in the timing of open purchase orders
- Activating special procedures for the handling of engineering changes affecting items under the planner's control
- Approving requests for unplanned stock disbursements

- Monitoring inventory for inactivity or obsolescence and recommending disposition
- Investigating and correcting errors in inventory records
- Initiating physical inventory counts
- Analyzing discrepancies or misalignments between item requirements and coverage and taking appropriate corrective action
- Requesting changes in the MPS

Most of these are routine and require no further elaboration, but a few of the inventory planner's duties warrant a more detailed review. Transactions continually modify inventory status, which, in turn, provides the clues to inventory action. The principal types of action are related to orders, that is, the releasing of planned orders and the changing of the quantity and/or timing of open orders. The inventory planner is constrained by the fact that it may be difficult or costly to change the quantity of an open purchase order and usually impossible (other than by splitting the lot) to change the quantity of an open shop order. His or her field of order-related action is, in practice, generally limited to:

- Releasing the order in the right quantity at the right time
- Rescheduling the due date of an open order if and as required to make it coincide with the date of actual need

In both these functions, the inventory planner is fully supported by the MRP system, which determines both the quantity and timing of planned-order releases and which also constantly monitors the validity of all open-order due dates. The following examples illustrate how the MRP system determines when to generate the two basic outputs or messages: "Release the order" and "Reschedule the order."

## Releasing a Planned Order

A planned order is mature (for release) when the planned release quantity appears in the current-period bucket. This happens either as a result of offsetting for lead time in the course of the requirements explosion or by passage of time, which gradually brings a planned-order release toward the first or current period. The current-period bucket in the planned-order release schedule is known as the *action bucket*. The (MRP) computer program tests the contents of this bucket, and when it exceeds zero, the system generates a message (report) to the inventory planner that the order is due for release. Figure 14-3 shows such a condition.

The planner, who reviews the request and takes the actual action, normally has the privilege of overriding the system in terms of changing the quantity of the order at this point. For example, a shortage of raw material may not allow the order to be released in the full quantity planned by the MRP system, and the planner may decide to reduce the order quantity rather than delay the order release.

Conversely, the planner may wish to increase the order quantity for some reason. He or she should not do this without assuring himself or herself that the action will not

**FIGURE 14-3**

A planned order mature for release.

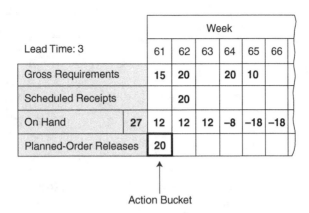

| Lead Time: 3 | | Week | | | | | |
|---|---|---|---|---|---|---|---|
| | | 61 | 62 | 63 | 64 | 65 | 66 |
| Gross Requirements | | 15 | 20 | | 20 | 10 | |
| Scheduled Receipts | | | 20 | | | | |
| On Hand | 27 | 12 | 12 | 12 | -8 | -18 | -18 |
| Planned-Order Releases | | 20 | | | | | |

Action Bucket

cause a problem at the level of component material that has been planned by the MRP system to cover requirements of perhaps several parent planned orders in quantities that the system planned for these orders. If the planner exceeds one of these quantities, he or she may short another one.

In net-change MRP systems (see Chapter 7), changing planned-order quantities at the time of shop-order release means upsetting the interlevel equilibrium that the system continuously strives to maintain. Such changes in effect introduce errors into the system's records that the planner should correct immediately (through special transactions) if he or she must override the system's recommendation. Note that the preceding comments pertain to shop orders; planned purchase orders are not as critical because they do not involve component materials within the system. When the planned order shown in Figure 14-3 as mature is released, the transaction reporting this will change the status of the item to that shown in Figure 14-4.

At this point, both open orders are scheduled correctly, that is, aligned against the dates of need. If the gross requirements change subsequently, however, the dates of need

**FIGURE 14-4**

Status change following planned-order release.

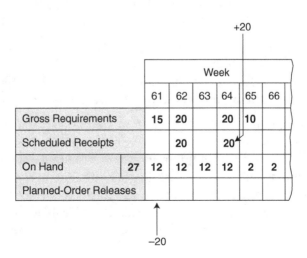

+20

| | | Week | | | | | |
|---|---|---|---|---|---|---|---|
| | | 61 | 62 | 63 | 64 | 65 | 66 |
| Gross Requirements | | 15 | 20 | | 20 | 10 | |
| Scheduled Receipts | | | 20 | | 20 | | |
| On Hand | 27 | 12 | 12 | 12 | 12 | 2 | 2 |
| Planned-Order Releases | | | | | | | |

-20

**FIGURE 14-5**

Open orders do
not coincide with
dates of need.

| | | Week | | | | | |
|---|---|---|---|---|---|---|---|
| | | 61 | 62 | 63 | 64 | 65 | 66 |
| Gross Requirements | | 30 | 5 | | 10 | 10 | 10 |
| Scheduled Receipts | | ←–20 | | | 20–→ | | |
| On Hand | 27 | –3 | 12 | 12 | 22 | 12 | 2 |
| Planned-Order Releases | | | | | | | |

may no longer coincide with the order-completion times (due dates). Such a condition is illustrated in Figure 14-5. Note that in this example only the timing, not the quantity, of the total requirements has changed. The coverage therefore is still adequate—no additional orders are required—but the timing of the open orders is now wrong.

## Rescheduling an Open Order

A net-change MRP system detects this condition immediately on processing the transaction that caused the gross requirements to change. A regenerative MRP system detects this during the requirements planning run. A changed gross requirements schedule necessitates a recomputation of the projected on-hand schedule, and the new schedule contains the clues to the action required. In Figure 14-5, there is a net requirement (for three) in the first period, followed by open orders in subsequent periods. The system is programmed not to generate a new planned order to cover the net requirement (regular logic) but to request rescheduling of the closest open order.

In the third period (week 63 in the example), there are 12 on hand, and the gross requirement in the subsequent period is only 10. This requirement can be covered in full by the quantity on hand, and there is clearly no need for the open order to arrive in week 64, as originally scheduled. This order should be rescheduled (its due date changed) to week 65, when it is actually needed. The two tests for open-order misalignment are as follows:

1. Are there any open orders scheduled for periods following the period in which a net requirement appears?
2. Is there an open order scheduled for a period in which the gross requirement equals or is less than the on-hand quantity at the end of the preceding period?

The MRP program carries out these two simple tests whenever the projected on-hand/net requirements schedule is being recomputed. If a test is positive, the system generates the appropriate rescheduling message. Note that extension of the second test to subsequent periods will indicate that an open order should be canceled when the on-hand quantity in the period preceding the scheduled receipt of the order is sufficient to cover all remaining gross requirements. This is equivalent to rescheduling the order beyond the far edge of the planning horizon. Note also that the system recomputes the planned-order schedule so as to align it properly with net requirements, which means that planned orders are being rescheduled automatically, without action on the part of the inventory planner.

If the relative priorities in the shop (and those of open purchase orders) are to be kept valid, the planner must reschedule due dates not only for orders needed earlier than originally planned but also for those needed later. The tendency is to concentrate on orders that need to be completed early to prevent shortages and to delay action or ignore the others. It is understood that following the rescheduling action by the inventory planner, which is reflected in the MRP system, the revised shop order due dates will be input to the operations scheduling system and to the dispatching (operations sequencing) system. The former reschedules all remaining operations of the affected shop orders in accordance with the new order due dates; the new operation start (or finish) dates then may serve as a basis for dispatching and are also used to recompute work load. When priority ratios (rather than operation start dates) are used for purposes of dispatching, the new order due dates enter into the preparation of the daily dispatch list. As far as revising purchase-order due dates is concerned, the inventory planner recommends action, by submitting the revised dates of need, to the purchasing department. Only when the latter actually acts does the MRP system reflect the rescheduling.

The planner may decide not to advance the order due date (contrary to an indication by the MRP system) when there is safety stock or when the new date would be impossible to meet. In the latter case, the proper course of action is to peg upward in an effort to solve the problem, possibly all the way to the MPS, which may have to be changed.

## Problems of Net Requirements Coverage

Probably the most serious problems that the inventory planner must cope with are discrepancies or misalignments between net requirements and coverage, resulting from unplanned events or increases in gross requirements. The planner has limited means at his or her disposal when trying to rebalance the status of a given inventory item. He or she cannot change the quantity on hand, nor can he or she change gross requirements by direct intervention. The planner can only change orders, that is, the timing of an open order and both the quantity and timing of a planned order, as discussed previously. Thus to change the gross requirements for a given item, the planner must change the planned order schedule of its parent item(s).

To be able to do so, the inventory planner depends on the two special capabilities of the MRP system that were mentioned previously, that is, pegged requirements and firm planned order. How the planner would use these is illustrated in Figures 14-6, 14-7, and 14-8.

Following the original status of item $Y$, a purchased material with five weeks' lead time (see Figure 14-6), gross requirements in week 31 increase from 20 to 30. As a result, the MRP system requests an immediate order release for 35 units (see Figure 14-7). The

**FIGURE 14-6**

Original status of item Y.

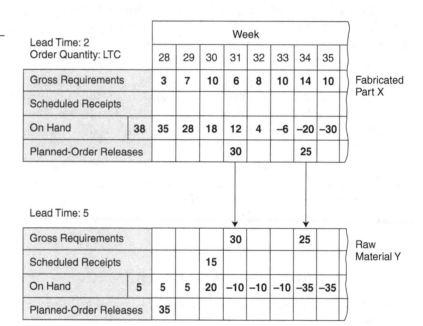

| Lead Time: 5 | | Week | | | | | | | | |
|---|---|---|---|---|---|---|---|---|---|---|
| | | 28 | 29 | 30 | 31 | 32 | 33 | 34 | 35 | |
| Gross Requirements | | | | | 20 | | | 25 | | Item Y |
| Scheduled Receipts | | | | 15 | | | | | | |
| On Hand | 5 | 5 | 5 | 20 | 0 | 0 | 0 | -25 | | |
| Planned-Order Releases | | | 25 | | | | | | | |

**FIGURE 14-7**

A problem of coverage.

| Lead Time: 2 Order Quantity: LTC | | Week | | | | | | | | |
|---|---|---|---|---|---|---|---|---|---|---|
| | | 28 | 29 | 30 | 31 | 32 | 33 | 34 | 35 | |
| Gross Requirements | | 3 | 7 | 10 | 6 | 8 | 10 | 14 | 10 | Fabricated Part X |
| Scheduled Receipts | | | | | | | | | | |
| On Hand | 38 | 35 | 28 | 18 | 12 | 4 | -6 | -20 | -30 | |
| Planned-Order Releases | | | | | 30 | | | 25 | | |

| Lead Time: 5 | | | | | | | | | | |
|---|---|---|---|---|---|---|---|---|---|---|
| Gross Requirements | | | | | 30 | | | 25 | | Raw Material Y |
| Scheduled Receipts | | | | 15 | | | | | | |
| On Hand | 5 | 5 | 5 | 20 | -10 | -10 | -10 | -35 | -35 | |
| Planned-Order Releases | | 35 | | | | | | | | |

inventory planner reviews this request against the current status of the item and detects a problem: The item, which has a procurement lead time of five weeks, is needed in three weeks.

Before placing a purchase requisition on a rush basis, the planner decides to peg to parent items to see if the problem might be solved some other way. The record of fabricated part X (see Figure 14-7), from which the gross requirement stems, indicates that the planned order scheduled for release in week 31 covers net requirements of weeks 33, 34, and 35 and is computed by (let us say) a least-total-cost (LTC) lot-sizing algorithm. The solution is evident: The parent planned order can be reduced without causing a problem in the status of item X.

The planner reduces the planned order in question to 20 via a transaction and designates it as a firm planned order by means of a special command input to the system. This is necessary to prevent the system from increasing the planned order back up to 30 during the next replanning cycle. This planned order is now "frozen," and after replan-

**FIGURE 14-8**

Solution of
problem of
coverage.

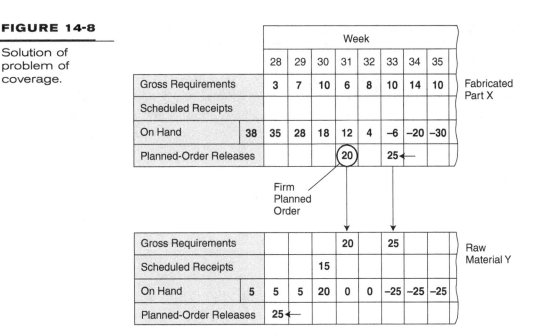

Firm
Planned
Order

| | | Week | | | | | | | |
|---|---|---|---|---|---|---|---|---|---|
| | | 28 | 29 | 30 | 31 | 32 | 33 | 34 | 35 |
| Gross Requirements | | 3 | 7 | 10 | 6 | 8 | 10 | 14 | 10 | Fabricated Part X |
| Scheduled Receipts | | | | | | | | | |
| On Hand | 38 | 35 | 28 | 18 | 12 | 4 | –6 | –20 | –30 |
| Planned-Order Releases | | | | | (20) | | 25← | | |

| | | | | | 31 | 32 | | | |
|---|---|---|---|---|---|---|---|---|---|
| Gross Requirements | | | | | 20 | | 25 | | | Raw Material Y |
| Scheduled Receipts | | | | 15 | | | | | |
| On Hand | 5 | 5 | 5 | 20 | 0 | 0 | –25 | –25 | –25 |
| Planned-Order Releases | 25← | | | | | | | | |

ning, the two records appear as in Figure 14-8. The problem is solved, and an order for 25 units of item *Y* will be released under normal lead time. Note that because the planned order for item *X* has been reduced, the MRP system has compensated by moving the next planned order in the parent record forward. In the real situation, the quantity of the second planned order (recomputed under LTC) also might be affected.

The preceding example illustrates a problem of coverage caused by an increase in component-item gross requirements, but the same type of problem would have arisen if the vendor of the open order for 15 had indicated that it was unable to ship on time. If item *Y* were a fabricated part, the scrapping of 10 of the 15 in process would have the same effect. In our example, the inventory planner was able to reduce the parent planned order because lot sizing covered multiple periods' net requirements. Had the planned order covered a single period, it still could have been reduced by an amount within its scrap allowance or other excess over the quantity of the net requirement. Safety stock at the parent level also would allow a reduction in the planned order.

The type of problem illustrated in the preceding examples sometimes can be solved without having to reduce the quantity of the parent planned order; instead, only its timing is changed. If the parent item's planned lead time can be compressed (as it often can—see discussion of lead-time flexibility in Chapters 7 and 12), the respective planned-order release can be rescheduled for a later period and held in place as a firm planned order.

Rescheduling a parent planned-order release consequently (after the next explosion) will reschedule the corresponding component gross requirement and thus solve the problem of net requirements coverage. In Figure 14-7, for example, if the lead time of parent item *X* were reduced by one week (for purposes of the first planned-order release only), it would result in the net requirement for component item $Y$ occurring one week later. This would alleviate the problem of net requirements coverage by allowing an extra week for procurement of the material.

If the inventory planner is unable to solve a problem of coverage by pegging to the next-higher level and manipulating planned-order or open-order data, he or she may peg from the parent record to its parent records in an attempt to find a solution. Successive pegs may lead to the MPS, which may have to be changed to solve a particular component-item problem.

## CHAPTER 15

# Industry Effect on MRP

To better understand the competitive position of an enterprise, the relationship between the volumes produced by the company and the variety produced is compared. An interesting diagonal has evolved where most companies are clustered in order to compete effectively. This position on the diagonal is the optimal position in terms of production cost and responsiveness. Movements from that competitive diagonal can be either a competitive advantage for the company or a disaster. According to Wheelwright and Hayes, this can be represented in the produce/process matrix shown in Figure 15-1.

This figure shows an inverse relationship between variety and volume. In general, as the product volume increases, the variety tends to decrease. Attempting to compete off this diagonal is not sustainable.

## PROJECT MANUFACTURING COMPANY

The far upper-left corner of the volume/variety matrix is a company that produces a very high variety of products but in very low volume. This can mean that a single product may be developed, planned, and produced once and never produced again. These products or deliverables typically are managed as unique projects. A project-driven company competes in the market based on the wide variety of products that it can produce using the same resources. This type of company uses some material requirements planning (MRP) for determining what needs to be ordered and when. In addition, this organization normally will use a project management system to determine the critical path for the activities of the project. The tools of the project management company also include a program-evaluation review technique (PERT) and Gantt charts. These tools can provide the expected finish date once the start date is determined using forward scheduling. For each task in the scheduling network, by using forward scheduling, the earliest a task can start is calculated. Or, given a desired completion date, the suggested start date can be calcu-

**FIGURE 15-1**

Wheelwright and Hayes product/process matrix.

| Process Structure Process Life Cycle Stage | Product Structure Product Life Cycle Stage | Low Volume Unique (one of a kind) | Low Volume Multiple Products | High Volume Standardized Product | Very High Volume Commodity Product |
|---|---|---|---|---|---|
| | (Project) | | | | |
| Jumbled flow (job shop) | | Job shop | | | |
| Disconnected line flow (batch) | | | Batch | | |
| Connected line flow (assembly line) | | | | Assembly line | |
| Continuous flow (continuous) | | | | | Continuous |

lated using backward scheduling. Backward scheduling determines the latest each task can start. In the process of scheduling all the required tasks, some tasks have a difference between early and late start or early and late finish.

When planning material to be available to begin a task, the difference between an early and late start is significant. The question quickly arises about when the material should be available: in time for the earliest possible start or hold off investing that capital in inventory until the last possible minute? The project-type company must decide and establish the material policy for ordering needed materials, choosing them to be available at the early start, late start, or average start date. In most companies of this type, the policy is to have the materials available at the earliest possible start date because a project-driven company's cost typically is driven most by the resources used rather than by the materials. These resources are usually the constraint to the company delivering a higher level of output. Having a resource idle because materials are not ready or available can cause a great financial loss because this resource's capacity cannot be regained once it is lost.

To manage these scarce resources more effectively and improve the overall time and expense required to complete a project, a scheduling methodology has been developed recently in the project scheduling area called *critical-chain scheduling*. This scheduling method pulls all the individual slack times from each operation and provides a schedule buffer for significant paths within the overall project. The traditional project scheduling method has this slack-time buffer broken up at each operation. Project management reality is that since the most critical resources are people, and since people are driven by deadlines, getting a task completed early is virtually impossible. The natural tendency is to wait until the last possible finish date to focus on the work and accomplish the tasks because the resources typically have more to do than there is capacity available. Figure 15-2 shows how the individual activity buffers are moved to the end of the project so that real requirement dates for each activity can be identified.

**FIGURE 15-2**

Critical-chain
schedule.

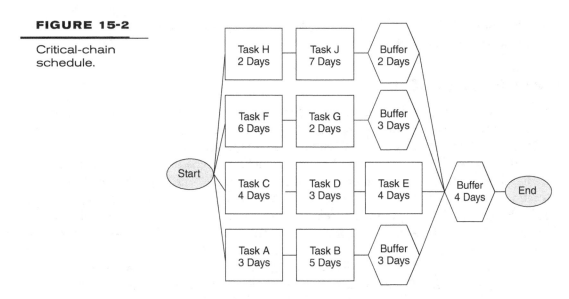

The critical-chain method moves the slack time from each task into a time buffer for a sequence of tasks. This provides more accurate information so that everyone can work to the real deadline but still have some buffer time in the overall project schedule in the event an emergency arises. For more information on this scheduling approach, the book, *Critical Chain* (North River Press, 1997), by Eli Goldratt, is available.

These new tools were developed in response to the overall pressure to reduce lead times for a project. A competitive strategy for a project-driven company is lead time. A company that can complete a quality project in less lead time than the competition typically can demand a premium for this performance. Having inventory available at the earliest expected moment that the operation could begin can enhance lead-time response. This is not to say that all materials should be purchased at the beginning of the project. However, the better alternative to guessing when the activities with slack time will really start is to use critical-chain scheduling. By using these resource-schedule buffers, more accurate start dates are calculated with which to determine real material need. The more agile a company is in the execution of a project, the less opportunity there is for cost overruns owing to unexpected crises. Or, in the immortal words of Marine Colonel William Scott (retired) in the management of a large aviation remanufacturing depot, "The longer the cow is in the pasture, the more grass it can eat!" A direct relationship exists between lead time and cost. When the business processes are capable and agile, costs reduce dramatically. However, if the quick response time is accomplished by expediting and manual intervention, short lead times can be financially disastrous. Anyone who has worked on a project that was behind schedule will attest to this fact. Schedule and cost never can be recovered simultaneously. Effective project management requires up-to-date information with which to make quality decisions. Having a closed-loop information system that provides feedback to the plan based on actual performance is essential for success in this environment. Implementation of enterprise resource planning (ERP) in this environment has its unique challenges. An implementation success strategy for project-driven companies is discussed in Chapter 16.

## MAKE TO STOCK

Make-to-stock companies typically ship to customers on demand. The customers are not willing to wait very long for their needs to be fulfilled. They expect the products they want to be on the shelf, typically in a retail environment. Since manufacturing has to build products in advance of customer demand, the manufacturing schedule typically is driven by a demand forecast. Actual customer demand then consumes this forecast. Ideally, the sales force will sell to the *available to promise* (ATP). ATP is the uncommitted portion of inventory. ATP assumes that the plan will be executed as designed and provides visibility on how much inventory will be available for customer orders. Figure 15-3 provides an example of ATP.

Available to promise is calculated only in the first time period and whenever there is an expected receipt. The demand time fence is the expected time within which no additional customer orders are expected. The demand time fence typically is the length of time that the customer is expecting to wait for the product he or she has ordered to be shipped. Within the demand time fence, the actual customer demand is used to calculate the projected available balance. Once the planning horizon goes beyond the demand time fence, the projected available balance uses the greater of the forecast or the customer order. The available-to-promise line uses only confirmed customer demand. Available to promise is the uncommitted portion of inventory. This is why it is calculated only for the first time period and whenever there is an expected receipt. Another way to think about it is how long does the inventory need to last given the current customer backlog.

Cumulative ATP shows how many pieces are available between receipts that have not been committed already. In some cases, there may be insufficient inventory to cover the demands that are already known in period 2 until the next receipt in period 4. This is when a process known as *backward ATP* is used to reserve that inventory to ensure that the known customer orders will be covered. In Figure 15-3, the real ATP in period 1 should be 47 pieces. If all 92 pieces are committed to a customer, then the customer's

**FIGURE 15-3**

Available to promise.

Part A
Demand Time Fence: 3
Beginning On Hand – 172

|                              | 1   | 2   | 3  | 4   | 5   | 6   | 7   | 8   | 9   | 10  |
|------------------------------|-----|-----|----|-----|-----|-----|-----|-----|-----|-----|
| **Forecast**                 | 100 | 90  | 80 | 75  | 80  | 90  | 100 | 100 | 120 | 130 |
| **Actual Customer Demand**   | 80  | 120 | 75 | 30  | 20  | 10  | 0   | 0   | 0   | 0   |
| **Project Available Balance**| 92  | 122 | 47 | 122 | 42  | 102 | 2   | 52  | 82  | 102 |
| **Available to Promise**     | 92  | –45 |    | 120 |     |     |     |     |     |     |
| **Cumulative ATP**           | 92  | 47  | 47 | 167 | 167 | 167 | 167 | 167 | 167 | 167 |
| **Master Production Schedule**|    | 150 |    | 150 |     | 150 |     | 150 | 150 | 150 |

order in period 3 will be short by 45 pieces. If the order-entry personnel use the project-ed available balance to promise customer orders, customers who have provided the com-pany with sufficient lead time may be penalized, whereas customers who just called in may get product immediately. Just comparing the two lines in Figure 15-3 clearly demon-strates the errors that can be made by using the project available balance.

However, remember the three rules of forecasting: Forecasts are always wrong, they are worse projecting further in the future, and worse in more detail. What a terrible situ-ation for this type of company! This company depends entirely on forecasts to drive its planning for material and capacity. If the forecasts are incorrect, then the resulting plan-ning is incorrect. If the planning is incorrect, then the inventory that is purchased may be exactly the wrong material in exactly the wrong amount. This has a terrible impact on cash flow and can affect overall profitability. The make-to-stock company is the most dif-ficult kind of company to reduce inventory and still provide good customer service with-out change to integrated business planning and supply-chain management processes. One proven strategy for reducing inventory is to have a relatively small number of fin-ished goods compared with a make-to-order or assemble-to-order enterprise. Also, remember that the faster a company can respond means that there is less need to have a detailed forecast far into the future. Improved agility relates directly to improved inven-tory performance.

A make-to-stock company must fight the urge to try to be everything to everybody if its processes are not flexible and agile. The marketing department will try to add small variations to meet the needs of unique market channels. This can wreak havoc on a com-pany if the processes are not capable of handling these changes easily. Many make-to-stock companies have deferred the final configuration of products successfully until the customer order is received. The advantage is that the forecast error is reduced signifi-cantly because resources to complete finished goods are not committed until an actual order is received. This competitive strategy is covered in the section on assemble to order.

This strategy may not be possible for the company if it is delivering products such as mayonnaise or canned green beans to a retail market. The customer is not willing to wait for the company to package on his or her demand. The customer expects to have immediate delivery from the shelf. This is why the make-to-stock company typically will exercise a strategy of carrying a safety stock of finished goods to buffer against the vari-ability of demands from the market. The key to success is where this buffer is located. However, understanding what is currently in the entire supply-chain pipeline can mean the difference between profits and loss for this enterprise. What may be perceived as con-sumer variability is nothing more than normal supply-chain replenishment cycles. Figure 15-4 shows how demand from just a few different retailers through a few distributors can wreak havoc on the manufacturer. This is not because the customer had wide variability in demand but rather because the batch ordering of products from the retailer to the dis-tributor and the distributor to the manufacturer has set the manufacturer up for guaran-teed failure. Having a collaborative approach with customers can provide win-win value because the distributor needs to carry less inventory to achieve the same fill rates because the manufacturer can build more closely to actual consumption in the supply chain. This approach was introduced in Chapter 2. The benefit for the manufacturer is that the forecasts are more accurate, and production resources are allocated to items that will really sell without a major reduction in price.

If safety stocks were calculated using the traditional statistical methods, the required safety stock to provide 95 percent customer service would be 805 units. If this is applied to the planning from Figure 15-4, the result would be as shown in Figure 15-5.

The combination of the safety stock and fixed order quantity yields an amazing average inventory of 1,640 units. This is over twice the expected safety stock. When this is considered on an enterprise-wide level, the assets dedicated to this hidden inventory could be significant. As an alternative, if visibility existed of the entire inventory in the

**FIGURE 15-4**

Lumpy demand from distribution to manufacturing.

**Distributor 1**
Beginning On Hand: 145

|  | 1 | 2 | 3 | 4 | 5 | 6 | 7 | 8 | 9 | 10 |
|---|---|---|---|---|---|---|---|---|---|---|
| Forecast | 100 | 90 | 80 | 75 | 80 | 90 | 100 | 100 | 120 | 130 |
| Projected Available Balance | 45 | 105 | 25 | 100 | 20 | 80 | 130 | 30 | 60 | 80 |
| Planned Receipt |  | 150 |  | 150 |  | 150 | 150 |  | 150 | 150 |
| Planned Order Release | 150 |  | 150 |  | 150 | 150 |  | 150 | 150 |  |

**Distributor 2**
Beginning On Hand: 190

|  | 1 | 2 | 3 | 4 | 5 | 6 | 7 | 8 | 9 | 10 |
|---|---|---|---|---|---|---|---|---|---|---|
| Forecast | 135 | 100 | 90 | 80 | 75 | 82 | 125 | 110 | 150 | 50 |
| Projected Available Balance | 55 | 255 | 165 | 85 | 10 | 228 | 103 | 293 | 143 | 93 |
| Planned Receipt |  | 300 |  |  |  | 300 |  | 300 |  |  |
| Planned Order Release | 300 |  |  |  | 300 |  | 300 |  |  |  |

**Distributor 3**
Beginning On Hand: 272

|  | 1 | 2 | 3 | 4 | 5 | 6 | 7 | 8 | 9 | 10 |
|---|---|---|---|---|---|---|---|---|---|---|
| Forecast | 80 | 75 | 80 | 90 | 100 | 125 | 130 | 120 | 50 | 85 |
| Projected Available Balance | 192 | 117 | 37 | 197 | 97 | 222 | 92 | 222 | 172 | 87 |
| Planned Receipt |  |  |  | 250 |  | 250 |  | 250 |  |  |
| Planned Order Release | 250 |  | 250 |  | 250 |  |  |  |  |  |

**Manufacturer**
Beginning On Hand: 200

|  | 1 | 2 | 3 | 4 | 5 | 6 | 7 | 8 | 9 | 10 |
|---|---|---|---|---|---|---|---|---|---|---|
| Total Demand | 700 | 0 | 400 | 0 | 700 | 150 | 300 | 150 | 150 | 0 |

**FIGURE 15-5**

Manufacturing
plan with safety
stock.

**Manufacturer**
Beginning On Hand: 200, Safety Stock: 805

|  | 1 | 2 | 3 | 4 | 5 | 6 | 7 | 8 | 9 | 10 |
|---|---|---|---|---|---|---|---|---|---|---|
| **Total Demand** | 700 | 0 | 400 | 0 | 700 | 150 | 300 | 150 | 150 | 0 |
| **Projected Available Balance** | 1000 | 1000 | 2100 | 2100 | 1400 | 1250 | 950 | 2300 | 2150 | 2150 |
| **Planned Receipt** | 1500 |  | 1500 |  |  |  |  | 1500 |  |  |

**FIGURE 15-6**

Supply-chain
solution.

**Manufacturer**
Beginning On Hand: 200, Safety Chain Solution

|  | 1 | 2 | 3 | 4 | 5 | 6 | 7 | 8 | 9 | 10 |
|---|---|---|---|---|---|---|---|---|---|---|
| **Total Demand** | 700 | 0 | 400 | 0 | 700 | 150 | 300 | 150 | 150 | 0 |
| **Projected Available Balance** | 0 | 0 | 0 | 0 | 0 | 0 | 0 | 0 | 0 | 0 |
| **Planned Receipt** | 500 |  | 400 |  | 700 | 150 | 300 | 150 | 150 |  |

supply chain, proactive planning as shown in Figure 15-6 could be used to provide a better level of customer service with no safety stock.

The availability of information about intended demand from distribution centers can result in drastically reduced inventory while providing higher levels of customer service. This is the motivation behind the implementation of supply-chain strategies. Having visibility of inventory in the supply chain from the manufacturer to the end consumer should be a priority for a make-to-stock company. This visibility provides the opportunity for the manufacturer to be proactive in its planning rather than reactive. Having a more responsive manufacturer means that the distributor can carry less inventory and still maintain high fill rates.

The next level to consider is that even when the plan looks like everything is in balance, excess inventory still can be in the supply chain. Very simple lot sizing can result in excess inventory. Fixed lot sizing is a very common practice in many companies. This excess inventory can hide in the system because no system currently highlights its existence. Only if sophisticated analytics are used that are capable of identifying this hidden inventory would the enterprise realize what potential there is for significant savings in inventory and agility. If the inventory investment in the company seems to be high, try this test:

1. Run a query on the current database for the parts for which safety stock has been authorized. Typically, safety stock is set in the item master record as a fixed order quantity based on a general rule of thumb.
2. Multiply this authorized safety stock by the cost of the product according to the accounting department.
3. Add these individual numbers to get the total authorized investment in safety stock. If this number does not shock you, then continue to the next step.
4. For the parts with authorized safety stocks, multiply the current on-hand balance by the accounting cost for each part. Add these individual numbers to get

the total real investment in safety stock. Don't be surprised if the actual invest-ment is significantly higher than the authorized investment.

When this test has been run at many companies, the difference between the autho-rized and real investments has been as high as a factor of 4 to 6 times! The frustration of the senior management team is that after all the money that has been poured into com-puter system implementation, this fact is so well hidden. Make-to-stock companies typi-cally will order according to some economic lot-sizing rule and buffer with a fixed safe-ty stock. The net result is inventory—and lots of it!

## MAKE TO ORDER

A make-to-order company competes in the market by providing a wide variety of prod-ucts in the shortest lead time possible. In addition to common raw materials, all products in a make-to-order company tend to go through similar operations. This type of manu-facturing facility generally is capital-intensive with general-purpose equipment that can accomplish a wide range of processes. An example of this kind of business is a machine shop making sheet-metal parts for many customers. The operations used can include punching, forming, deburring, plating, and assembling. Almost an infinite number of fin-ished goods can be produced from these basic operations. To compete effectively in the market, the make-to-order manufacturer tends to focus its marketing in one type of industry, such as aerospace and defense, medical devices, computer parts, and so on. The constraint for growth in this type of company typically is knowledge of the market, the unique customer demands, and potential distribution channels and other routes to mar-ket rather than production capability. The cost of adding additional distribution channels is significantly more expensive than adding production capability.

The inventory strategy in this organization typically is to purchase a safety stock of the commonly used raw materials so that the overall response time to the customer can be reduced. Customers tend to order what they want at the last possible minute. Design changes after the order is placed are not unusual. Normally, relatively few raw materials are used in the normal course of business. The investment required in safety stock to shorten the response time is not all that significant. Another competitive strategy is to standardize the manufacturing processes to use common sizes of raw materials. Rather than using the size of material that provides the best material utilization, the company may standardize on sizes of raw material that are easily obtainable. Purchasing standard-size stock material prevents having to maintain a safety-stock inventory because these standard sizes are normally in stock at the supplier. In addition, these standard-size mate-rials typically are less expensive on a square-foot basis. However, it is true that more of the material will be wasted than if the best-fit material were purchased. The savings in material cost and inventory cost can far outweigh the material utilization benefit.

Understanding what the best solution is for the company as a whole must consider the cost of the wasted material, the inventory carrying cost to stock special material, the less expensive stock material, the competitive position of the company with respect to lead-time response, and many other factors that affect the overall cost. This final decision depends on many factors and must be considered from an overall competitive position for the enterprise. The capacity strategy usually focuses on maintaining aggressive cross-training with the operators so that they can operate a number of machines. This enhances the overall flexibility of the enterprise and could lead to a market advantage.

## ASSEMBLE TO ORDER

In an assemble-to-order company, the customer is provided with more product variety than in the make-to-stock company if he or she is willing to wait a small amount of time. Dell Computer has embraced this competitive strategy with documented success. Only the semi-finished subassemblies are forecast, built, and inventoried. When the customer orders a finished product, these items are assembled on demand to provide a custom product. The MRP system for an assemble-to-order company should contain a linear finite configurator. With a linear finite configurator, order-entry personnel can select from a pre-established list of options to build a finished part. Each of the choices that can be promised in a finished product must exist in inventory. This is very different from a dynamic parametric configurator. A linear finite configurator usually creates a temporary part number to represent and track the end item. If exactly that set of options is ordered again, the demand will be added to the previous order. This provides the enterprise with the visibility of the configurations that are the most popular and therefore may warrant a move to a make-to-stock strategy.

A success strategy for setting up this type of configurator is to make the choices that distinguish the product at the very end of the manufacturing process. Having the common parts early in the selection process also helps to make the assembly line run more smoothly. Capable-to-promise (CTP) functionality is still in its early stages in most MRP systems. Available to promise (ATP) is the process used to commit make-to-stock products. ATP works best on finished-goods items that are forecast, and then customer orders are received directly against that forecast. CTP matches the promise capability with how the assemble-to-order company actually plans its products. True CTP functionality requires visibility into the supplier's production schedules, inventory, and capacity. Material superbills (S-bills) are used for planning (see Chapter 11). Forecasts in the assemble-to-order company are accomplished at the semi-finished-goods level using percentage product mix as an indicator of relative need. Customer orders are received at the finished-goods level. CTP examines the material availability one level down and the available CTP delivery to the customer. A company using CTP also likely would be using a two-level master schedule.

Figure 15-7 shows an example of a superbill that can be used to plan this product. There is little need to know how many red heat shrink, type C connectors, 12-inch cable with strain relief were shipped. The important issue is to have sufficient wire, connectors, heat shrink, and strain reliefs available. These semi-finished goods are placed into inventory pending actual customer orders for the final assembled product.

**FIGURE 15-7**

S-bill for cable
assembly.

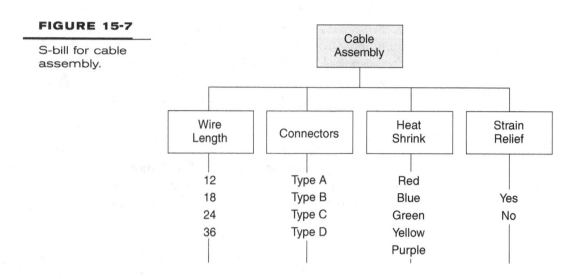

In an assemble-to-order environment, typically manufactured and purchased parts are common to many assemblies; each assembly is made up of a series of options. For this reason, when promising orders using CTP, it is essential that reservations are made against inventory that has been committed to a customer; otherwise, the same scarce sub-assembly may be promised to two different customers (or more). In addition, CTP also may take into account the capacity of the final assembly area when promising an order. Even though all the parts may be available, if there is no capacity with which to assemble them, the product will not be available when expected. The assemble-to-order company is well suited to the expectations of customers for mass customized products on demand. The challenge is having the right building blocks available from which to make the final product.

## MAKE TO STOCK/ASSEMBLE TO ORDER

Some companies may be a combination of types and encompass a number of positions on the volume/variety matrix. This company combines a number of different raw materials into significantly fewer semi-finished goods and then can explode those choices into many finished products.

An example of this kind of company would be a computer manufacturer that also assembles its own circuit boards and cables. Many individual parts need to be managed and planned to support the assembly of significantly fewer semi-finished goods. From these options, a wide variety of final products can be assembled. This company has all the challenges of both the make-to-stock and assemble-to-order companies. This company is driven from forecasts for semi-finished goods similar to the assemble-to-order company. The inventory strategy is to have these semi-finished goods available based on the overall sales and operations plan by product mix. Then this type of company has to plan and purchase the required raw materials to combine into subassemblies. These needed raw materials are planned using traditional gross-to-net logic. This company has the additional challenge of longer lead times because it must purchase all the separate components, assemble them into semi-finished goods, and then assemble the finished product to the customer's orders.

The assemble-to-order company has to worry only about the timely response to customer orders without the complication of ordering all the raw materials. The diversity of manufacturing processes also can be a real challenge in this type of company. A make-to-stock/assemble-to-order company can include processes as diverse as sheet-metal stamping to electronic assembly to painting. This diversity of operations required for fabrication normally requires batch manufacturing that results in on-hand inventory as a buffer between the fabrication and assembly operations because the assembly batches are significantly smaller than the fabrication batches. The assembly resource capacity is the buffer in this type of enterprise. Having a tight integration of design with fabrication is essential for overall success. Using standard parts in multiple models helps to reduce the complexity of the fabrication process and increases the volume of each individual part. The cost per part is usually reduced as the volume increases. The design strategy is a critical success factor for this type of company.

## CHAPTER 16

# Project Manufacturing

According to Ralph Currier Davis in *The Fundamentals of Top Management*, a project is "any undertaking that has definite, final objectives representing specified values to be used in the satisfaction of some need or desire." For a project manufacturer, this final objective is a physical product that meets the requirements of the customer. The project manufacturer builds products that are very low in volume and extremely high in variety. It is not uncommon to build items that are one of a kind. At the other end of this production spectrum are the process and repetitive manufacturers. The project manufacturers also usually have a requirement to track costs to a top-level project. Projects may span multiple years, and the cost collection must continue to aggregate at the very top project level. Managing scope, time, and resources, including costs, is essential for the enterprise.

Project management adds a great deal of complexity for the material requirements planning (MRP) system because the basic assumptions under all MRP engines is that all materials and activities are scheduled on the critical path. The fundamentals behind project management are that there is one critical path and that the slack time on the parallel paths is used to balance capacity. The project manufacturer's dilemma is, How should the system plan the materials to support the activities not on the project's critical path? Should they be available at the early or late start date of the activity? When are resources really required?

The standard logic of an MRP system assumes that if the part is the same for fit, form, and function as is designated by the part number, it can be used anywhere that part number needs to go. The project manufacturer may purchase materials specific to a single project that potentially could be used for multiple projects. The desire is to track the actual costs of these materials to the top-level project. However, this is in direct conflict with the standard costing process that is the default of most accounting systems. These complexities of material planning, scheduling, and overall costing highlight some of the project manufacturer's uniqueness. This chapter is not intended as a comprehensive reference in project management but rather highlights the application of MRP to this environment.

## PROJECT LIFE CYCLES

Too many project managers believe that the phases of a project are

1. Wild enthusiasm.
2. Mass disillusionment.
3. Search for the guilty.
4. Punish the innocent.
5. Promote the non-involved.

This belief has evolved from the personal experience of many project managers as they have been striving to do a good job managing the project. Effective project management requires effective management skills and techniques supported by appropriate technological tools. Unlike some projects, where the overall statement of work can be descoped to bring the project in on schedule and on budget and declare accomplishment, the project manager for a manufacturing company usually cannot descope the final product to be shipped and still declare victory. The product still must meet all the requirements set by the customer. This is a very difficult industry for a project manager.

According to David I. Cleland and William Richard King, the phases of effective project management include

1. Conceptual phase
2. Definition phase
3. Production phase
4. Operational phase
5. Divestment phase

### Conceptual Phase

The *conceptual phase* occurs when the design team is working with the customer to determine the overall requirements for the product and potential deficiencies of the existing processes and products. The initial feasibility of the technical, environmental, and economic reality is also examined during this phase. One example would be the recent development of superfast MRP systems. These systems were conceived many years ago, but only recently have they become a reality. The reason is that the available technology could not support them. Now that computers have multiple-gigabyte memory available, these calculations can be completed directly in the active memory core without incurring the slowdown of reading and writing to the hard drive.

In the conceptual phase, the project team provides the answers to

- What will the product cost?
- When will the product be available?
- What will the product do?
- How can the product be integrated into the existing systems?

The overall design and production approach is determined during this phase, and an initial statement of work is prepared for further detail in the definition phase.

## Definition Phase

The definition phase occurs when the detailed plan is prepared. This determines the realistic cost, schedule, performance requirements, quantity, and timing required for human and other critical resources. A good project manager also will identify the areas that are risky or cause for concern. These areas then are detailed further for recovery and contingency plans. This is very different from the traditional MRP environment, where the master schedule is entered, supported by a bill of material and routings. In the traditional MRP environment, the assumption is that everything will work exactly as planned. Any variability or unexpected events typically are covered with safety-stock inventory or available surge capacity. In the project manufacturing industry, the likelihood of the same inventory being used again is very small. Capacity must be scheduled carefully to provide optimal cost performance.

In the definition phase, a detailed statement of work is developed and broken down to the necessary level for control purposes. An effective statement of work (SOW) should clearly define the objective for the project and how success will be measured. The SOW should include cost and schedule targets as well as quality targets and usually becomes the contractual SOW. Since revenue is directly related to progress against the SOW, it is very important to have agreement with the customer on the definition of key words. For example, a customer may desire to have a product tested in water. Your intent is to test the product in a local lake. The customer meant water as in the Pacific Ocean. The cost implications can be significant. Even establishing the documentation format for the product manuals can be important because manuals and drawings created using new versions of software may not be readable by previous versions. The project manufacturer also may be paid at certain points in the completion of the product, so having measurable completion criteria is essential for each significant step in the work breakdown structure.

Once the SOW has been defined and approved, the work begins on the work breakdown structure (WBS). A graphic representation of the WBS looks like a bill of material (BOM) that has been laid on one side. The duration of each task and the relationship of one task to another can be clearly visualized in this WBS, as shown in Figure 16-1.

This breakdown of activities allows scheduling materials and capacity based on the timing that is required. When this WBS is phased against a time line, the timing for each task can be calculated.

## Production Phase

The production phase of a project begins with verification of the product production specifications and the beginning of unit production. The final preparation and dissemination of documents is incorporated in this phase, including the development of technical and service manuals and other traceability that is required for the product. Most companies have moved to releasing this information electronically owing to the cost of

**FIGURE 16-1**

Project task structure.

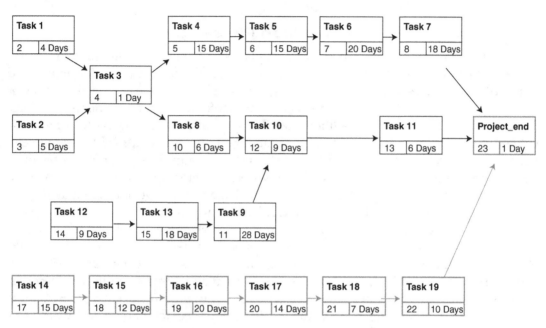

preparing and maintaining this essential document in a paper format. The product is tested to ensure that the specifications defined in the definition phase have been achieved. MRP can be used to bridge the transition from the planning to the production phase by supporting the project with detailed capacity and material planning. Timely status feedback is essential during this phase to provide an early warning of any part of the project that may not support the required completion date.

## Operational Phase

The operational phase of the project occurs when the product has been delivered to the customer. This can be a time when the field service personnel may be intimately involved to aid in installation or training for the product. During the operational phase, the realized costs are compared with the quoted costs to determine whether or not the product has been delivered at a profit. The actual cost of the product should not be a surprise at the very end of the project. Effective project management tools provide running feedback on completion status through a process called *earned-value analysis* or $CS^2$. This compares cost and schedule through a series of measures such as:

- Budgeted cost of work scheduled (BCWS)
- Actual cost of work scheduled (ACWS)
- Budgeted cost of work performed (BCWP)

- Actual cost of work performed (ACWP)
- Schedule variance (SV) = BCWP – BCWS or expressed as a percentage SVP = SV/BCWS
- Cost variance (CV) = BCWP – ACWP or expressed as a percentage CVP = CV/BCWP

No matter what measures are used, it is critically important to evaluate the overall performance of the project to learn what worked well and what could be improved for the next product. Without this postmortem on a project, the lessons learned on one project are likely to be learned over and over and over. Not only can the team learn from things that have gone wrong and should be avoided in the future, but learning what went well also helps the team to determine how to repeat the success. Identifying what could be improved provides the opportunity to possibly avoid that situation in the future.

## Divestment Phase

The divestment phase of a project manufacturer is the phase where the project team is transitioned from one project to the next. Some companies will dedicate personnel 100 percent of the time to a particular project. When that project is completed, it is essential to recognize the organizational and cost implications of having another project ready to start. If the project team does not see another product for development available on the horizon, there could be a tendency to stretch the current project out as long as possible. This can contribute to cost overruns on the current project and adversely affect the profits for this product. The divestment phase is a critical time that must be managed effectively so that the organization can take from the lessons learned on the previous product and apply them to the next product.

# PROJECTS IN MRP

The role of MRP in project management is to provide the common information that supports detailed planning and integration from product design through manufacturing. Product data management (PDM) is used frequently for design-cycle control, including the initial design release, distribution of the desired design to suppliers and potential multiple manufacturing sites, and managing changes that occur after the design release. Incorporation of the PDM tool has been shown to provide a 15 to 20 percent improvement in production cost by its linkages to the production engine. A full product lifecycle management (PLM) approach facilitates tapping into innovation. Here, the improvement to the enterprise can be several orders of magnitude. PDM and all its facets are beyond the scope of this book. Technology at this writing is still evolving in this area, but this promises to be a major focus for systems development in the next decade.

Linking the project management statement of work to MRP can be a challenge. The normal process is to perform a one-way download of the project into the MRP system and then begin to manage the project by exception only through MRP. However, very typical questions for this environment are, "How is my project doing?" and "What is the item that is extending the critical path, and if it is reduced, what is the next item?" These questions are very difficult to answer with standard MRP functionality. Use of a different approach to project scheduling, critical-chain scheduling, for a project provides improved visibility of lead time and other resource buffers. Tasks are scheduled with realistic expected times, and the buffers are moved to the appropriate spot in the critical chain.

If traditional project planning is used, with slack-time potential at each step except those on the critical path, then decisions must be made for planning material. The usual strategy is to order the material at lead time prior to the early start date. The advantage is that if the preceding task is completed early, the materials will be ready, so the project can continue. The disadvantage is that critical cash resources may be expended before they really need to be. Also, when materials have to sit around for a long time, they tend to get lost.

## CAPACITY DEPLOYMENT

Capacity deployment in the project manufacturing environment also can be a challenge. The normal response is to attempt to have all projects in work simultaneously. This allows personnel to easily move from one project to another when a problem arises. The downside of this approach is that typically projects will take much longer, and the company may develop another constraint, such as space or cash. To demonstrate this point, Hal Mather presents a wonderful riddle on how to deploy cranes in a ship unloading exercise. Assume that you have six ships. They will arrive in port at exactly the same time at six different piers. You own six movable cranes that are each capable of unloading a single ship in six days. If you put two cranes on one ship, then that ship can be unloaded in three days while the other ship waits. If you are in the shipping business and you make money by having ships on the sea and not in port, how should you deploy your cranes? Assume no interference between the cranes. The traditional approach to project management is to have all six ships in work at the same time. All projects would have some resource sprinkled on it. This is rather like trying to play poker with 52 people. Each person gets one card, and nobody can get a winning hand. The enterprise as a whole loses. Back to the ships, assuming that you make money by having ships on the water, your desire is to minimize the amount of time the ships are in port. Here are the options available:

### Turn Days for Each Ship

| Cranes/ship | Ship 1 | Ship 2 | Ship 3 | Ship 4 | Ship 5 | Ship 6 | Average Days in port |
|:-----------:|:------:|:------:|:------:|:------:|:------:|:------:|:--------------------:|
| 1 | 6 | 6 | 6 | 6 | 6 | 6 | 6 |
| 2 | 3 | 6 | 3 | 6 | 3 | 6 | 4.5 |
| 3 | 2 | 4 | 6 | 2 | 4 | 6 | 4 |
| 4 | 1.5 | 3 | 4.5 | 6 | 3 | 6 | 4 |
| 5 | 1.2 | 2.4 | 3.6 | 4.8 | 6 | 6 | 4 |
| 6 | 1 | 2 | 3 | 4 | 5 | 6 | 3.5 |

Clearly, the best solution for the enterprise is to have six cranes all dedicated to one ship. The same is true for project management. Rather than attempting to have critical resources multitask across multiple projects, it can be shown that dedicating resources to a single project can reduce the project lead time drastically. Consider that one of the competitive factors for a project manufacturer is overall lead time. This could provide a competitive edge.

## MATERIAL ALLOCATION

Material allocation is also more complex in a project manufacturer. Material can be purchased for a specific project and quite possibly could be usable on multiple projects. MRP has an assumption that all materials move into and out of stock. An MRP system that effectively supports the project management must have the functionality that material can be purchased directly to the project and not force the development of a part number and stock transactions. The downside of this approach is when the exact material is desired again, and there is no traceability of it ever having been bought because it went directly to the project. Once again, the implementation approach must weigh the costs and benefits before deciding if and how to implement a part-numbering system for project materials.

In governmental projects, there are requirements to account in great detail the costs paid for the parts and to provide assurance that indeed they were used on that project. This is sometimes referred to as tracking the *color of money*. A company that does business with the American military, the British military, and the Australian military may find itself in a position of having a part purchased for the British that is needed on an Australian project. The transactions and controls required to track the costs and movements of the part could cost more than the part itself. This level of Department of Defense cost accounting easily can lead to high levels of obsolete parts and excess inventory and expensive processes to track them.

Material allocation depends on the BOM. This is the same for standard job shop producers. BOMs must be 100 percent accurate to allow the integrated planning systems to work effectively. This is covered in more detail in Chapter 11. The use of PDM systems helps to keep the engineering BOM (EBOM) in alignment with the manufacturing BOM (MBOM). Where the project manufacturer differs is in the desire or need to track effectivity by serial number. Since the products that are made are typically large capital projects or machines, the effectivity of material usage is desired by the production unit number. Imagine that 12 units will be produced. An engineering change is required on the third unit. The desire would be to have a BOM with effectivity set by serial number, as in the following example:

### Parent 185726 Top Assembly

| Component | In | Out |
| --- | --- | --- |
| 136609 SST, housing plates | 1 | 3 |
| 142058 Titanium, housing plates | 4 | 12 |

Although this looks like a great idea, the material planning functionality plans by date. Somewhere in the MRP system there must be a conversion to a date. Consider that this date can be the expected receipt date, the expected release date, or the expected ship date. Be very sure when purchasing a system that claims to have serial effectivity to fully understand the requirements of making the material planning function work as desired.

Another challenge for the MRP system is dealing with as-required and zero-quantity items. *As-required items* typically are things such as shims, fasteners, lubricants, and hardware. The engineer is resistant to spending time counting up and specifying all the fasteners required for an assembly. This is further complicated by the ability to substitute fasteners of comparable quality, material, and grip length. In the event of a production problem, oversize fasteners also may be used. As anyone knows who has ever tried to get an assembly put together without all the fasteners, the fasteners may look small and inconsequential, but they can stop a major project from shipping. The most prevalent strategy is to standardize on the fasteners that will be used in all assemblies. These components then are kept on an order-point system, where the bins are refilled when they get low. This is an excellent area for vendor-managed inventory. Getting a supplier to take care of the fasteners and negotiating so that the enterprise pays only for what is actually used from the bins can reduce the overall cost dramatically and improve availability of these critical yet low-cost parts.

The same process can be used for "zero per" items. Zero-per items are things such as epoxy, LockTite™, and paint, where the amount used is so small that it does not bear quantifying on the BOM. However, similar to fasteners, these items can stop an assembly in its tracks. The risk of the vendor-managed inventory approach is that future orders may not require what was used in the past. Provided that product mix in the future is similar to that product mix in the past, this approach is very effective. As products change dramatically or require a unique material or configuration, these materials must be controlled through the formal material planning systems.

For consistency, some organizations will attempt to control everything through the formal enterprise resource planning (ERP) system and find out that the transactions required quickly become too cumbersome to manage. Remember to use the right tool for the right job, and the job will be much easier. Understand the process and the desired results before developing and implementing systems to achieve that goal.

## SUMMARY

The project manufacturer is a different and difficult type of enterprise to manage. Its ability to compete effectively in the market rests on its ability to be quicker than the competition to design and build products that are very low in volume and extremely high in variety. One of a kind is not unusual. Project management and project control costing adds a whole level of complexity to the MRP implementation. Projects may span well over a year, and the cost collection must continue at the very top project level. Effectively managing scope, time, and resources, including costs, is essential for the enterprise.

Part of the basic MRP functionality helps this management process, and other parts of the typical MRP system do not support the desired business model. This includes the effectivity on the BOM, scheduling activities, and allocating material to projects. Since the overall project is controlled through a separate project management system, part status synchronization is desired between the project system and the MRP system. This has become possible only recently through the development of open databases and middle-ware linkages. Recent developments in project management such as critical-chain scheduling also have provided new resources and insights into a well-established body of knowledge and toolkit. Since customers increasingly are demanding higher product variety in lower volumes, the pressure is on design and manufacturing to deliver these products more quickly in a profitable way.

As Heraclitus of Greece said in 513 B.C., "There is nothing permanent except change." Effective integration of MRP, PDM, and project management tools allows the project manufacturer to profitably manage in the face on these increasing changes.

## PROJECT MANAGEMENT RESOURCES

Cleland, David I. *Project Management: Strategic Design and Implementation.* New York: McGraw-Hill, 2002.

Goldratt, Eli. *Critical Chain* (APICS No. 03203). Great Barrington MA: North River Press, 1997 (in Spanish).

Kerzner, Harold. *Project Management: A Systems Approach to Planning, Scheduling and Controlling.* New York: Van Nostrand Reinhold, 2003.

Newbold, Robert C. *Project Management in the Fast Lane: Applying the Theory of Constraints.* Hoboken, NJ: St. Lucie Press, 1998.

ProChain Solutions, Inc., 12910 Harbor Drive, Lake Ridge, VA 22192, provides scheduling software supporting critical-chain scheduling; www.prochain.com.

Project Management Institute (PMI), Four Campus Blvd., Newtown Square, PA 19073, establishes project management standards and provides seminars, educational programs, and professional certification that organizations desire for their project leaders; (610) 356-4600; www.pmi.org.

# CHAPTER 17

# Remanufacturing

**R**emanufacturing is an industrial process in which worn-out products are restored to like-new condition. In contrast, a repaired product normally retains its identity, and only parts that have failed or are badly worn are replaced or serviced. Integrated planning and control systems help to coordinate and schedule the difficult job of having the right part in the right place at the right time to support this industry. As difficult as standard manufacturing is, the remanufacturer has a task that is several orders of magnitude more difficult. The first step in the remanufacturer's planning process is to take a product that is no longer usable and, based on past replacement or repair history, have the right components available to return the product to usefulness. This requires very sophisticated planning tools. Not only is there a statistical probability attached to which parts may be replaced, but there are also many different routings that the part could take to be put in "as new" condition. To further complicate the issue, the planning of materials and capacities in the remanufacturing environment greatly depends on the quality and availability of the carcass assets from which the process starts.

Another challenge to the remanufacturer is the simple process of tracking inventory. The product likely will have the same part number before and after the remanufacturing process. This requires some unique identification of part condition to plan the supply and demand accurately. A standard material planning system usually considers the status codes as information only and will combine the rebuilt and core product quantities together. The most common solution is to apply a different part number to the different-status parts so that they can be kept separate in the planning system. See Chapter 6 for material requirements planning (MRP) system requirements. Bills of materials (BOMs) then are used to link the parts usage together.

Another complexity becomes apparent when the components of a particular asset must be tracked directly back to that particular asset and cannot be used interchangeably in other products with the same item identification code. Add to this complex web of inventory tracking the desire to track the costs of the parts back to the parent item, and one can only begin to imagine how to plan and manage this enterprise effectively. Effective integrated systems can help to provide clarity in this web of complexity.

With the advent of environmental responsibility and the ISO14000 standard, it is not surprising to see remanufacturing processes used in an increasingly wide variety of industries. These industries include automotive, electronics, defense, communications, education, electrical, health care, food, furniture, glass, graphic arts, mining, transportation, retail, metal fabrication, pharmaceutical/chemical, plastics/rubber, lumber/paper, textiles, and apparel. As remanufacturing processes and planning become better defined and understood, it is not surprising to find that remanufacturing is good business and good for business. No longer is this operation one that consumes profits from a company; remanufacturing actually can produce profits for a company.

## REMANUFACTURING SIMILARITIES AND DIFFERENCES

Remanufacturing companies have long held the belief that they are very different from the traditional manufacturing enterprise. George W. Plossl, CFPIM addressed the question of just how different remanufacturing and manufacturing are. His conclusion was that there are more similarities than differences:

### Similarities:

1. Both involve suppliers, plants, and customers.
2. Both have two fundamental questions: Are we making enough in total, and are we working on the right items now?
3. Both have the same basic logic guides.
   - What will we make?
   - What resources are required?
   - Which are now available and adequate?
   - Which are on the way and should arrive soon?
   - What else must be procured and when?
4. All activities fall into one of two categories: planning and execution.
5. The same system framework is common to all manufacturing, including aerospace and defense and remanufacturing.

### Differences:

1. Disassembly is required of cores in remanufacturing but not in new product manufacturing.
2. Capacity planning involves less predictable rework, classic capacity requirements planning is not justified, and rough-cut capacity techniques generally are better.
3. MRP programs must handle decimal fractions and negative lead times, showing when components from disassembly will be available for final assembly.

Plossl's conclusion: There are many apparent differences between repair/remanufacturing and manufacturing, but the real differences are few and far outweighed by the similarities. This chapter will concentrate on the differences.

## MANAGING REMANUFACTURING MATERIAL

The starting point for a remanufacturing company is receipt of the core or carcass. This provides the basic working material for the process to begin. The core or carcass is an item that is intended for remanufacturing or repair. The finished remanufactured part may or may not go back to the same customer. If you have ever purchased brakes, a carburetor, or radiator for a car, there is a core charge that is commonly levied when you purchase the part. This cost is refunded when you bring in the nonfunctioning part. This rebate is to motivate the consumer of the rebuilt part to return sufficient raw material for remanufacture. Even with this financial incentive, a significant problem for remanufacturers is to ensure sufficient quantity of high-quality cores that can be remanufactured cost-effectively.

Once the part has been received, the first operation in every remanufacturing company is the carcass *assessment process*. This process is also known as *inspection and evaluation*. A highly skilled person is required to examine the received core and first determine if the part can be repaired or refurbished economically. Parts that cannot be returned to a usable condition economically are scrapped at this point. Some carcasses can no longer be salvaged economically. There is a fine balance between using all material that is returned and accomplishing the process in a manner that can be profitable. Some companies will set aside material that cannot be reclaimed currently until a later time when the part's value may increase as the cores become more difficult to obtain. This assessment process could be as simple as a visual inspection, or it may require disassembly and testing to determine the part's internal condition. Care must be taken in this initial operation not to create a disorganized junkyard where the assessment process is done.

## REMANUFACTURING BILLS OF MATERIALS

Once the part has been assessed as having economic value for salvage, it is time to begin the remanufacturing process. However, parts and capacity are needed to support that process. Similar to how a traditional manufacturing company plans to have the appropriate level of material and capacity available through the use of bills of material (BOMs) and routings, the remanufacturing enterprise develops the disassembly or teardown BOM, followed by a reassembly BOM. Figure 17-1 shows a graphic example of what this complete BOM looks like.

Remanufacturing BOMs typically have the shape of a diamond. A single part is disassembled, component parts are repaired or replaced, and then the parts are reassembled into a single parent part. The ability to plan for this type of disassembly process requires a very special BOM function. Lately, some MRP systems have begun to develop and introduce functionality that specifically meets the needs of the remanufacturer.

**FIGURE 17-1**

Remanufacturing
bill of material.

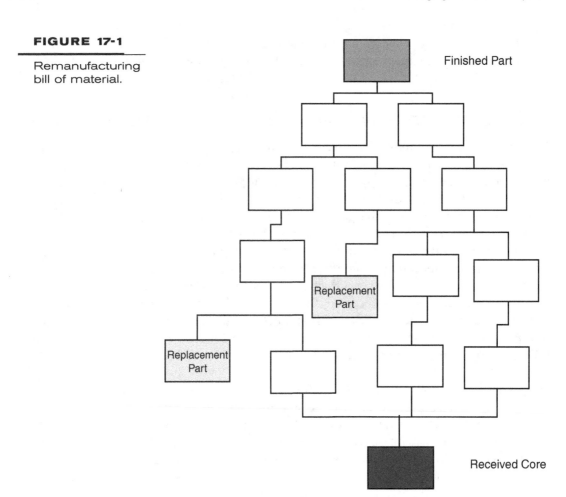

**142095 Clock, disassembly**

| | | |
|---|---|---|
| 142503 Inner works | | −1 each |
| 123291 Hands | | −2 each |
| 123032 Face | | −0.5 each |
| 136958 Clock | | 1 each |

Notice that this BOM contains a negative amount for three of the items. This allows standard material planning systems to be used to control the remanufacturing operation. Since MRP was designed originally to support a one-to-one or many-to-one relationship between components and parents parts, attempting to plan multiple parent parts on a single order is not possible. Having a negative amount in the BOM allows a standard system to handle the fact that one item is made into many in the initial remanufacturing disassembly phase.

The first step in a standard work-order release is to create a work-order number and then issue the materials to that order. In this case, the order would be created for the disassembly clock (part number 142095). A partial issue then is done to remove the 136958 clock from inventory. The 136958 clock is the carcass that has been returned for refurbishment. After physically disassembling this clock, the issue transaction then can be done for the inner works, hands, and face. When a negative quantity is issued, the result is that these parts actually will be placed into inventory as usable parts. The parent part is received as normal. The costs also will be credited to the work order correctly. Labor time is charged to the order just as in any other manufacturing order. Notice that the expected by-products of this clock assembly include

| | |
|---|---|
| 142095 Clock, disassembly | 1 each |
| 142503 Inner works | 1 each |
| 123291 Hands | 2 each |
| 123032 Face | 0.5 each |

You cannot physically have one-half of a clock face. This fractional amount represents the probability that the face will be usable after assembly. In this case, 50 percent of the faces will be usable for the reassembled units. The issue transaction is performed at the end of the process, with the actual amounts realized from the disassembly process. The parent part for this order, the 142095 clock disassembly, usually represents the largest part of the carcass that is tracked through the process. This could be the frame of the radiator, the main airframe of a plane, or the chassis of a computer.

The BOM is used as a guide for the inspector/evaluator in the teardown and evaluation process of an item intended for remanufacturing. Once the process has been performed a number of times, the quantities for the material can be updated. Depending on what comes off the assembly that can be reused, the BOM is modified after the evaluation to create a bill of repair that represents the required scope of work. A repair BOM is a BOM that has been created to define the actual scope of work required to return an item to serviceable condition. This BOM results from the actual examination and evaluation of an item intended for repair and is used for master scheduling and MRP explosion purposes.

For the example given previously, the bill of repair could be

**242095 Clock, refurbished**

| | |
|---|---|
| 242503 Inner works | 1 each |
| 223291 Hands | 2 each |
| 123032 Face | 1 each |
| 142095 Clock | 1 each |
| 239853 Box | 1 each |

Notice that in this repair BOM there is a new part number for the inner works and the hands, but the part number for the face has stayed the same. This is because the inner works and the hands had to be processed through other processes to make them ready for the re-assembly process. This is very difficult to track on a single work order because different parts are going many different ways. One way to handle this problem is to have a mini-BOM to process these parts with the supporting routings. These BOMs and routings then can be used in the standard fashion to take components and process them into parent parts. The amount of processing to bring the face to a usable condition is very small, so the same part number is used to track this item. Also, this bill of repair shows a new part, a gift box, that is required to complete the product for sale.

Recall that for this particular item, the face had a 50 percent chance of being replaced or reused. This is called an *occurrence factor*. Within the repair/remanufacturing environment, some repair operations do not occur 100 percent of the time. The occurrence factor is associated with how often a repair is required to bring the average part to a serviceable condition and is expressed at the operation level in the router. The planning system must have full visibility of the repair BOM, including the occurrence factors, to plan material and capacity effectively. For the example used here, the BOM entered in the planning system would be

**242095 Clock, refurbished**

|  |  |
|---|---|
| 242503 Inner works | 1 each |
| 142503 Inner works | 1 each |
| 223291 Hands | 2 each |
| 123291 Hands | 2 each |
| 123032 Face | 0.5 each |
| 223032 Face | 0.5 each |
| 142095 Clock disassembly | 1 each |
| 142503 Inner works | −1 each |
| 123291 Hands | −2 each |
| 123032 Face | −0.5 each |
| 136958 Clock | 1 each |
| 239853 Gift box | 1 each |

This can quickly become very confusing. "A picture is worth a thousand words," so Figure 17-2 is the same BOM represented in a graphic form.

In this representation of the BOM, it is much clearer that the new face (223032) and the gift box (239853) are parts that are purchased from the outside and used in the assembly. The core clock that is returned (136958) is expected to provide the chassis (142095), the hands (123291), the inner works (142503), and the face (123032) required for the reassembly process. However, this disassembly process yields the face only half the time. The other half of the time a new face (223032) is required to complete the finished product. The inner works and the hands are expected to require additional processing after disassembly before they can be reassembled.

**FIGURE 17-2**

Remanufacturing
bill of material.

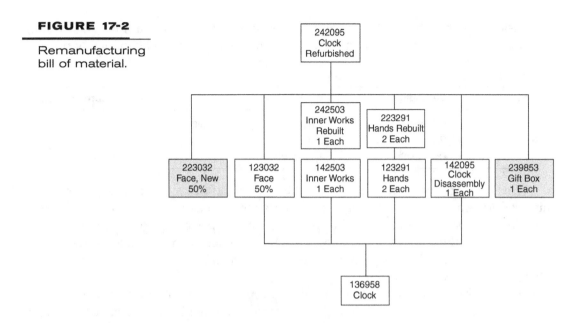

## REMANUFACTURING ROUTINGS

Similar to a standard material planning requirement, routings must be prepared that describe the process by which the components are transformed into the parents. In this example, routings would be required for:

242095 Clock, refurbished

242503 Inner works

223291 Hands

142095 Clock disassembly

Notice that no routing is required for the face (123032) because this is a by-product of the disassembly process of the 136958 clock. These routings may have probabilities attached to them, similar to the occurrence factors in the BOM. A standard material planning system requires that the probabilities be reflected in the routing input to the system. A system that has been built specifically for the remanufacturing industry allows the entry of occurrence factors separately.

The routing below could represent the planning routing for the processing of the hands.

**223291 Hands**

| Step | Description | Work center | Hours |
|------|-------------|-------------|-------|
| 10 | Cleaning | Plating | 2.0 |
| 20 | Polish | Buffing | 3.5 |
| 30 | Reassembly | Assembly | 1.0 |

Assuming that this routing is accurate, it still could be misleading. This routing actually could be the planning for a product that really is requiring the following process:

**223291 Hands**

| Step | Description | Work center | Hours | Occurrence factor |
|------|-------------|-------------|-------|-------------------|
| 10   | Cleaning    | Plating     | 8.0   | 25%               |
| 20   | Polish      | Buffing     | 7.0   | 50%               |
| 30   | Reassembly  | Assembly    | 1.0   | 100%              |

This routing more accurately shows the factors and times that are used to plan for this product. Another level of complexity occurs when sometimes the parts can be used as is, and they only need to follow this process a certain percentage of the time. This factor is known as the *repair factor*. The repair factor is sometimes referred to as the *frequency of repair*. This defines the percentage of time an average item must be repaired for return to a serviceable condition.

The other option is to use the part as is or replace it with a new part. The example in this chapter used the face of the clock as a demonstration. Planning to have these parts available requires definition of the *replacement factor*. The replacement factor defines the percentage of time an average item will require replacement. In a remanufacturing industry, not only do the times and quantities need to be correct, but the occurrence, repair, and replacement factors also must be maintained to provide accurate inputs for planning. Even with all these factors and planning, sometimes unexpected things happen, and a part needs to be replaced or repaired that wasn't planned. This is called *unplanned repair*. Unplanned repair is unknown until remanufacturing teardown and inspection.

## REMANUFACTURING INVENTORY MANAGEMENT

Not surprisingly, inventory management in a remanufacturing company is quite similar to that in any other manufacturing company. However, it does have its unique twists. When receiving a carcass for repair or remanufacture, the part number on the product is the same number as that on a good unit. Some systems will have status codes as an active field in the database such that it becomes part of a unique identifier and is traceable through the system. For example:

**136604 Clock**

| A code: | 145 |
|---------|-----|
| F code: | 53  |
| G code: | 97  |

This could represent a single part with three different usability statuses. The A-code parts may be the parts that are available for customer orders. F-code parts could be those which have been evaluated as repairable but are waiting for parts. G-code parts may be carcasses that have not been examined yet. Each business will have different codes for different inventory states, but the advantage of this system is that the same parent part number is used throughout the process, and only the status code changes. However, beware that some systems will have grades or status codes that are informational only and do not become part of a unique identifier. In this case, the MRP system would see a total of 295 pieces of 136604 clocks in inventory from the preceding example with no distinction for usable grade.

Traceability is another requirement to many remanufacturers. Owing to product liability and configuration control, the enterprise may wish to link the parts that come off an assembly directly back to the original assembly. As the disassembly parts move through the refurbishment process, the intended destination product may be part of the record. This becomes very difficult to track and is at best transaction-intensive. The end item to which the part is destined must become a significant part of the item identifier. The impact is that each part in the shop must be uniquely identified. The system required to support this is very complex.

The other risk is that the part that is intended for use on the end item somehow falls behind in its processing owing to quality or material problems and a part for a later unit comes to the assembly line. Normal behavior would be to use the available part and document it on the assembly paperwork. This now leaves another unit missing its legitimate component, and the process can get out of hand quickly. This process is known as *back-robbing*. This also occurs when a part is taken directly from another unit that cannot be completed and is used on a unit that is close to completion and is missing that part. When backrobbing starts, it is terribly complex to track. In addition, it is virtually impossible to do so accurately. The short-term benefit of using the part is far outweighed by the system complexity necessary to track it. The probability of getting the parts' traceability to balance out in the end is close to zero.

In other ways, managing the inventory assets in a remanufacturing enterprise is quite similar to that for other manufacturing companies. Sufficient investment in inventory assets is needed to be able to support the ongoing business. However, too much inventory and cash is wasted. Too little inventory and the production operation cannot complete its work.

In addition to the inventory strategy choices described in Chapter 4, the remanufacturing enterprise also must consider at what threshold replacement factor should the inventory investment be funded. If a replacement factor is less than 30 percent, it is very likely that too much cash may be tied up in slow-moving inventory. At lower replacement factors, the material planning system may require only a fraction of a part based on the scheduled volumes. However, the smallest quantity that any part can be purchased in is one piece. This one piece may never be used and may remain forever as obsolete inventory. Another strategy for these parts is to wait until the part is required and then order. Clearly the tradeoff is response time. If the part has a sufficiently low replacement factor, this process still may be capable of providing acceptable customer service. As in all decisions, the consequences must be weighed against the possible alternatives and the best overall decision made for the enterprise.

# TERMS RELATED TO THE REMANUFACTURING INDUSTRY[1]

**carcass:** A nonserviceable item obtained from a customer that is intended for use in remanufacturing.

**disassembly or teardown bill of material:** The bill of material (BOM) used as a guide for the inspector/evaluator in the teardown and evaluation process of an item intended for remanufacturing. This bill is subsequently modified after evaluation to create a bill of repair that represents the required scope of work.

**MRO:** Maintenance, repair, and overhaul (an alternative definition for this acronym used in the remanufacturing industry).

**occurrence factor:** Within the repair/remanufacturing environment, some repair operations do not occur 100 percent of the time. The occurrence factor is associated with how often a repair is required to bring the average part to a serviceable condition and is expressed at the operation level in the router.

**remanufactured parts:** Components or assemblies that are refurbished or rebuilt to perform the original function. Syn: refurbished goods, refurbished parts.

**remanufacturing:** An industrial process in which worn-out products are restored to like-new condition. In contrast, a repaired product normally retains its identity, and only those parts that have failed or are badly worn are replaced or serviced. In general, the remanufacturing environment is where worn-out products are restored to like-new condition.

**remanufacturing resource planning (RMRP):** A manufacturing resource planning (MRP II) application in the remanufacturing sector.

**repair bill of material:** A BOM that has been created to define the actual scope of work required to return an item to serviceable condition. This bill results from the actual examination and evaluation of an item intended for repair and is used for master scheduling and MRP explosion purposes.

**repair factor:** The repair factor, sometimes referred to as the *frequency of repair*, defines the percentage of time an average item must be repaired for return to a serviceable condition. This factor is also expressed as a percentage applied to the quantity per assembly on the bill of material assembly/component relationship. It is used for forecasting material and required capacity in advance of carcass receipt.

**replacement factor:** The replacement factor defines the percentage of time an average item will require replacement. This factor is expressed as a percentage applied to the quantity per assembly on the bill of material assembly/component relationship. It is used for forecasting material and required capacity in advance of carcass receipt.

**unplanned repair:** Repair and replacement data that are unknown until remanufacturing teardown and inspection.

---

[1] *APICS Dictionary*, 12th ed. (New York: Blackstone, 2008).

## CHAPTER 18

# Process Industry Application

**P**rocess-flow industries comprise about half of manufacturers worldwide, with the proportions much higher in Australia, New Zealand, and South Africa. The process industry has always been a challenge and a poor fit for traditional material requirements planning (MRP) systems. In the first edition of this book, Joe Orlicky posed the question of whether MRP would ever be used in process industry.

Process industries are typically highly automated plants with a large capital investment. Examples of process industries include food processors, refining, pulp and paper mills, beverage, primary metals mills, and plastics and chemical manufacturers. To realize the best return on investment and the lowest product costs, these plants generally run 24 hours a day, 7 days a week. The changeover of the line from one product to another typically is quite expensive. The whole plant is usually down during the changeover. The entire production work force is idle, and the expensive capital assets are not producing revenue. Costs, however, continue. For this reason, the main focus for any enterprise planning system in the process industry is effective capacity management, including product sequencing and optimization of orders through the plant, rather than material planning. The two main tools are called *process-flow scheduling* (PFS) and *advanced planning and scheduling* (APS).

## PROCESS INDUSTRY OVERVIEW

Process-flow scheduling provides the highest utilization level possible in the plant by sequencing changeovers and scheduling by-products and coproducts to minimize downtime. This scheduling process is also known as *block* or *campaign scheduling*. Once the capacity is planned, then the specific production output is confirmed against the order book and optimized for profitability given the capacity constraint—exactly the opposite of the process for discrete manufacturing. The traditional manufacturing business model and the basis on which MRP was developed is to first plan the priorities of the enterprise and then to confirm the capacity availability.

Figure 18-1 shows the difference between the bills of material (BOMs) from a traditional MRP plant and those from the process industry. The traditional MRP scheduling logic is used in the "A" plant, whereas the process plant is exactly the opposite, or a "V" plant.

The MRP logic would have a difficult time managing all the by-products and co-products if all process plants were only "V" type plants. By-products are materials of value that are produced as a residue of the production process. For example, in an injection-molding operation, the material used to hold the part in the die is called a *gate*. These gates are removed as the part is removed from the injection mold and collected. Customarily, this material then is reground and stocked as regrind. The regrind is blended with the virgin material to mold more parts. The percentage of regrind that can be used optimally by part and by specific component material will change. However, the ability to use regrind dramatically reduces the cost of manufacturing these parts.

Co-products are products that are typically manufactured together or sequentially because of product or process similarities. This could include grouping parts of many different sizes and shapes to be cut from one piece of material to optimize material use. This is called *nesting* in discrete production. In corrugated-box plants, different customer orders may be run as co-products to maximize use of the corrugated web. In process plants, such as chemical and refining plants, operations are the result of very complex stoichiometric models. These models use process conditions or are controlled by raw material quality,

**FIGURE 18-1**

Standard MRP versus PFS bills of material.

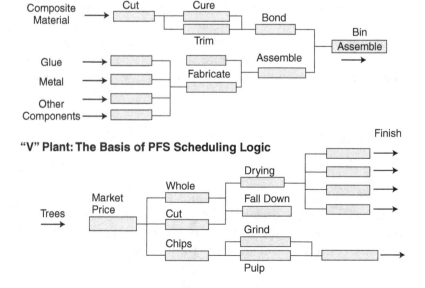

temperature, pressure, and flow to determine how production should be scheduled. This provides an additional level of complexity to the material planning process.

The process plant is rarely a single "V" type process. It is not unusual to find a blend of manufacturing types inside a process plant. The initial processing of the main product may occur in a continuous flow, but it is very common to find that final packaging is done in lots or batches. Food processing is a good example. The product may be produced continuously but then must be packaged into a finite number of different package sizes. The final BOM resembles a "T."

The scheduling interface between the long, continuous runs of product must be integrated with the final product configuration so that the product meets all manufacturing requirements. This can include how long the work-in-process can be held in bulk before packaging. The forecast market demands must be met and the best profit achieved. This schedule is further detailed into the individual retail pack. The retail pack then usually has some kind of shipping or case overpack for sale to the distributor. These distribution packs can be simple, where one product is put in each case, or a complex pack. A complex pack is a mix of products or an assortment put into one case pack. These distribution packs then must be consolidated onto a pallet for shipping purposes. These pallets are tracked through the system, but visibility to the individual retail pack must be maintained. Having appropriate scheduling tools is essential in this environment so that the very best schedule can be developed.

In addition to this type of bulk-to-pack manufacturing processes, process industries use different quality measures. The discrete plant typically inspects parts with a binary result. The part is either good or not. Good parts are used, and bad parts are scrapped or reworked. A process plant can test a batch or a part to determine its potency or grade. The result is not as simple as good or bad. The same production costs may be incurred to produce a batch of product with a low- or high-grade result. The low grade only can be sold for a low price—if it can be sold at all. The high grade demands a premium in the market. The high grade also may be affected by storage time. If the product is stored too long, the grade may deteriorate into a lower-grade salable product or may deteriorate into unsalable product. Each run will produce many different products, and the end result is not known until the run is complete and the testing accomplished.

In refining, during the summer season, gasoline is in higher demand and commands a higher price than other products. In the winter season, heating oil has the priority. The profitability of the operation depends on being able to shift the production schedule between these two seasons, taking into consideration market demand, the quality of the crude being received, the physical constraints of the process equipment, major maintenance, overhaul requirements, season of the year owing to regulations regarding pollution, and storage capacity for building inventory in the chosen market location(s). Traditional manufacturing only has to worry about where a part was placed in storage so that it can be retrieved. Many process plants can go back to the same location, and the product has changed by virtue of sitting on the shelf. This is yet another complication in scheduling the process plant.

The most common return on investment for an MRP implementation in a discrete manufacturing plant is through the effective planning of component materials, as measured by inventory reduction. The process industry is unique in that it generally uses few raw materials. Raw materials frequently are purchased in very large bulk quantities such as rail cars or ships. These materials commonly are purchased at a commodity price that can fluctuate widely in the market. Detailed material planning by exploding requirements through BOMs does not provide the same benefit for the process industry. The detailed material planning capability of the discrete MRP system must be replaced with a process flow scheduling approach.

Adding to the difficulty caused by the conversion process, the inherent characteristics of the process-flow business cause challenges for traditional MRP implementation. Process-plant businesses are continuously shuffling brands and product lines. These businesses are constantly selling and acquiring lines of product to align with the desired strategic plan. The MRP system must be sufficiently flexible to adapt to these changes after the initial implementation. This can be a very different approach than some implementations take, where the configuration of the system is locked down and difficult to change once the system has been configured. Providing the ability to reconfigure quickly based on business changes, mergers, and acquisitions is essential for overall success of the process business.

Another area that is unique to the process industry is the inherent nature of supply-chain management. Owing to the purchase and sale of plants, one week a supplier could be a feeder plant, and then quickly this plant could be a sister plant owned by the same corporate entity. A process business requires a well-designed business model to understand the impact and consequences of customer and supplier locations so that a holistic solution can be developed that benefits the entire supply chain. Since the total operating cost of the plant equipment is relatively fixed, overall profitability must be ensured by best using capacity and the people operating the plant with minimal overhead functions. In some process industries, tolling and exchange contracts with complementary or competitive suppliers also are used to balance supply and demand for multisite operations that may have a significant geographic distribution.

## PROCESS-FLOW SCHEDULING

The process industry is characterized by having relatively few raw materials that can explode into a variety of end products, coproducts, and by-products. A proven method for scheduling this type of output has been developing over the past 25 years under the guidance of Dr. Sam Taylor and Dr. Steve Bolander. This type of scheduling is known as *process-flow scheduling* (PFS). This book is not intended to be a complete discussion of this scheduling technique. We intend just to introduce the concept. This chapter attempts to create an awareness of these tools and their differences from the MRP body of knowledge already presented. The close integration of the internal schedules, the reliability of constrained capacity usage, and the impact of outside events have positioned the process industry as the leader in advanced planning and scheduling (APS) systems.

APS engines include the business application of simulation, heuristics (best-of-business rules), linear programming (LP), and constraint-controlled intuitive modeling such as fuzzy logic. These sophisticated mathematical modeling tools embrace all aspects of the business that can be affected by a supply-chain management implementation. Integrating demand information from the customer's customer to the supplier's supplier is enhanced by also incorporating additional dimensions of the business such as the revenue chain. This allows more comprehensive "what if" scenario planning in evaluating new market potentials, corporate takeover return-on-investment analysis, price-sensitivity analysis, and logistics configuration analysis.

The first process flow scheduling tools were implemented in the late 1980s and focused on the consumer goods industry. The calculation models were rather simplistic and attempted to provide the single "best" answer based on the model input. As computers have become more sophisticated, the models representing the business have become more sophisticated as well. It is now possible to create almost "virtual reality" for the business to evaluate alternate plans. As in running a real business, a single, large model is insufficient to represent all the integrated functions within an enterprise. Effective PFS systems have the ability to define multiple models to best describe the business.

The three main approaches for solving the process industry scheduling problem include simulation, heuristics, and optimization. Simulation attempts to represent in a computer the interrelationships of a business. An effective simulator directly relates to the business and allows the iteration of many different decisions to determine the impact on the enterprise. The simulator can be as simple as a spreadsheet that provides "what if" capability for different production schedules. The openness of current technology supporting MRP systems provides the opportunity to download information from the main system into spreadsheets for seamless manipulation in "what if" analysis. The simulator can handle important tradeoffs analytically and clearly identify the impact of certain decisions. This quantifiable analysis allows decisions to be made on the basis of fact, and fewer decisions are based on intuition. At times, the amount of data becomes overwhelming, causing difficulty in building a mental picture of a particular problem or solution.

The simulator can be used to develop a more sophisticated model to reflect a particular enterprise. These simulators can provide an excellent teaching tool or can help in managing a real business by allowing management to see the consequences of decisions before implementation. With the continuing growth of computer processing power and the decline in computer processing costs, simulations are sure to become more widely used in day-to-day operations.

Heuristics are simplifying rules or rules of thumb that are used to develop a feasible schedule. These rules are based on intuition or experience instead of mathematical optimization. These rules may be required because the simulation and optimization may not be able to provide a feasible solution without them. Another use for heuristic rules is to develop an initial solution from which improvements can be made. An example may be that production cannot be increased or decreased more than 10,000 units for each major schedule change. Another rule may be that major schedule changes can be incorporated only once a quarter. Reasons exist for these business rules, but they cannot be quantified sufficiently for incorporation into the simulator. Remember that the computer tools cannot take over for good management of the company. Materials, capacity, and other resources have been managed well in the past. The computer tools should not replace effective management but rather should augment it.

Optimization attempts to calculate the best solution given the bottleneck to achieve the desired results. The focus of optimization could include most profits, shortest total lead time, best customer service for a preferred customer, smallest total changeover time, or making whatever measure that is selected the best that it can be. For optimization to be effective, a system must be defined for which the demand exceeds the possible supply so that a constraint can be determined. Optimization then provides the best possible solution to the problem in terms of this specified objective function. In the process industry, with its dependent setups and variant production batch sizes, optimization modelers enable the scheduler to consider a variety of inputs when developing the schedule.

## MRP SYSTEM REQUIREMENTS

This section lists the typical requirements and a brief explanation for an effective MRP system in a process industry. These requirements are not listed in any specific order of importance.

*Supply-chain management.* Collaborative forecasting and planning are a real must for the process industry. Having visibility of customer inventory and channel sell-out data patterns in addition to the traditional channel sell-in information can provide proactive requirements information. Promotion, sale price, and competitive impact are integral in this supply-chain solution. Hard and soft allocation should be allowed in the system. Soft allocation promises product based on overall volumes. The reality of execution is that the process for filling orders is first come, first served. A preferred customer may find itself without any product. Hard allocation makes an assignment of inventory to a particular customer order. The hard allocation process allows preferred or more profitable customers first availability of the product rather than first come, first served. This hard allocation also may be supported by contract agreements and pricing. Tolling and exchanges also can be used for balancing supply and demand within the supply chain. These data are also necessary to determine where best to place inventory, as discussed in Chapter 4.

*Multiple plants, warehouses, and branches.* Most process industries must control inventory at a number of different physical locations, including multiple factory and packaging locations, distribution warehouses, and possibly branch retail locations. Some of the process-industry products are in liquid form and therefore are distributed via public or private pipelines. The control parameters for this mode is very different from standard MRP discrete distribution approaches.

***Multiple units of measure for each item.*** The units of measure can be very complex for a process industry. A piece may be tracked in inventory as an "each," but the part may be sold by the pound. The system must have the capability to recognize the part as both units of measure and convert between these units of measure automatically. The unit-of-measure conversion also should include tracking retail packs to case packs to pallet packs. These multiple units of measure also require very sophisticated and complex pricing models to support the business as well. The same product may have different prices depending on the location to which it will be shipped. Volume and customer discounts are also normal. The product can change composition during transit owing to external influences, such as pressure or temperature, or internal influences, such as product mixing in common pipelines or storage tanks. The planning system must be able to take these variations into account.

***Formula management.*** Similar to how a cookbook is written, a different formula is required for different production batches in a process plant. Unlike the discrete business, these formulas are different from the BOM in that if a double batch is required, each component in the formula may not be doubled. Any good cook can relate failed attempts at doubling a recipe. A big issue in the process industry is how exactly a recipe is scaled to a desired batch size. These different recipes should be able to be tracked in the enterprise resource planning (ERP) system for different batch sizes. Also, component lots need to be tracked to meet government regulations for processing, particularly in food and pharmaceutical operations.

***Quality tests and specifications.*** Since quality is not a binary function for the process industry, the quality tests and specification should be traceable in a computerized system. This is frequently an external execution system. The tolerances for specification also should be integrated into this process so that automatic grading is possible. Variances on shop orders are considered as a statistical process control chart over time rather than on each order, as in discrete manufacturing.

Orders then can be allocated automatically based on grading. Assume that grade A is superior to grade B and grade B is superior to grade C. If a customer orders grade C and no grade C inventory is in stock, the system should follow predefined business rules for allocating available B grade material or A grade material to the order. This type of functionality is very unique and normally is handled through a suffix at the end of the part number to denote grade. Sophisticated substitution logic then is required to handle this process.

***Pack bill of materials.*** Another unique BOM function is the pack BOM. This defines which items are packed and shipped together to the customer. This can include packing multiple items into a single package for sale, such as an assortment of the same product, or packing complementary products together for sale as a complete unit. One example would be to pack a television, VCR, and stereo speakers together for shipment as an individual entertainment unit. Typically, this configure-to-order strategy is completed at the distribution center.

*Flexible planning solutions.* The planning solution must be able to incorporate and manipulate pack quantities. This includes substitution of component items that are not available with component items of equal or higher value. Defining these business rules can take an extensive amount of time during the implementation but can ensure that customer demand will be filled most expeditiously with little manual effort. Part of this planning process also should include the recognition of shelf life. If a part will expire on the first of next month, the system should recognize that the lot will not be available for an order that is scheduled to start or ship on the fifteenth. Any established priorities, such as customer or profitable products, also should be recognized in the planning process. Multiple variant allocations also should be supported by the planning system. Examples could include product grade, colors, or size.

*Lot tracking and product genealogy.* This tracking includes knowing which lots of raw material were used to manufacture the finished goods. The ultimate destination for each lot also must be traced in the event a recall is required. All lots shipped to a certain customer should be easily identifiable in addition to all customers who are shipped a certain lot. This traceability is essential for risk management and is a governmental requirement in industries such as pharmaceuticals, medical devices, and food processing.

*Process costing.* The costing system should be able to accurately allocate costs across multiple parents on the same work order based on established business rules. These parent parts could include co-products, by-products, and grade variants. Fall-down in grade targets also must be handled. Fall-downs usually are sold at a significantly reduced price or at an almost nominal cost. The cost to produce the fall-down product is exactly the same as that for first-run product. The variance between the production cost and the product's value must be handled within the costing system as a process run chart rather than discrete shop order variances.

Large process plants are required to operate at or near capacity for 24 hours a day, with preference being for only planned maintenance outages. The budget for maintenance typically is 20 to 30 percent of the total operating budget or higher depending on larger process-unit turnarounds. The requirement for robust asset management that considers planned, emergency, preventive, predictive, and turnaround or outage maintenance requirements is a priority in the planning system. Key areas that must be integrated include material procurement, stores inventory, and maintenance systems.

## SUMMARY

The process industry is very different from the traditional job shop. High-volume products made in relatively few varieties characterize this industry. Capacity is the critical constraint in the process industry and is planned first in the planning cycle. The first MRP systems were developed for the job shop to manage the complex flow of raw materials through variable capacity. The priority plan initially assumes that there is infinite capacity. Capacity planning is used to validate the priority plan in the discrete enterprise.

The process industry accomplishes its planning exactly in reverse. Capacity is planned first, and then the priority plan is determined. Materials are ordered based on the availability of capacity rather than the other way around. Therefore, it is not surprising that these traditional systems do not fit the process industry well and Orlicky posed his question in the first edition of this book. Unique scheduling tools are used for process industries, including process-flow scheduling. The MRP system must have these unique functions to allow this integrated system to best fit the process enterprise. Using the right tools for the right job makes the job much easier. Attempting to use the same tools to solve every problem is asking for disaster. Successful MRP implementation in the process industry is the ability to know the difference.

## CHAPTER 19

# Repetitive Manufacturing Application

**U**nlike the traditional process of a master schedule netting gross to net requirements through to a detailed material plan, the repetitive manufacturer needs to perform rate-based material planning supported by rate-based parts delivery schedules. In some companies, this rate-based schedule must be tied to a mixed-model-sequenced line for items going to finished goods or may require traceability to a customer order. Processing individual work orders either by the practitioner or under the software covers by the computer does not add value. Rather than detailed job-order costing, the variance in expected versus actual output rates over time is a key performance measure. A key measure is linearity, which expects that each day has a consistent output. Some companies will measure each hour of output. This linearity of output provides key insight into the overall effectiveness of the process. Product costing changes to a four-wall-period approach rather than performing detailed issue transactions to a work order. This process control approach is quite similar to the process industry, as discussed in Chapter 18.

## GENERAL REPETITIVE APPLICATION

A repetitive manufacturer makes high-volume products in low variety. Most commonly, this type of manufacturer competes in the market based on price and/or lead-time response. The manufacturing strategy used to meet the market is usually make to stock, configure to order, or assemble to order. The ability to promise delivery to the customer accurately is very important for the repetitive manufacturer. Less important is the ability to track costs to a specific production unit. Costs are considered over a period of time rather than for a particular unit, similar to process industry. Bills of material (BOMs) have relatively few levels, and routings are fixed and reliable. This is a very different environment from the traditional discrete job shop for which material requirements planning (MRP) was developed. Discrete job shops have a wide variety of potential routings, capacity planning is a major challenge, and costing is accomplished by job. Clearly, these two different environments require different tools.

In a repetitive manufacturing operation, the conversion process is accomplished through a very predictable series of sequential operations. Work-in-process is relatively low, and these sequential operations are highly dependent on each other. Forcing repetitive manufacturing approaches into job shop–oriented computer systems is possible, but one barrier to a successful fit is the amount of paperwork and transactions. Since the lot sizes are so small, the amount of paperwork is quite large because a unique order is still expected for each production lot that is released in a discrete manufacturing system. If the same process of paperwork and transactions is used to build repetitive product as discrete product in a job shop, the production work force is soon buried under a mountain of paper.

A look at the economic order quantity (EOQ) formula in Chapter 8 can provide insight here. Although this formula is very old, and many consider it to be obsolete, it helps to make the point of why different management processes are needed to support a repetitive operation. Even though this formula may not be used explicitly to calculate the lot size in a repetitive manufacturer, an effective management process does this kind of analysis to determine optimal lot size. A repetitive manufacturer wants to have very small lot sizes with the eventual goal of producing in lot sizes of one as close as possible to market demand so that the manufacturer can react as the market reacts. This is different from the commodity manufacturing company at the far right of the volume/variety matrix. This formula also can be analyzed in reverse to identify factors that must be changed to allow these small lot sizes to be achieved in a cost-effective manner.

Changes in lot sizes should have no significant impact on annual usage. The total demand does not depend on the production lot size. Given that the inventory carrying cost typically is a fixed overhead cost divided by production volume, this factor also should not be affected by a change in lot size. The same is true for the cost of the item. The case could be made that one of the expected benefits of reducing lot size is that quality should improve because there is less time between the production and use of material. Quality improvement eventually should benefit product cost. In the short term, a decision to reduce lot size will not have a major impact on product cost. Therefore, the only factor that can be adjusted to reduce the cost impact of the decision to produce in smaller lot sizes is the cost of setup. A significant part of the setup cost is the fixed time and expense required to issue work orders, transact parts, close paperwork, and address all the other routine expectations found in a job shop. This cost does not depend on the number of parts on each work order but rather is related directly to the number of orders processed. The repetitive manufacturer has no time for all this non-value-added activity for each individual part in lot sizes of one. The process must be reengineered to provide the required information without the non-value-added cost.

Since product is built in high volume through a repetitive process, there is no requirement for a detailed level of progress reporting because this feedback does not add value, only cost. However, in a repetitive manufacturer, the manufacturing process can be thought of as a river that flows continuously at a relatively stable level. When the flow of inputs is balanced with the flow of outputs, the overall level in the river stays constant. The water level in the river reflects the lead time through the plant and the inventory in the operation. Since the lead time for a repetitive manufacturer is so short, the work-in-process is very low. Critical feedback processes for the repetitive enterprise include the ability to report production-rate variances in addition to the expected cost variances.

The repetitive sequential operations are directly related to each other, so if one operation stops, the balance of the line soon stops. Piles of inventory are not allowed to build up between operations. This stable level of work-in-process simplifies the control and reporting systems. True repetitive systems have rate-based production scheduling and backflushing capability. This backflushing can occur at a pay point partially through the routing or simply at the end of the process. The system does not create work orders in the background. When a job shop system attempts to masquerade as a repetitive system, this background creation of work orders is common practice. The creation and processing of work orders in the background can consume a great deal of computer power and processing time.

High volumes and very low variety characterize the repetitive product. The raw materials needed are repetitious in quantity and timing. The end items are still discrete but are produced in a very short cycle time, typically less than one day. Management of this type of operation turns to balancing the capacity along the line rather than planning detailed routings. The BOMs tend to be very flat. This means that they have very few levels. Routings are simple, with only one or two steps, because the operations are closely coupled with each other. The entire line is either running or not. The option of having some work centers working while others are idle is not common in this environment. Costs for this type of operation are easy to allocate directly to these focused lines.

The *four-wall approach* is used to track inventory. Receipts are transacted when the product arrives, and deductions are made to the inventory when the final product leaves. Intermediate tracking does not add value to the process, only cost, so such tracking is not done. The traditional intermediate tracking is no longer required because the materials are in process only for a very short period of time.

## KANBAN

Just-in-time (JIT) execution tools are used to bring materials to the line very close to the time of need, as in kanban. In actuality, *kanban* is a Japanese word that literally means "sign board." According to the American Production and Inventory Control Society (APICS), a kanban is a "method of just-in-time production that uses standard containers or lot sizes with a single card attached to each. It is a pull system in which work centers signal with a card that they wish to withdraw parts from feeding operations or suppliers. This is also known as a *move card*, *production card*, or *synchronized production*." Figure 19-1 shows the actual Japanese kanji characters for a kanban system.

A kanban can be a light that signals replenishment, a card that is moved, an empty container that is sent to the supplier to be filled when the active one is empty, or even a

**FIGURE 19-1**

Kanji characters
for a kanban
system.
(Courtesy of
Toshiyuki Okai.)

KAN BAN SI SU TE MU

看板システム

KANBAN System

fax or e-mail message that authorizes movement of material. The supplier then quickly refills the container and sends it back to the line directly to the point of use. This replenishment can happen many times a day to bring parts to the production area as needed. The primary factors for a kanban system are lead time, item cost, and consumption rate. In addition, user variables include the desired frequency to receive the material and the desired level of certainty that a particular item will be on the shelf at any given time. Kanban is really very similar to the order-point system that had been in use for many years prior to the development of MRP. The important thing to remember when sizing a kanban is to consider the potential demand and supply variability for the part and the total cost to the enterprise.

The order-point system sends out a signal for parts when the inventory has fallen to the level defined as the order point. The expected lead time to replenish parts should be equal to the inventory of parts left. If the lead time were two weeks, then the order point would be set at approximately two weeks' worth of parts. The delivery is expected to arrive just as the parts run out. The main difference between kanban and order point is the time elapsed between signal and replenishment. The order-point signal may take days or weeks to replenish. The kanban typically is replenished within minutes or hours. The kanban process works extremely well when the volume and variety stay constant over a period of time. Provided that the future looks like the immediate past, a supplier, either internal or external, experiences stable demand.

The approach works well when the demand for the parts is relatively stable. This stable demand provides a predictable run-out for the inventory. The kanban average inventory is significantly less than that of the order-point system because the replenishment lead time is also significantly less. Kanbans typically can be replenished several times in the same day. The small amount of inventory stored in the kanban also translates to relatively stable demand to the supplier.

Given the very quick replenishment, the demand to the supplier in a kanban environment looks stable. This is very different from the spikes in demand found in an order-point process. See Chapter 4 for a more complete discussion of order point.

The scheduling task for the kanban supplier is relatively easy, provided that no configuration changes or wide variability in volume occurs. Theoretically, this may look very good, but the real world is rarely so easy to manage. Variability and volatility are increasing at an increasing rate. The risk in using kanban only, without MRP to plan materials, is what happens when a change in configuration or quantity is required. Since the kanban is sent to the internal or external supplier only minutes or hours ahead of requirement, the supplier may be unable to respond in time, and the whole line can be disrupted or stopped until the supplier is able to provide the needed part. There is insufficient response time for the supplier to react to a requirement that blindsides him or her. If a supplier has been delivering blue parts at a constant rate and then, all of a sudden, a kanban shows up with a demand for a striped part, the chance that the striped part will be available is very small without having some forewarning that the demand is coming. This is why an MRP system still has an integral place in a repetitive manufacturing business. MRP can very effectively plan to the day exactly what is required and when. Practically speaking, most facilities find a daily schedule to be sufficient to support production. Rarely can an enterprise schedule and react to the hour or exact time. The kanban then can become the execution tool of less than one day's duration with the exact timing of the replenishment when production is ready.

## RATE-BASED SCHEDULING

Rate-based scheduling occurs when a production schedule is entered directly into the system as a date range with volumes. The following could be an example of a rate-based schedule:

| Product 135948 | Production per Day |
|---|---|
| June 10–August 15 | 12,000 |
| August 16–September 22 | 15,000 |
| September 23–October 31 | 20,000 |
| November 1–November 30 | 30,000 |
| December 1–March 30 | 20,000 |
| March 31–June 9 | 10,000 |

The rate-based schedule also can be entered as rate per shift if that level of control is desired. This higher level of detail is required to plan the staffing for each shift. The increased visibility prevents surprises and allows shop supervisors to adjust the affected resources to match the desired production rate.

**Production Schedule**

| Product 135948 | Shift 1 | Shift 2 | Shift 3 |
|---|---|---|---|
| June 10–August 15 | 6,000 | 3,000 | 3,000 |
| August 16–September 22 | 10,000 | 3,000 | 2,000 |
| September 23–October 31 | 10,000 | 5,000 | 5,000 |
| November 1–November 30 | 10,000 | 10,000 | 10,000 |
| December 1–March 30 | 10,000 | 10,000 | 0 |
| March 31–June 9 | 10,000 | 0 | 0 |

This example provides the necessary visibility required to identify when additional shifts are required. Since there is usually a premium paid to the off-shifts, the strategy that is usually followed is to maximize production on first shift before adding additional shifts.

No work orders are created to track these requirements. A true repetitive system will have the function to input the expected schedule in a way very similar to the preceding chart. This makes schedule changes and "what if" analysis much simpler to accomplish.

## PRODUCTION-SALES-INVENTORY ANALYSIS

In a repetitive manufacturing company, the product usually is built to stock. Customers expect their orders to be fulfilled very quickly from a finite number of possible configurations. Forecasting is used to project future demand so that production rates can be scheduled. The production rate is compared with the projected sales rate to determine the impact on inventory. If the production rate exceeds the sales rate, then inventory will rise. If the sales rate exceeds the production rate, inventory will fall. Inventory can consume significant cash and physical space in the enterprise. These resources should be planned in advance to ensure that sufficient capacity exists to support the overall plan rather than being reactive if cash or space runs short. This production-sales-inventory (PSI) report compares the planned production rate with the sales rate (Figure 19-2).

In the figure, the cumulative production rate climbs at a steady slope. This relatively stable production rate is used to fulfill the more variable demands. This strategy is followed commonly when the capacity flexibility is insufficient to track to actual demand. The risk is that expected demand will not be realized and excess or obsolete inventory may result. The following table reflects the raw data that were used to generate the PSI table below. Using the PSI graph is far easier to identify problems or issues than attempting to use the raw data.

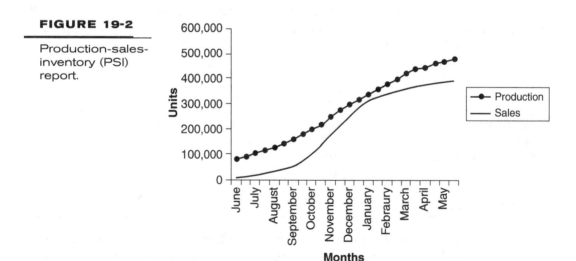

**FIGURE 19-2**

Production-sales-inventory (PSI) report.

|            | Production | Sales   | Inventory |
|------------|------------|---------|-----------|
| June       | 100,000    | 50,000  | 75,000    |
|            | 120,000    | 50,000  | 82,000    |
| July       | 120,000    | 60,000  | 88,000    |
|            | 120,000    | 60,000  | 94,000    |
| August     | 120,000    | 100,000 | 96,000    |
|            | 150,000    | 100,000 | 101,000   |
| September  | 150,000    | 100,000 | 106,000   |
|            | 200,000    | 200,000 | 106,000   |
| October    | 200,000    | 200,000 | 106,000   |
|            | 200,000    | 400,000 | 86,000    |
| November   | 300,000    | 400,000 | 76,000    |
|            | 300,000    | 400,000 | 66,000    |
| December   | 200,000    | 400,000 | 46,000    |
|            | 200,000    | 400,000 | 26,000    |
| January    | 200,000    | 200,000 | 26,000    |
|            | 200,000    | 200,000 | 26,000    |
| February   | 200,000    | 100,000 | 36,000    |
|            | 200,000    | 100,000 | 46,000    |
| March      | 200,000    | 100,000 | 56,000    |
|            | 200,000    | 80,000  | 68,000    |
| April      | 100,000    | 80,000  | 70,000    |
|            | 100,000    | 50,000  | 75,000    |
| May        | 100,000    | 50,000  | 80,000    |
|            | 100,000    | 50,000  | 85,000    |

In a PSI graph, the distance between the production and sales lines represents the amount of expected inventory. Notice in Figure 19-2 that the system started with some inventory on hand. The expected surge in demand during the months of December to February is covered by the production rate from previous months. This PSI report shows that all demands are expected to be covered at the desired timing. No customer should expect to wait to have his or her demands filled. Figure 19-3 shows an expected back-order situation.

Since the sales line crosses the production line, backorders to customers can be expected. There will be insufficient inventory from late December until early April to fulfill expected demand. This is where knowledge of the business is required to determine if this is a feasible plan. Provided that customers are willing to wait for this production output, all customer demands should be filled by May. However, if this is an item that will not tolerate a backorder situation, such as a gift item that is desired during the holiday season, the unfilled sales then would be lost sales. Including these sales as part of a revenue forecast would be unwise. The PSI report provides visibility to this risk and allows proactive management and alternative identification.

The comparison of production and sales also provides the capability to calculate available to promise (ATP). ATP is the uncommitted portion of inventory and is used fre-

**FIGURE 19-3**

Production-sales-inventory (PSI) report with shortage projection.

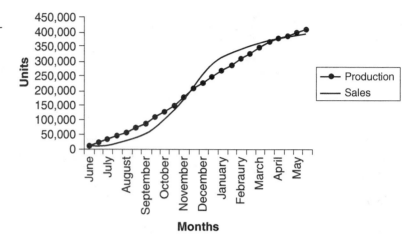

quently in make-to-stock companies. ATP includes only actual customer orders and planned and expected inventory receipts. The forecast is not used to calculate ATP. As supply-chain systems have become more powerful and sophisticated, this concept has been expanded to include allocated available to promise (AATP). Even though a customer's order may not be placed in the MRP system, the product is allocated to that preferred customer and is not available for general consumption, as shown in Figure 19-4.

When the actual customer order arrives matching the allocation, the allocation is reduced by the quantity the customer actually has ordered. If the quantity exceeds the level that has been reserved, the overall ATP is reduced. Once the production schedule is developed and approved, the stage is set for putting the plan into motion.

**FIGURE 19-4**

Allocated available to promise (AATP).

**Part A**
Demand Time Fence: 3
Beginning On Hand: 172

|                              | 1   | 2    | 3   | 4   | 5   | 6   | 7   | 8   | 9   | 10  |
|------------------------------|-----|------|-----|-----|-----|-----|-----|-----|-----|-----|
| Forecast                     | 100 | 90   | 80  | 75  | 80  | 90  | 100 | 100 | 120 | 130 |
| Actual Customer Demand       | 80  | 120  | 75  | 30  | 20  | 10  | 0   | 0   | 0   | 0   |
| Projected Available Balance  | 92  | 122  | 47  | 122 | 42  | 102 | 2   | 52  | 82  | 102 |
| Allocated for Customer       |     |      | 30  | 20  |     | 20  |     |     |     |     |
| Available to Promise         | 92  | −45  |     | 100 |     | 140 |     | 150 | 150 | 150 |
| Cumulative ATP               | 92  | 47   | 47  | 147 | 147 | 287 | 287 | 437 | 587 | 737 |
| Allocated ATP                | 92  | 17   | 17  | 80  | 80  | 120 | 120 | 150 | 150 | 150 |
| Master Production Schedule   |     | 150  |     | 150 |     | 150 |     | 150 | 150 | 150 |

## BACKFLUSH

This postdeduction records the component parts used in an assembly or subassembly by exploding the BOMs by the production count of assemblies produced. In the low-volume, high-variety production process (job shop), a work order is used to collect the job costs for material and labor. Since repetitive manufacturing usually determines individual product cost by collecting direct costs over time and dividing them by the number of units completed, the work-order transactions specific to an individual part or group of parts are not necessary. In addition, since the BOMs are stable in a repetitive industry and the product cannot be released to the line without availability of all the parts, the detailed tracking of material to a work order is not needed. This is why the process of backflushing is used to deduct components based on the number of finished goods completed.

If 200 lamps were produced, then 200 bulbs, 200 yokes, 200 bases, 200 shades, and 200 electrical cords had to have been used. Only exceptions to the BOM need to be managed and reported to the system. These exceptions could be parts that are scrapped, rejected, destroyed, substituted, or otherwise different from the BOM. Having a good process for capturing and reporting these exceptions is essential to having an effective backflush. Attempting to implement backflushing with significant errors in the BOMs or inventory is inviting disaster. Effective inventory control is not possible, and the information used by the ordering process will be incorrect. This incorrect information likely will result in shortages to the line that will quickly shut down the whole operation. Backflushing may look very easy, and theoretically, as a process, it really is. With one transaction, the finished-good part is received into inventory, and all components are relieved. This almost looks like "automagic." However, without good control of exceptional usage or inaccuracy in the BOM, this magic can become a nightmare.

A more advanced process of backflushing is called *pay-point backflushing*. Pay-point or count-point backflushing is used when the manufacturing process extends for more than one day. Documenting material movements with transactions more than one day apart makes controlling inventory very difficult. Daily cycle counts must take into account the inventory that is on the line semi-completed and not yet transacted. This problem does not exist in the job shop because all materials are transacted at the beginning of the order when the material is issued from stock.

### Product 136593 Widget

| | |
|---|---|
| 142604 Component A | Routing step 10 |
| 142496 Component B | Routing step 10 |
| 148375 Component C | Routing step 20 |
| 193857 Component D | Routing step 30 |
| 103857 Component E | Routing step 50 |

For this product, if routing step 10 is considered a count point (pay point), a transaction would be done that would complete that operation, and the two parts, 142604 and 142496, would be consumed from the line inventory and moved to work-in-process. The next pay point may not be until routing step 30. When this transaction is done, both component $C$ and $D$ will be consumed. If the transaction was missed at routing step 10, the system also should consume the materials assigned to routing step 10. Component $E$ is not consumed until the very end, when the product is transacted as completed and put into inventory. At this time, the MRP system will transact all parts that have not already been consumed, relieve work-in-process of the temporary staging, and receive the part into finished goods. This is all accomplished by a single completion transaction. By using the pay-point backflush method, validation of inventory on the line is a much simpler process because less work-in-process needs to be considered when reconciling inventory.

A good inventory management process to support backflush is to perform a transaction when moving a bin of parts to the line location and then backflush from the line location. This does create an extra inventory transaction for the bin of parts, but the control may be worth it, especially in the early stages of implementation. This provides a clear demarcation between the active inventory on the production line and any inventory that should be in the stockroom. If the inventory in the line location goes negative, it is much easier to detect and resolve the process problem that is causing it rather than allowing the error to continue and affect the main stockroom. Of course, the ideal would be that there is no inventory in a main stockroom as a backup for the line. Unfortunately, most companies do not have sufficient reliability in supply-chain performance for this to be true for all parts.

Backflushing in an enterprise that requires lot or serial number traceability, which can appear like a mission impossible. MRP does not support backflush capability for parts requiring lot or serial number traceability. This is because the lot number of the component parts must be identified when material transactions are done. The answer to this problem has been the process of issuing material to work orders including detailed lot numbers of component parts. However, this does not mean that a facility with lot or serial number traceability is doomed to a life with work orders. Creatively applying the backflush process is still possible. Rather than using the standard backflush feature, the standard work order issue function is used at the end of the manufacturing process. This is known as a *postdeduct process*. Rather than attempting to predict which lot of parts will go onto a particular end item through the traditional job order release and parts issue function, the information on what lot numbers of parts were really used is collected as the part flows through manufacturing. This is usually tracked through a piece of paper traveling with the part and then transacted at the end of the process. Another option is to use bar-coding technology to collect that data and automatically deduct the parts from inventory as the parent part is flowing through the process. At the end of the process, a hard copy can be printed for the device master record. This process has all the benefits of backflushing but still provides the required traceability.

In addition, some MRP systems provide the option of backflushing labor as part of the transaction. The backflush transaction creates a general ledger transaction for labor at the same time the material is relieved. Another option could be to consider labor a period expense and simply divide it by the number of units completed. This is an excellent practice to follow in a repetitive or process industry. Rarely is labor a variable cost for the product. Labor in such organization tends to be fixed in the short term.

## PERIOD COSTING

A steady level of work-in-process and a reliably short throughput time make period costing possible. Work-in-process is directly related to lead time. When lead time stays constant, then so does the level of work-in-process. Keeping track of every detailed transaction costs more than the value of the information received. Statistical process control charts can be used to track costs over time. Using the same rules as would be used when managing part dimensions, costs can be expected to vary within a certain tolerance. When the trend begins to move out of control or a special occurrence happens to cause a one-time spike in cost, this can be identified quickly. This kind of reporting is now available through the use of data analytics tools. Figure 19-5 shows an example of how this cost analysis may look. This can track either total costs or just the variances against standard cost.

As would be expected in a new-product development, the costs are more widely variable at the beginning of the analysis. Once the product has become stabilized, the reliability of cost is very high over time. One exception can be seen around time period 40, when the cost spikes sharply. This out-of-control condition can be readily identified and resolved. This would not be possible in a regular costing process. This spike would be averaged in over time and likely would pass by without being noticed.

## HIGH-VOLUME MIXED-MODEL MANUFACTURING

The repetitive approach to MRP implementation also can be leveraged into a mixed-model discrete manufacturer, where the processes are repetitive. Examples of this type of manufacturer are automobiles, consumer electronics, computers, and airplane overhead bins. Every end item could be absolutely unique, but each one uses the same process. The various differences could be color, hole patterns, attached parts, and name plates.

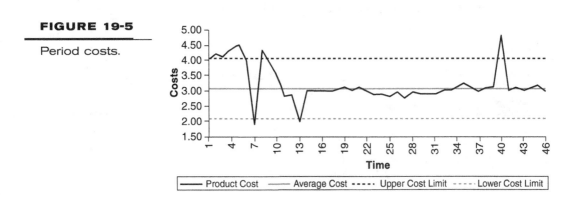

**FIGURE 19-5**

Period costs.

Material and process traceability is also a very common requirement in this type of environment. This means that the end product needs to track the lot number of the materials used and the operator who worked on the units to an end-item batch or serial number. High-volume, high-variety repetitive production has one foot in the discrete work order world and the other in the rate-based world. Managing the material and capacity planning can be a real challenge. The same repetitive tools can be used for a final assembly operation, where many different configurations of parts are running down the same line. MRP can be linked into the execution on the shop floor to determine the material configuration and the day required. Normally, the smallest planning time bucket for MRP is a single day. A kanban process still can authorize replenishment to the minute supported by MRP scheduling the configuration of the parts to the day.

Some industries require that the component parts come to the line in a specific sequence, and the response time from the supplier is longer than is manageable using a simple kanban. Functionality tying less than daily-planning buckets to a serial number is rarely found in ERP. In this environment, it is very common that line sequencing of the desired models also must be netted through to the planned purchase order so that the supplier can sequence the incoming materials to the needs of the line. This process is known as *seiban*. The literal translation of *seiban* comes from three Japanese words that have been shortened into one term. *Seizou* is production. *Bango* is number. *Kan Ri* is management. Therefore, seiban can be translated to be "management by lot number." The kanji characters for this term are shown in Figure 19-6.

This type of line sequencing is found commonly in automotive and truck assembly. Replenishments from suppliers are received multiple times each day, and having the right part on the truck in the right order is essential for overall productivity of the assembly line.

## CONFIGURATORS

Repetitive manufacturing frequently uses a configurator tool. *Configurators* are software systems that create, maintain, and use product models that allow complete definition of all possible product options and variations with a minimum of entries. Things to ask about configurators in a repetitive environment include

**FIGURE 19-6**

Kanji characters
for seiban.
(Courtesy of
Toshiyuki Okai.)

SEI   ZOU   BAN   GO   KAN   RI

製 造 番 号 管 理

SEI   BAN

- Is the configurator a linear finite model, or does it have dynamic or parametric capability? Linear finite models are models in which a limited number of choices can be made. Dynamic or parametric models are models in which measurements and sizes can be added. Dynamic configurators are used for the fabrication of windows, doors, and wire cable lengths. The parametric configurator knows that once a product gets to a certain size or complexity, the configuration may not be feasible or additional parts may be required to make the configuration feasible.
- Is the configurator an integral part of the MRP system, or is it a bolt-on that must be interfaced? An integrated configurator will save countless hours of testing and integration during implementation and through maintenance.
- Does the configurator interface or integrate with the front-office tools to provide seamless pricing and quoting capability?
- Is the configurator Web-enabled to allow your customers to configure their own products and determine prices before the order is placed?

## SUMMARY

The repetitive operation is the most simple to comprehend, yet it can be the one of the most difficult to manage. Any unexpected breakdown can cause the entire plant to shut down. Material is closely synchronized with the production process. Traceability may be required to the end-item level. If the BOMs have been flattened too much, critical stocking points that would compress lead time may be missed. The goal to eliminate inventory could end up making the entire supply chain too fragile.

The positive side of the repetitive manufacturing approach is that there are no piles of inventory covering up the problems. Any disruption is quickly identified. A sense of urgency is common in identification and resolution of the problems encountered in this type of facility.

Capacity planning is more straightforward because products move down a highly predictable series of machines or operations either on a transfer line or through a manufacturing cell. The routing in a repetitive operation is usually only one step. Lead times through the process are less than one day. Processes that take more than one day are supported with tools such as pay-point backflushing.

A number of unique tools are available to the repetitive manufacturer. Kanban, seiban, configurators, AATP, backflushing, postdeduct, and period costing all can be used at the appropriate time and place in the process. When these tools are understood, including when they are applicable, management of the repetitive enterprise can be simplified. As William Milliron said, "Use the right tool for the right job, and the job is easy. Use the wrong tool and you will fight the job every step of the way."

## CHAPTER 20

# Sales and Operations Planning

## Evolution of Sales and Operations Planning: From Production Planning to Dynamic Business Performance Management

*By Andy Coldrick and Dick Ling*

In the face of the Sarbanes-Oxley Act–mandated increased requirements for timely reporting to Wall Street and other investors, a significant limitation facing every senior-level executive is the lack of integration between what sales plans to sell, what operations plans to make, and what the financial plan expects to return. Each of these plans is typically a silo within each function. The sales team develops and manages to the sales forecast. The operations team develops and manages to a master production schedule (MPS). Finance develops and manages to the budget. In reality, each of these three plans depends on the other two. Overall company success depends on the holistic integration of these different functions. Sales and operations planning (S&OP) provides that big picture.

S&OP has been used extensively since its creation in the late 1980s. The process evolved during the 1990s, and numerous companies have gained tangible business benefits in improved customer service and reduced inventories and have used the process to facilitate growth and sustained profitability. Other organizations are struggling with supply-chain planning, collecting data for a monthly meeting in which sales, marketing, and finance are less than enthusiastic attendees.

This chapter traces the evolution of S&OP and its application to companies of different industries and sizes and suggests reasons why some companies have maximized the benefits and others have just become stuck in old paradigms. Our intent is that, by providing you with an understanding the evolution, we provide you a context and frame of reference to allow you to identify where you are today and gain some insights into how to improve or change what you have. Evolution takes too long and does not guarantee success. Achievement of fast, sustainable benefits is through a challenging plan that identifies behaviors, activities, and functions that need to change so that we can align S&OP with the executive management agenda. For those who are just about to start an implementation or reimplementation, we suggest taking the learning from the 1990s and insist on business leadership from the start and avoiding the pain a supply-driven process brings. We call this innovative right-to-left approach *breakthrough S&OP*.

# WHAT'S IN A NAME?

What started out as sales and operations planning has taken many different forms, now being referred to by a multitude of names. These include *integrated business planning, integrated business management, integrated performance management, rolling business planning, regional business management,* and *sales inventory and operations planning* (SIOP) to name a few. Several organizations continue to use the acronym S&OP, although stages of maturity are quite different from company to company. Unfortunately, for many, a change in name does not guarantee a corresponding change in the underlying process, behaviors, and results. We still call this *sales and operations planning*. Some still use S&OP for demand-supply balancing; others use it as a powerful cross-functional business process.

## What Is Sales and Operations Planning?

Early versions of S&OP were driven by manufacturing to make material planning and manufacturing planning more stable. These manufacturing origins show through in many applications of S&OP and are characterized in some telling ways, including:

- *The timetable set around supply meetings.* This involves representatives from sales, marketing, and commercial having to attend a set of S&OP meetings established around manufacturing locations and planning routine, rather than category/ range, reviews or customer- and market-driven events and calendars.
- *The view that sales and marketing are one homogeneous organization with "a view."* This misses the very fact that sales and marketing have different drivers and objectives and potentially conflicting views that need to be understood and reconciled.
- *The obsession with a single set of numbers.* This is based on the naive belief that it is possible to create one number that will represent all views of the business and that all uncertainty can be eliminated to the degree that you can plan and control the future this rigidly through the entire planning horizon. This is a relic of what was called "best practices manufacturing resource planning (MRPII)." At the time, many people were fixed on the idea that S&OP was a new part of MRP II designed to give one number on which MRP II depended. In fact, the truth was, and still is, that S&OP is the big picture. Previously, MRP II and now enterprise resource planning (ERP) provide the detail planning and execution to support S&OP.

Today, we see S&OP as a process for cross-functional decision making. Ultimately, the process enables a business to accomplish monitoring and updating of its strategic intent using the monthly operating plan as a robust foundation. However, through its evolution, it has taken different forms in different applications and in many cases has fallen short of its full potential.

S&OP is a forward-looking process with a minimum horizon of 18 months or 6 quarters that integrates and aligns strategic and tactical views and decisions and directs operational planning and execution. It is not a short-term scheduling tool with only a four- to six-month horizon (Figure 20-1).

**FIGURE 20-1**

Traditional S&OP model.

There are two key points in Figure 20-1. First, S&OP as a powerful decision-making process has to be the driver of tactical and operational planning and execution, with the financial view from S&OP credibly supporting the business plan. Second, the planning horizon must be a minimum of 18 months to ensure that decisions are made about year end in the context of the following year. A simple way of visualizing this is to see operational planning as the short-term day-to-day flawless execution. Tactical planning is about delivering this year's budget, and strategic planning is delivering future years performance. S&OP, a monthly process looking both inward and outward, enables changes in assumptions to be evaluated and is used to monitor progress forward and update strategies when needed.

The principal focus of S&OP during the 1980s and 1990s was how to get a good operational foundation in place. This foundation provides the ability to evaluate demand and to ensure that sufficient resources are in place across the business to meet that demand. Changes are assessed monthly, and plans are updated and communicated. The first impetus was provided by Dick Ling with the creation of S&OP, which we now call *traditional sales and operations planning*.

## TRADITIONAL SALES AND OPERATIONS PLANNING

S&OP was created in the late 1980s by Dick Ling (his book, *Orchestrating Success*, coauthored with Walter Goddard, was published in 1982). At the time, MRP II was in vogue, and S&OP started to be seen as a driver whose principal focus was to make MRP II work in a single manufacturing plant within a business.

At the time, S&OP was a breakthrough because in many businesses annual business planning, sales planning, and production planning were completely separate exercises. There were one-way hand-offs and massive disconnects; finance as the neutral function often was used as the referee in disputes among sales, marketing, and manufacturing. Multiple sets of numbers existed, and all improvements were functional and therefore disconnected. For example, an inventory-reduction project to improve cash flow would be initiated by finance and supported by manufacturing. Sales and marketing would make no contribution until customer service suffered. The inventory-reduction project then would be followed by a customer-first project led by sales, until inventory or the cost of complexity again became the focal point.

The premise of traditional S&OP is that customer service and inventory are *resultants*. To manage them effectively, we must manage the drivers, that is, demand and supply.

S&OP was a breakthrough in that it forced sales, marketing, and manufacturing to agree once a month to *one set of numbers* for sales, production, and inventory. Within the month, there would be a sales planning meeting, chaired by the sales director, agreeing on the volumes at family level predicted for sale for the next 12 to 18 months, called *demand planning*. The manufacturing director then would run a meeting called *supply planning* to respond, using resource capacity management, with the corresponding production and inventory plans. This would be followed by a pre-S&OP meeting where sales, marketing, and manufacturing agreed with each other for one day in the month to prepare for an S&OP meeting with the general manager/managing director and other board members. Following the S&OP meeting, or just before, some reconciliation of volumes with financials would be done as a check against the budget. Is this revolutionary? No! It is merely *organized common sense*. The process is shown in Figure 20-2.

The focus on managing demand and supply as drivers (seeing inventory as a resultant) gave many businesses improved customer service and lower inventories. These operational benefits often stemmed from an attention to detail, and the *S&OP spreadsheet* provided the data that helped to spot the results of independently managed events.

Many early applications of this focused on manufacturing views of product families rather than external views of the business, for example, channel or brand. An example of the output from traditional S&OP meetings for an assembled product is shown in Figure 20-3.

During the late 1980s and early 1990s, we saw many people struggling because they saw basic demand and supply planning as an end in itself. The attention to detail and desire for stability that drove early benefits were pursued to extreme lengths, creating a new set of problems.

**FIGURE 20-2**

Organized
common sense.

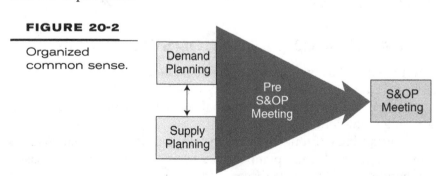

**FIGURE 20-3**

Product family volume plan.

| Past Months | | | | Future Months | | | | | | | | | |
|---|---|---|---|---|---|---|---|---|---|---|---|---|---|

| Sales | -3 | -2 | -1 | 1 | 2 | 3 | 4 | 5 | 6 | 7 | 8 | 9 | Ann |
|---|---|---|---|---|---|---|---|---|---|---|---|---|---|
| Planned | 1200 | 1200 | 1200 | 1200 | 1200 | 1200 | 1200 | 1200 | 1200 | 1200 | 1200 | 1200 | 14400 |
| Previous | 1211 | 1197 | 1200 | 1200 | 1200 | 1200 | 1200 | 1200 | 1200 | 1200 | 1200 | 1200 | 14408 |
| Act/Proj | 1211 | 1197 | 1251 | 1125 | 1250 | 1250 | 1250 | 1250 | 1250 | 1250 | 1250 | 1250 | 14890 |
| Diff | 11 | -3 | 57 | 25 | 50 | 50 | 50 | 50 | 50 | 50 | 50 | 50 | 490 |
| Cum Diff | 11 | 8 | 65 | | | | | | | | | | |
| **Production** | | | | | | | | | | | | | |
| Planned | 1200 | 1200 | 1200 | 1200 | 1200 | 1200 | 1200 | 1200 | 1200 | 1200 | 1200 | 1200 | 14400 |
| Previous | 1195 | 1202 | 1200 | 1200 | 1200 | 1200 | 1200 | 1200 | 1200 | 1200 | 1200 | 1200 | 14397 |
| Act/Proj | 1195 | 1202 | 1197 | 1200 | 1200 | 1400 | 1250 | 1250 | 1250 | 1250 | 1250 | 1250 | 14894 |
| Diff | -5 | 2 | -3 | 0 | 0 | 200 | 50 | 50 | 50 | 50 | 50 | 50 | 494 |
| Cum Diff | -5 | -3 | -6 | | | | | | | | | | |
| **Inventory** | | | | | | | | | | | | | |
| Planned | 600 | 600 | 600 | 600 | 600 | 600 | 600 | 600 | 600 | 600 | 600 | 600 | 600 |
| Previous | 584 | 589 | 589 | 589 | 589 | 589 | 589 | 589 | 589 | 589 | 589 | 589 | 589 |
| Act/Proj | 584 | 589 | 529 | 504 | 454 | 604 | 604 | 604 | 604 | 604 | 604 | 604 | 604 |
| Diff | -16 | -11 | -71 | -96 | -146 | 4 | 4 | 4 | 4 | 4 | 4 | 4 | 4 |
| Cover (Wks) | 2.06 | 2.07 | 1.83 | 1.75 | 1.56 | 2.08 | 2.08 | 2.08 | 2.08 | 2.08 | 2.08 | 2.08 | 2.08 |

# S&OP, THE UNIFIER—TRADITIONAL S&OP CHALLENGED

Following the initial euphoria resulting from getting control, enthusiasm waned, and traditional S&OP started to be seen as a logistics project, merely demand and supply volume planning focused on year end only, with too much detail (stock keeping unit/pack level/line item forecasts going out for 12 months). The dream that S&OP was the unifier faded, and it started to be seen as a middle-management logistics responsibility. Demand planners, often reporting to the supply organization, owned the numbers rather than sales and marketing management, and the process was designed not to cope with the impact of increased innovation and customer responsiveness that many organizations were driving. It appeared at the time that S&OP was relevant only in organizations with limited innovations committed to a cost leadership strategy.

A single set of numbers was a supply-chain dream, but it was an obstacle to other functions. Sales, marketing, and finance were more interested in a *range*, doing their own financial scenario planning in separate activities from supply. Executive management manages uncertainties, probability odds on events planning, and ranges of numbers.

Without robust financial linking, volume forecasting became a lower priority than financial forecasting. Sales, marketing, and general management were measured on financial results, and manufacturing and the supply chain were measured on operational targets based on volume predictions, where new activities were not well forecast (Figure 20-4).

The two vertical arrows illustrate the point: Whatever the output is from S&OP (*thin upward arrow*), the weight carried by the budget number (*thick downward arrow*) takes precedence, overriding any decisions made in S&OP. Because the operational number for the supply chain was lower in priority than the financial number and very often was different, the S&OP meeting became the forum where supply people grumbled about forecast accuracy against their single set of numbers—the impossible dream. It was becoming apparent that getting a single number from a pre-S&OP meeting where people had

**FIGURE 20-4**

S&OP change
model.

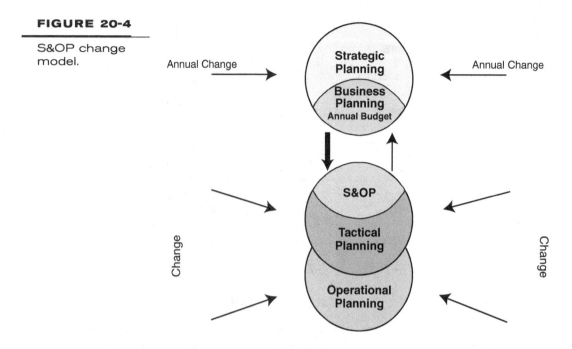

their own functional agenda was virtually impossible. Very often the one who shouted loudest got his or her way, but if finance did not agree, the number was questionable.

In fact, this obsession with one number was seen as the Holy Grail and S&OP as the great unifier—the way to get a single number. Why? Because multiple sets of numbers create confusion. They do, but the antidote is not a single number. What is needed is an agreed-on latest view that comes as a result of reconciliation of different views (see next section). In a traditional pre-S&OP meeting in an organization with a silo culture, there was no recognition that different views add value—they were considered an obstacle to a decision. In such organizations, one can see the politics. What is said in a meeting is influenced by functional positioning. It influences where and when things are said, by whom, to whom, and against whom. It is impossible to harness all the ability and knowledge of all participants to obtain the optimal agreed-on latest view in such an environment.

This culture is often reinforced by the *A* versus *B* syndrome. The symptoms of this are polarity and argument around two different options when creation of a third or fourth option may well be the best solution. Here, verbal dexterity in support of one or criticizing the other absorbs all the talent and time. Often whoever is fastest on his or her feet with the best information or examples in support wins for *A* versus *B*—even if more considered thought would produce the opposite and conclude that *B* is better and even a solution where an undeveloped *C* or *D* would be better still.

Some reconciliation of different views must take place before there is agreement on what a number should be. We should have seen that S&OP must be the great reconciler before it could be a unifier.

The early 1990s saw an additional complication, the advent of the single market in Europe. This heralded the regional business concept where the business unit or category, strategic marketing, finance, and supply chain were to be managed regionally. Sales, tactical marketing, and financial management of the legal entity were to be managed in the countries. This added complexity and ambiguity to the traditional S&OP process. Multiple sales and marketing units interfacing with multiple sourcing units raise questions. How many meetings do you have? Do you have a meeting in a sourcing unit or in the market or both? Should you have meetings, or should you focus on an integrated process?

We know some people wedded to the traditional S&OP concept who religiously hold S&OP meetings at their European plants and ensure that sales and marketing attend monthly meetings at all the plants that supply their products. Imagine a sales director/manager who sells product supplied from five different plants in Europe! This poor soul wastes five days a month of precious time with customers by being trapped in some outdated S&OP concept. S&OP not only requires a reconciler—but a cross-functional integrator also is mandatory.

## S&OP, THE RECONCILER AND INTEGRATOR

With this background and with some early work on Europeanization with several multinational businesses, we developed a more robust model that is often called the *five-step process*. Since then, the process has continued to evolve into a model that we use to frame the key steps of S&OP (Figure 20-5).

The relevance and significance of the five steps and how they must be integrated have been tested over the last decade. They have been adapted to meet the pressures of

**FIGURE 20-5**

S&OP–
Ling/Coldrick
model.

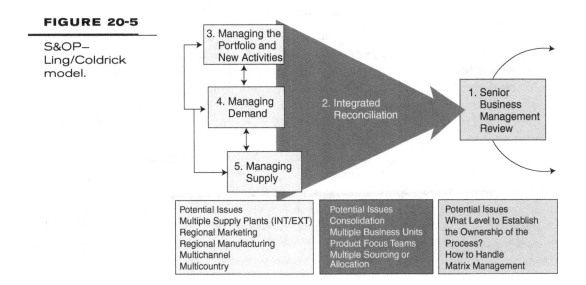

different business strategies and handle the needs of different industry sectors within and beyond manufacturing, including retail and services. A number of key themes have emerged.

## Integrated Reconciliation

The development of integrated reconciliation has highlighted the importance of financial involvement and leadership early in the process and has changed the agenda from a volume discussion to a business one. It is about reconciling different views. There is added value in discussing different views and the reasons for them. It increases the understanding of what the numbers mean, which focuses attention on the assumptions underpinning the numbers, along with opportunities and vulnerabilities. The conversation is principally about what has changed since the last review and why. Without assumptions, the conversation is why have the numbers changed. When the focus changes from just numbers to assumptions, the need for marrying forecasting with foresight becomes even more apparent.

Here are some of the questions involved in integrated reconciliation:

1. What is the impact of integrating new activities, demand and supply, on the business (not just the supply chain)? What are the emerging issues and gaps? What are the opportunities and risks? You must have volume and value information and assumption changes to answer these questions. Understanding these questions leads to an imperative that finance is an integral part of all five steps, whereas in many examples of S&OP, finance is added at the pre-S&OP meeting and the S&OP meeting.

2. What scenarios are important to make better decisions in the future?

3. What decisions should we make and which ones should be escalated to the senior management review?

The step is not a meeting as such but an iterative process run by a senior cross-functional team in the business. The team highlights key issues and decisions required for the senior management team. In fact, the team determines the agenda for senior business management review. Participants in integrated reconciliation are the future executives in the business. This exercise is seen as a key training ground for the next generation of presidents and vice presidents. This is fundamentally different from the pre-S&OP meeting in traditional S&OP, where the main focus was on volume and its impact on resources. Understanding integrated reconciliation has directed a broadening of the scope of new activities, demand and supply management. Integrated reconciliation as a process leads directly into the senior business management review, which focuses on understanding change: What is our current performance? What decisions are still outstanding? And what decisions have been made already in integrated reconciliation? The business agenda also raises further questions: Are we on track with the business plan? And are we still on track with our strategic intent?

You'll note that in this model there is no such step as a pre-S&OP meeting or executive S&OP meeting. Our thought is that the aligned and integrated five steps define the entire S&OP process, and we have deliberately omitted reference to an S&OP meeting. We have seen too many examples in large companies of the S&OP meeting being the focus, and within large multinationals, S&OP meetings were springing up everywhere by country, by cluster, by business unit, by manufacturing plant, and so on. Our view is that the senior business management review would be similar to an executive S&OP meeting; however, if this is preceded by a traditional pre-S&OP meeting, the agenda would be more operational. Our view of a senior business management review is colored by how different the integrated reconciliation process is from the traditional pre-S&OP meeting.

## Managing the Portfolio and New Activities

New activities have developed significantly over the last 10 years. As with traditional S&OP in the 1980s, the notion of integrating new-product planning with supply and demand planning of the existing portfolio was something of a breakthrough despite being common sense. With increasing focus on innovation and the use of stage and gate decision processes and innovation funnel management, some companies took the opportunity to integrate these emerging approaches and developing them in parallel to S&OP. Early attempts at integration often focused on the commercialization/introduction stage of the funnel. The aims were to ensure preparation for launch and phase-in/phase-out and to see that cannibalization effects were understood, motivated by helping production not to be caught out when introducing a new product.

Progressive organizations, often those driving very aggressive innovation agendas, realized that connecting only the back end of the process missed opportunities to manage the innovation funnel in an integrated way.

The scope also was broadened in another direction by those who saw the need to manage new activities beyond the narrower definition of product. Although the list is different in every application, a common theme in opening up this step beyond just product is identification of the activities that:

- Have a significant impact on demand and/or supply (volume and value) and any other support resources
- Need to be managed across the business in a joined-up way, with decisions driven through a structured review process
- Require visibility and management across a portfolio of activities, leading to better prioritization, resource allocation, and decisions

This redefinition from new product to new activities opened the scope outside a business-as-usual situation and in so doing opened up the appeal to a much broader audience in the business. This changing context had a dramatic effect on the view and understanding of the demand and supply steps of the process.

## Managing Demand

Demand management, including the accountability for forecasting, has developed significantly. In the early years of S&OP, a lot of effort went into agreeing to a volume forecast emphasizing a single set of numbers. Demand forecasting was very often part of the supply organization, and forecast accuracy was seen as the principal measure rather than customer service. Some organizations even went so far as saying, "You did not forecast this; therefore, we cannot make it!" obviously alienating sales and marketing. The thinking that sales and marketing form one homogeneous organization with a single view of the numbers misses the fact that these two functions have different drivers and objectives.

By the mid 1990s, people realized the importance of sales and marketing inclusion in forecasting. At the same time, "Customer, Customer, Customer!" was fashionable, and sales became the focal point for forecasting and the one-size-fits-all solution, ignoring the importance of marketing input.

In many businesses following a "customer relationship" strategy, sales leadership is appropriate, but in organizations with product/service differentiation strategies, marketing is the principal driver of medium- to long-term demand prediction. This distinction is covered in more depth later in this chapter.

Giving sales single accountability for the forecast led to some organizations spending too much time analyzing detailed history in an attempt to get the forecast accurate instead of being with the customers gaining knowledge about future trends. Against this background of trying to get the forecast accurate, there was a growing realization that there is a different inherent uncertainty in different markets, channels, and sectors, as well as with different products and customers. After years of complaining about forecast accuracy and trying to crank the handle faster on the same old detailed forecasting machine, companies began to wake up to forecasting for what it is—predicting the future! By no means does this remove the responsibility for forecasting, but it did lead to new and innovative ways of making a more educated prediction. In agreeing with a forecast, an important piece of knowledge is to understand the range (high and low), and providing numbers without documented supporting assumptions is unhelpful. In some companies, the rule is that a forecast number cannot be changed unless an assumption is also changed. A summary example is shown later in this chapter in Figure 20-6.

Today, we understand that a robust demand plan over a minimum of 18 months is possible only by reconciling cross-functional views; volume and value must be integrated. Finance and logistics/supply chain are committed to this output. In general, sales input by major customer (with input from account managers) and channel is important in the short term, typically the first four to six months. Marketing provides information beyond four months based on market share, goals, and brand/product health and marketing investment. Strategic marketing and research and development (R&D) in many cases have a role beyond 12 months, particularly in new activities. There must be reconciliation between foresight (i.e., strategic marketing) and forecasting (i.e., tactical marketing and sales). These are guidelines only to illustrate the collaborative approach and will vary depending on the business. The responsibility of finance and supply chain/logistics is to ensure that the volume and financial forecast are reconciled and aligned. The demand plan is at an aggregate level, and the aggregate families are chosen, understood, and used by all functions. Simulations at the aggregate level are more helpful than trying to do "what ifs" at the stock keeping unit (SKU). Choose a software solution that facilitates this capability.

Traditional S&OP tended to use manufacturing families. In demand-driven environments and product differentiation businesses today, we see the aggregate product family being brand or brand/technology. Why would one choose an aggregate group with little relevance to marketing and sales?

## Managing Supply

Supply management also has broadened in its scope. Traditionally, it applied to just manufacturing, but now it is extended from manufacturing to a wider view of sourcing that includes other resources, including external ones. In multinational organizations, it has been extended to supply-chain optimization, making the best sourcing decisions from scenario planning. This has challenged the planning capability of many organizations in that traditionally many planners have been used to management and execution in detail at single supply points; supply-chain optimization is a wider role calling for the ability to test different scenarios and recommend and make the right choices.

## Monthly Timing and Integration

The widening scope of new activities, demand, and supply has put even greater pressure on a more joined-up approach. As businesses have become bigger and more complex, S&OP has become a bigger challenge. We have noticed that without an integrated process, organizations attempt to solve complexity by being more functional.

Highlighting the importance of new activities and their impact on demand and supply, and forcing volume and financial integration through the integrated reconciliation process, we facilitate a business management understanding based on a robust operational view (volume and value) across the organization. This integration, combined with the increased future horizon, emphasizes the connection between the steps as the most important element in success. It also stresses the need for this to be seen as an iterative process, normally run on a monthly cycle, so that decisions taken during the process and confirmed in the senior management review in the month are fully communicated into the organization and executed through the process in month 2 and beyond (Figure 20-6).

The arrows from the senior business management review into the next cycle show the importance of making, committing to, and communicating decisions and taking action. These changes to the process had a major impact on the leadership of the implementation and its use in the organization.

Traditional S&OP normally was led from the supply side of the organization. The importance of new activity integration and early volume/financial reconciliation will not

**FIGURE 20-6**

Iterative business
management.

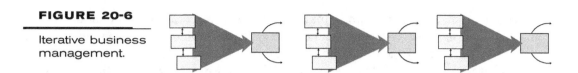

be recognized if supply people lead the implementation. Early engagement of finance is crucial, and strategic influencing of sales and marketing is vital from the beginning to ensure success. The paradox is that supply-chain people are normally the first to recognize a need for S&OP, but they should not be seen as the principal leaders.

## Managing Paradox

Used properly, the S&OP process provides a means of managing the choices and trade-offs across the business and taking decisions to keep on track with strategic direction. Inevitably, these decisions will need to cut through the conflict caused by opposing pressures of seemingly contradictory views. The implementation, as well as the use, of S&OP is riddled with paradoxical choices.

Reconciling these paradoxes is a key to unlocking the real potential of S&OP. To illustrate, here are a few of the repeat offenders:

- *The leadership paradox.* In many companies, the supply organization initiates the S&OP process as a reaction/response to the pain caused by changing forecasts and the apparent disregard from the demand side of the business for stability, efficiency, and reduced complexity. By biasing the S&OP agenda with a supply perspective, typically those functions that need to be actively involved and driving the process are often alienated. These situations pose a challenge: Those who have initiated the process need to relinquish their leadership of it if they are to realize the total business benefits.

- *The horizon paradox.* There is a constant tension between the need to take decisions to protect long-term value and the need to hit the numbers now. There are compelling arguments for each of these objectives, and the only sustainable proposition is to do *both*! Some have made the error of defining S&OP with a horizon of 3 to 18 months; the intention was good in attempting to focus management attention beyond the immediate short term. Unfortunately, it led to disconnected processes—a short-term sales and operations scheduling process (0 to 3 months) with weekly or daily review and a separate S&OP process (3 to 18 months) reviewed monthly and, sadly, no link between the two. A major benefit of S&OP is that decisions taken in the medium to long term will decrease the number of surprises in the short term. For example, a demand peak in months 9 through 11 must be solved by outsourcing; S&OP would make this decision proactively, knowing ahead of time the consequences of the decision, so that when months 9 to 11 become months 2 to 4 months, we already know how to manage this peak. If S&OP is disconnected, planners in short-term scheduling will behave reactively when they see the demand peak in month 3, make decisions, and count the cost afterwards.

S&OP done properly would show a minimum of 0 to 18 months at an aggregate level over the whole horizon, providing an aggregate view of the short term in the context of the medium to long term. Quarterly and year-end targets are important, but decisions must be taken in the wider perspective of next year and so on. A year-end focus is pragmatic and appropriate; a year-end obsession is unhealthy for future sustainability. This is why a minimum of 18 months' visibility is recommended.

- *The consistency paradox.* One person's consistency is another's bureaucracy. A mix of personal preferences, functional bias, and national and company cultures all add up to a very specific reaction to a prescribed way of doing things. Ultimately, we would want to strike the appropriate balance between allowing space for people to be creative in addressing issues and opportunities, taking decisions close to the action, while providing a framework to direct that creative energy and allow the interdependent elements of the organization to mesh together effectively. The paradox is buried in the phrase appropriate balance. Why is it that in some environments a timetable is taken as an absolute deadline, while in others it's merely a suggestion or even an imposition? Why do some organizations see a consistent template/format as a necessary way to allow integration and aggregation of information, whereas others see it as a request or challenge for innovative ideas on how to lay things out differently?

In this chapter we have touched on some other potential dilemmas that will need to be reconciled. We work to develop the behaviors and capabilities to cope with and thrive on this ambiguity and confront the choices—allowing you to break through these paradoxes and establish solutions that get both and in so doing add more value!

The changes made from the traditional S&OP model (Figure 20-1) to our five-step process (Figure 20-5) have led many businesses toward a robust operational foundation (Figure 20-7).

The two arrows between S&OP and the business plan are of equal strength, which means that we have a robust and credible latest view of the business that may be different from the budget, but the two must be reconciled. Each is credible, and we must answer questions on what we need to do differently to meet the business plan. Building the ongoing reconciliation is the first step toward a cross-functional business planning process.

The next step is to update the strategies from reconciliation of the business and operating plans, which is the goal expressed in Figure 20-1. The ultimate test is whether the senior team has the commitment and confidence in the process to dismantle the incumbent budgeting process.

**FIGURE 20-7**

S&OP operational
model.

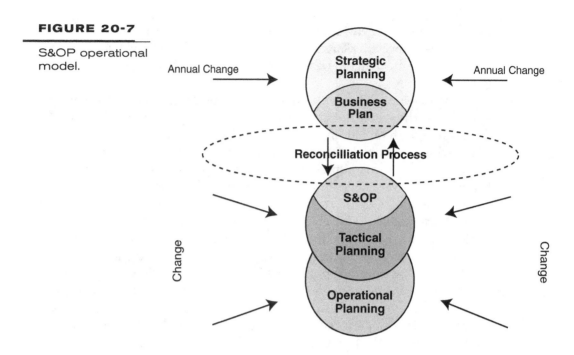

## KNOWLEDGE AND KNOW-HOW
## VERSUS DROWNING IN DATA

Any implementation puts pressure on the need for management information and not transactional detail. We designed a typical management information spreadsheet showing the importance of integrating volumes and financials with underpinning qualitative information on assumptions, changes to assumptions, and decision support. This format has been proven as a powerful means of communication and ensuring consistent understanding of the story behind the numbers. Enhanced by today's powerful information technology, this allows succinct management information to be available in any environment for fast, effective decision making (Figure 20-8).

Our experience in developing S&OP into a business management process has led to some interesting findings suggesting that aggregating data alone does not necessarily give good management information. This is sometimes called *drowning in data but starving for information.*

Our emphasis is on roughly right rather than precisely wrong to help businesses avoid the trap of projecting forward two years' worth of detail at the SKU level. The detailed approach commonly found in statistical forecasting software leads to an answer that looks about right but cannot be understood. Thousands of minor changes within the black box are not visible at a higher level. Management has no idea why the latest view has changed or whether the latest changes to plans have been incorporated. Instead, management is told that the system says, "This is the answer; you have to believe it!" A

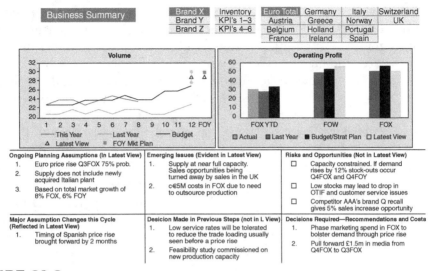

**FIGURE 20-8**

Volumes and financials in one sheet.

better approach is to build knowledge and know-how by having a business discussion around the following questions:

- What major assumptions is this forecast based on?
- What changes to assumptions have been built into this forecast since last cycle?
- What are the issues and gaps I should know about?
- What are the risks and opportunities around this latest view?
- What decisions have already been taken but are not yet reflected in this view?
- What decisions should we be taking now?

The statistical forecast is always a useful input to the S&OP process but is not as valuable as understanding changes to assumptions. For example, the assumption on low-volume growth in the budget to support increased margin now may have changed to a big-volume push to hit a year-end target at the expense of margin. Making a high-level adjustment on the roughly right principle will be easier to do and more accurate for the medium to long term than changing the forecast at customer or SKU detail level.

If businesses pour enough resources (i.e., salespeople, demand managers, and systems) into planning at the detailed level over a medium- to long-term horizon, they can get an answer, albeit an answer that will not be understandable to management because of a lack of high-level assumptions. The main danger, apart from diverting the sales force from selling, is in creating an illusion of accuracy. The huge effort and cost involved in creating the forecast mean that people will tend to believe in it, whatever external changes may be occurring.

A further problem with detailed planning over the medium to long term is that no businesses have enough resources to do scenario planning and optimization at the

SKU/customer level. Simplification by looking at the overall context (e.g., region, brand, channel, technology type, etc.) of the data that reflect major drivers of change in the marketplace and aligning the application of roughly right assumptions rather than precisely wrong tinkering are fundamental before trying to model even a few options.

Finally, any strategy that calls for fewer and faster innovations and shortened product life cycles will mean that more items will fall into the planning horizon, even though they have not been invented yet, let alone have an assigned product code!

The level of detail involved in planning should change as the time horizon changes (Figure 20-9). In the short term, businesses need accurate transactional data on supply, sales, stocks, and so on and will be focusing resources on collaboration with customers and the deployment of advanced supply-chain optimization tools. In the medium to long term, the roughly right, not precisely wrong level of detail is required. Aggregating data using brand families, average revenues by technology type, and so on with a focus on the development of scenarios using business planning/simulation tools becomes important. This can be achieved only if finance has a strong leadership role in implementation.

## UNCERTAINTY VERSUS A SINGLE SET OF NUMBERS

Although there has always been a recognition that forecasts are either wrong or lucky, the early S&OP movement followed a premise that with immense pressure and focus the quest for a single set of numbers could remove uncertainty. In the early years, that pressure and focus often created some benefit by putting a basic rigor in place in sales and marketing and raising the profile of the need for improved forecast accuracy for operations. For many, though, the operational ownership and quest for a single set of numbers

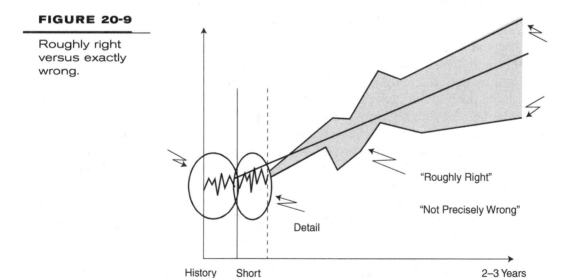

**FIGURE 20-9**

Roughly right versus exactly wrong.

"Roughly Right"

"Not Precisely Wrong"

Detail

History      Short                                                      2–3 Years
             Term

continued to dominate the conversation and became a mindless pursuit of accuracy beyond what was intrinsic in the markets in which a business operated.

Finance, meanwhile, had its own systems and realized that robust projections about the future relied on understanding scenarios. S&OP demands for a single set of numbers were perceived by many finance people to be naive.

In the context of decisive action, the role of managing and communicating assumptions played a huge part in enhancing the understanding and reconciliation of different views of the future. Supporting the numbers with the underlying assumptions, risks, and opportunities brings a richer dialogue about future projections and helps to focus on the greatest causes of uncertainty. Understanding the range (high/low) of opportunities, risks, and uncertainties is crucial to the discussion before agreeing to the latest view (see Figure 20-9).

In recent years, increased innovation, broader offerings to customers, and reduced product life cycles have made the old one-size-fits-all approach to crunching the numbers simply inadequate. Progressive organizations have switched the focus from numbers only to balancing numbers with the underlying assumptions to reduce the inherent uncertainty to a minimum and better managed risk.

However, uncertainty cannot be removed completely. There is still a range, and that range needs to be understood in the context of the decisions that will be taken on that information. In environments with extreme uncertainties (e.g., companies on the bleeding edge of technology breakthrough or movement into new markets and channels and pharmaceutical companies in the early stages of development), there is a need to go beyond understanding the range around a given projection and may be the need to run alternative scenarios based on unique sets of different assumptions.

This level of sophistication requires a different level of input from across the business and an understanding of how and when to use the output from different scenarios, but the acceptance of a range of uncertainty and the probable need for scenario planning typifies maturity in the S&OP process. Making decisions based on probability and/or odds of an event occurring with a particular result becomes very relevant. At what point should the odds trigger a scenario that we should include in the agreed latest view?

## S&OP AS THE ALIGNER TO SUCCESS AND FUTURE SUSTAINABILITY

In the last five years we have worked with several clients who had strong fundamental S&OP processes. They had a focus on a business agenda spanning 12 to 18 months in the future, aligned with business planning. However, to gear the S&OP process to future sustainability, we need to ensure that there is alignment with the first two years of the strategic view. Without this, S&OP becomes aligned with the business agenda only 18 to 24 months out, with a temptation to focus on the first 12 months.

With these clients, we have found great benefit in helping them align their S&OP process with the strategy of the business and the shape of the future product portfolio. It is a mistake to assume that when strategies are formulated, they do not change. For example, there are organizations that are formed from the merger of two companies that, having promised the stock market reductions in the cost structure, see this drive for efficiency as a necessary foundation for profitable growth. A common problem is that executives recognize that the strategic thrust has changed; some managers believe that the business is still following cost reduction as the primary target. S&OP done properly and aligned with the business strategy is a great tool for ensuring that the strategic and operations plans are in sync. We do not see the senior business management review as a forum to develop strategy, but it certainly is the proper forum to review strategy execution; it is there to ensure that the operational plans and latest view are in line with the strategic plan. Any major divergences are highlighted and discussed at a separate strategy review. This is an important distinction: S&OP in itself is not strategic, but it is the bridge between strategy and operations.

Earlier we highlighted the breakthrough of S&OP as the reconciler, but it must be reconciled to something. The choices are to reconcile to an 18-month business planning agenda with an emphasis on hitting the budget or, alternatively, seeing the next two years as a continuum toward future sustainability. The first year would be the budget, and the second year would be the early part of the strategic plan.

The two questions need to be posed to any business whose S&OP process needs to be aligned to future sustainability:

1. Of the three most well-known strategies—cost leadership, customer relationships, and product/service differentiation—which is the strategic direction in your business?
2. How much is new-product introduction part of the future portfolio? How much is really new as opposed to what is repackaged?

The type of executive leadership and the differing emphases of elements within S&OP depend on strategic intent and the future product portfolio. Understanding the difference in emphasis comes from aligning S&OP with the business agenda through right-to-left thinking.

## S&OP Executive Leadership Depends on Strategic Intent

Every business has a strategy; it is either crafted by executives or exists in the organization by default. The most common strategy default is operational excellence! Operational excellence, however, is not really a strategy; it is a necessary discipline and is a very important element of cost leadership, but the discipline is equally needed in any organization following strategies such as customer relationships and product/service differentiation. As an example, there were some businesses in the early twenty-first century that followed customer intimacy blindly and even called their S&OP process demand-driven, and became very responsive, but they ignored the discipline of operational excellence (sometimes called *operational effectiveness*). These businesses grew revenue by being responsive but became less profitable because they paid little attention to the cost of responsiveness. Improving operational effectiveness toward excellence is necessary to achieving superior profitability and is imperative but not sufficient. In itself, it is not a strategy.

A business with a cost leadership strategy aligns more to the traditional S&OP process. Managing the portfolio of new activities is a minor role; managing supply, because it is the biggest cost, has the major role, and the objective is to supply demand at the lowest cost. Therefore, supply feels the need for S&OP more than most. Eliminating forecast bias and improving forecast accuracy are the priorities. Other critical success factors include discipline, efficiency and effectiveness, clearly defined roles, waste elimination, continuous improvement, and reducing layers in the organization. A single set of numbers in this environment is appropriate, and executive leadership should be from finance and supply. Key measurements, in addition to forecast bias and accuracy, are customer service to promise, asset utilization, and cost (Figure 20-10).

Customer relationships is a strategy followed by businesses that believe that customer segmentation and providing a tailored service are the keys to growth and success. S&OP implementation is led by sales; with strong support from marketing, finance, and the supply chain, the decision-making process focuses on volume and revenue growth with an understanding of opportunities and risk—a range of numbers rather than a single set. Emphasis is on sales planning, with extensive involvement of account managers, promotional activity, and the timely introduction of product-line extensions. High levels of customer service and supply-chain responsiveness at minimum cost are standard expectations. Principal targets include customer retention, revenue, and profit by customer/channel.

In businesses that follow product/service differentiation as a strategy, leadership in S&OP implementation normally is strategic marketing, with support from research and development. Strong support is also needed form sales, finance, and the supply chain.

Decision making focuses on volume and margin growth, understanding opportunities and risks—again, a range of numbers rather than a single set—and in certain cases,

**FIGURE 20-10**

Impact of strategic focus on the S&OP process.

| Strategic Focus | Emphasis | Focus | Key Measurements |
|---|---|---|---|
| **Cost Leadership** | • One set of numbers for supply | • Volume<br>• Cost | • Customer service<br>• Forecast variability/stability<br>• Asset utilization<br>• Cost |
| **Customer Relationships** | • Bottom up view from account managers<br>• Sales planning<br>• Impact of promotions<br>• Macro overlay of customer segmentation | • Volume<br>• Revenue growth<br>• Opportunities (Hi)<br>• Risks (Lo) | • Customer retention/lifetime value<br>• Revenue by customer/channel<br>• Customer profitability |
| **Product Differentiation** | • Strong strategic marketing overlay on bottom up view<br>• Invest, growth vs. defense<br>• New activity prediction and risk managment<br>• Strong portfolio management | • Volume and profit<br>• Opportunities (Hi)<br>• Risks (Lo) | • New products as percent of profit/revenue<br>• New product time to profit<br>• Profit by brand/category<br>• Brand health<br>• Market share |

Forecast *Accuracy & Bias* need to be measured in all cases, however, the targets will be different—one size does not fit all.

scenario planning; uncertainty goes with the turf. Emphasis is strongly on strategic marketing, new-activity success, pipeline fill, minimizing obsolescence, and portfolio management. High levels of customer service and supply-chain responsiveness are expected, although there is forecast uncertainty. Primary targets include new products as a percent of profit, new-product/time to profit and profit by brand/segment, brand health, and market share.

Getting clarity in strategic intent is important before embarking on an S&OP implementation or reimplementation. Without this clarity, an S&OP implementation embarks on a one-size-fits-all approach; here, by default, the assumption is that operational excellence is a strategy. If the one-size-fits-all approach is followed for businesses using the strategy of customer relationships, product/service differentiation, there will be no enthusiasm from marketing, sales, finance, or business management. These key participants resent spending valuable time in a process that spends several hours in a month focused on volume and cost implications, one set of numbers for supply, and a set of measures that mainly interest the supply chain and only one measure (customer service) of interest to sales, and little to offer marketing.

## Different Portfolio Models and Their Impact on S&OP

How different the future will be from the past and present is important in understanding the business issues that connect to the S&OP process—S&OP is all about managing change and its consequences! Figure 20-11 shows five different portfolio models with their different emphases on S&OP.

From models 1 through 5 we go from a future devoid of new activity to one with a high degree of new-product introduction; in fact, in model 5, the new-activity impetus is coming from products that are new to the world.

An S&OP process in portfolio model 1 would be traditional, and since there is no new activity in the next five years, demand and supply balancing would be the emphasis in S&OP. Because forecasting standard products in markets that are not growing is rel-

**FIGURE 20-11**

Five portfolio models.

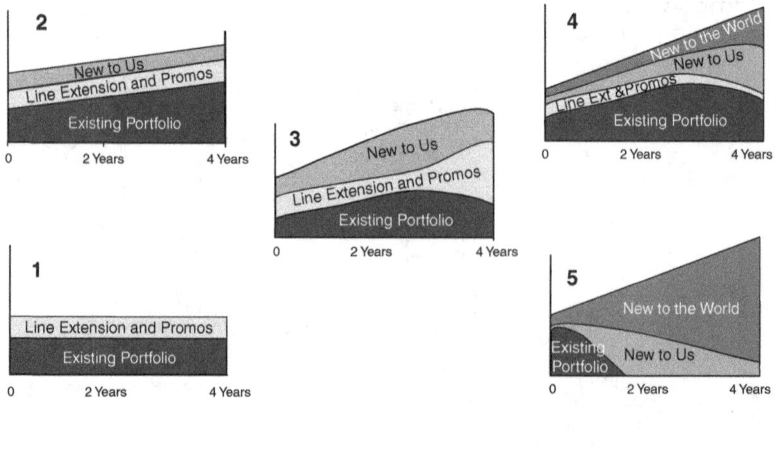

atively straightforward, there would be an emphasis on forecast accuracy and a single number in supply. A business of this kind typically would be following a strategy of cost leadership. An example of this type of business would be commodity chemicals.

In model 2, there is more new activity, but it is relatively straightforward, and the business appears to have linear growth. New activity would play a part, but it would be a minor role. An example of this would be an industrial chemicals organization whose main business is commodities but that is also looking at specialty chemicals and may be acquiring small businesses to augment the new-to-us category. The strategy here is primarily cost leadership, but the response in specialty chemicals could be differentiated service because of the higher margins on these products.

The most challenging business model for traditional S&OP is portfolio model 5, where the existing portfolio today will not be around in four years' time. These are businesses with a high degree of technology change and rapid implementation of new products. Portfolio management, including new products, is the single most important step in the S&OP process. The traditional S&OP model of demand and supply balancing would appear to be of little relevance to executives in this environment. Uncertainty and a range of numbers in the integrated reconciliation step and the importance of simulation and its impact on profitability have enormous consequences. Measurements such as time to market and time to profit are immensely important. Standard S&OP software that does not facilitate forecasting of new products before they are given a specific product code is an obstacle in this environment. Manufacturers of electronics, mobile phones, software, and computers are in this portfolio model. The strategy normally followed in these companies is product differentiation coupled with service differentiation.

Many food and drink companies and fast-moving consumer goods and pharmaceutical companies are examples of portfolio models 3 and 4. Typically, they would follow product/service differentiation or customer relationships.

If your business has a portfolio similar to models 3, 4, and 5, spending time only implementing a demand and supply process such as the traditional model in Figure 20-1 is really inappropriate.

## Strategic Intent and Future Product Portfolio and Their Impact on S&OP

Understanding the business strategy is essential to understanding the emphases on the way S&OP will work. In the preceding section we discussed how S&OP product portfolio models work and how these go hand in hand with understanding of strategic models. Strategies are about choices and tradeoffs, and each business needs to understand the principal strategy it is following. It is not unusual to find that an organization might have different business units following different strategies.

A one-size-fits-all universal checklist for S&OP is not helpful; a business guide showing that there are choices depending on strategy and product portfolio can be very helpful.

## DISCOVERIES LEADING TO BREAKTHROUGH S&OP

We have summarized the breakthroughs in S&OP knowledge and demonstrated the sequence in which they were discovered (Figure 20-12).

*Discovery 1: S&OP as the great unifier.* The desired outcome was one view or a single set of numbers, which was the considered antidote to multiple sets of numbers. This pursuit of the Holy Grail created dysfunction in the organization.

*Discovery 2: S&OP as the great reconciler and integrator.* It is not possible to get an agreed-on latest view unless the different views are reconciled. There must be recognition that different views add value to the process and provide a greater understanding of the latest business view. In addition, when multiple markets and supply points are involved, the S&OP process has to be integrated in a timely fashion. These two factors are the reason behind why we changed the pre-S&OP term to integrated reconciliation, that is, reconciling different views across functions, countries, and regions.

*Discovery 3: S&OP as the great aligner.* Reconciliation involves making apparently conflicting things compatible or consistent with each other. In the early days, we tried to align S&OP with the budget only. Unfortunately, this led to behavior supporting a 12-month view at the beginning of the fiscal year but became shorter term as the fiscal year progressed. The breakthrough came with S&OP as the great aligner; this was when we realized that it must be aligned beyond year end. The future involves the product portfolio and the strategic intent of the business.

**FIGURE 20-12**

Evolution of S&OP from left to right.

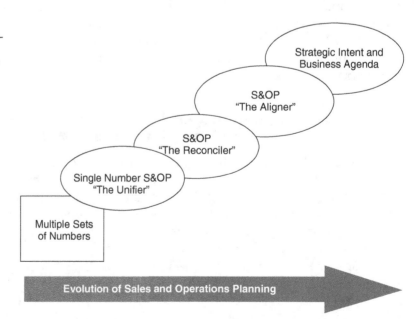

Although this is how S&OP has developed, we must reverse the sequence to implement successfully and quickly. Success to us means that you get real financial benefits from the process in six months. This implementation methodology is called right to left, which means that we start the implementation from the strategic intent and future product portfolio (Figure 20-13).

The implementation phases are

*Phase 1: S&OP as the great aligner.* The first step is the senior business management review and involves getting the business agenda around the strategic intent and future product portfolio.

*Phase 2: S&OP as the great reconciler and integrator.* This means that the business agenda in the senior business management review (step 1) drives the activities and discussion of integrated reconciliation (step 2). The different views adding value are explored and reconciled with the strategic intent of the business. In more complex companies involving global, regional, and country-wide activities, a mandatory step is to integrate and reconcile these views both top down and bottom up.

*Phase 3: S&OP as the great unifier.* This means that the operational steps of new activities, demand, and supply are directed from the needs of the integrated reconciliation process. Different views are welcomed because they add value to the knowledge of the business. Functional behavior is not tolerated in these steps, and cross-functional discussion is expected. All participants in the S&OP process are committed to a unified agreed-on latest view, having considered the different options.

**FIGURE 20-13**

Breakthrough
S&OP.

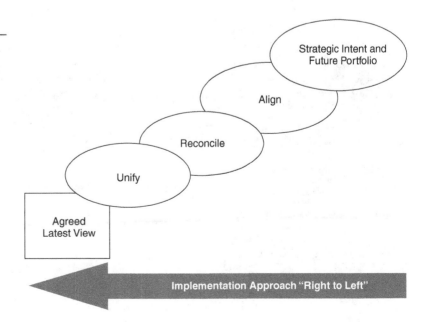

## Implementation Approach

This is a time-phased implementation plan showing the activities that should take place over an implementation time frame of six months. Phase 1 takes place in months 1 and 2, that is, alignment with senior management. Phase 2 takes place in months 2 and 3, that is, driving the senior management agenda through integrated reconciliation. Phase 3 takes place in months 4 and 5, that is, driving the needs of integrated reconciliation to be the outputs from new activities, demand, and supply (Figure 20-14).

   Although we believe that a customized approach to S&OP is important, there are certain principles that remain constant. The first and most critical one is that no business can afford to spend time on S&OP unless there is a return on S&OP in six months. Early successes and breakthroughs are mandatory. This implementation approach must be customized for your environment.

# APPLICATION OF S&OP TO VARIOUS ENVIRONMENTS

Does S&OP apply to any business? We believe that it does! Most S&OP use is in businesses that make and sell products. However, there are other businesses, such as small service companies, to which the principles of S&OP also apply. These businesses have a

**FIGURE 20-14**

S&OP leadership and understanding.

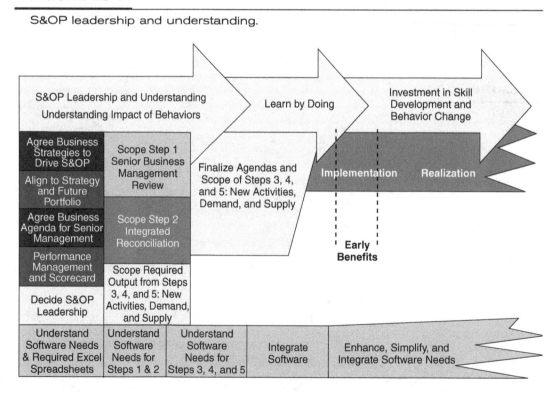

demand from the marketplace, and they need a process mechanism to evaluate the impact on their resources; they do not manufacture products, but they supply services through people. Some practitioners call this *sales and resource planning*, but the principles of S&OP are inherent in the process.

Are there some businesses that would not use S&OP properly? If a company does not want transparency, for whatever reason, S&OP is a threat. Strategic moves such as acquisitions or secretly preparing a business for a sale to other bidders, where the intention is only known to a select few, is not a fertile ground for S&OP. Businesses that are only interested in maximizing year end regardless of the effect on the following year are not good candidates either. S&OP treasures cross-functional strengths and medium- to long-term success.

## Global and Regional Considerations

This guide must be tailored to businesses that operate globally. The S&OP process works globally, regionally, and locally (in countries). It also must work in conditions where there is more than one business unit or routes to market.

Some companies have approached this issue with a traditional bottom-up approach, where country S&OP processes are aggregated to the regions, which then aggregate to a global S&OP meeting, as shown in Figure 20-15. This approach, called *old S&OP* on the left-hand side, results in a bottom-up aggregation of numbers facilitated by ERP, seen wrongly in our view, as the S&OP integration system. Global management sees its task first as, "Can I believe the data?" and second as, "How can we communicate that we do not like this number?" Disaggregation in these scenarios becomes a nightmare.

A better approach is shown on the right-hand side of Figure 20-15, where action can be taken at the global level, and specific changes can be made to regions and then countries. This entails understanding that each element in the three- or five-step process takes place globally, regionally, or locally and that these steps need to be aligned and integrated (Figure 20-16).

**FIGURE 20-15**

Old versus new S&OP.

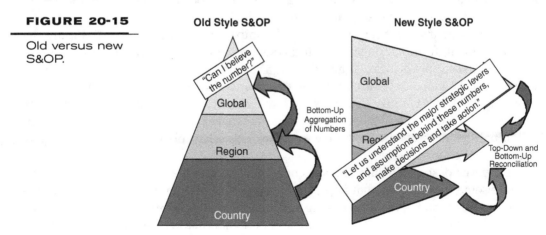

**FIGURE 20-16**

Aligning decision making with executive strategy.

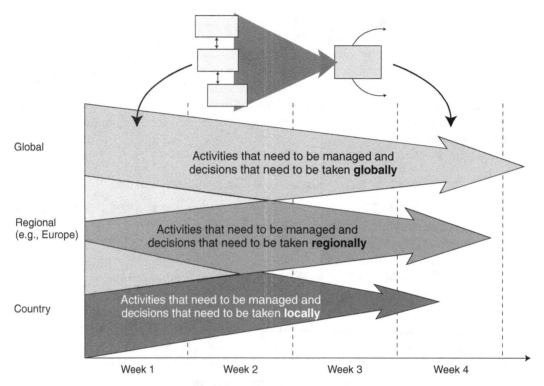

This is a subject in its own right and beyond the scope of this chapter. The design of the S&OP process needs to be specific to the company because the company has various degrees of complexity. For example, managing the portfolio and new activities may be directed globally, coordinated regionally, and executed locally. There is a managing-the-portfolio and new-activities step at all three levels.

In some pharmaceutical and chemical companies, active-ingredient supply decisions are global, formulation-supply decisions are regional, and end-item (packed and labeled SKUs) supply is local. Again, there is a managing-supply step at all levels, where decisions at the global level are more long term and strategic, regional issues are strategic and tactical, and country decisions are largely tactical. Our experience is that there is no one-size-fits-all guide in these more complex scenarios. This guide must be tailored to individual business needs, or it may be more helpful to obtain the guides to global, regional, and country S&OP.

However, despite the differences in complexity and company structure and operation, there are some subprocesses that apply universally and are essential to successful business operation. Assumption-based forecasting and scenario planning are two of these. It could be argued that if a company operates in a single country and market, assumptions on the business environment and competition are known to all, but this is certainly not true in global organizations where there can be vast differences between regions, countries, and markets. It is essential that the assumptions on which the strategy, longer-term forecasts, and budget are based are clearly identified and grounded. For larger companies, these should start with economic and political issues before considering market dynamics and competitive position. The hopes and aspirations for the product portfolio in terms of market share and the timing of and forecasts for new product introductions should feature strongly.

The business forecasts based on these assumptions would be a range, the extent of which would be underpinned by the positive and negative assumptions. There is no place here for a single set of numbers! The monthly regional and global S&OPs would review significant changes to the assumptions and modify forecasts accordingly while firming short-term forecasts.

Hand in hand with assumption-based forecasts is scenario planning. Each forecast should have a most likely outcome with upsides and downsides. The manufacturing and supply functions build their firm plan around the most likely outcome but have a plan of how to meet upsides and downsides should they arise together with decision points. These are the latest times when these alternative plans can be initiated successfully.

The S&OPs at the country, regional, and global levels are cross-functional groups that review the assumptions, forecasts, and scenario plans on an exception basis. Discussion is focused on either major changes to assumptions or a pending decision point for a scenario plan. Budget, cash flow, and strategic consequences of these changes, of course, are worked up by the finance function. With such a tight grip of the business, it is sometimes possible for the global or regional teams to switch resources to capture exceptional market opportunities because they have a clear, current view of markets. It is our view that the power of S&OP is unrivaled in these situations, and it also can be used extremely effectively in broader resource planning and allocation across companies including capital planning.

## Small to Medium-Sized Businesses and Developing Markets

Another question often posed is, "My business is small, for example, $2 million to $8 million revenue. Do I still need S&OP?" Our view is that any successful business uses S&OP principles formally or informally. A one-person company or a small group of entrepreneurs runs S&OP informally in their heads or around the coffee machine. They do not have the formal process steps outlined in the earlier sections, but they are continually evaluating predicted demand and balancing resources to meet it. They share assumptions and changes to assumptions informally on a daily basis.

This chapter has been written to apply mainly to medium to large businesses, but the challenge for smaller businesses is that as they grow and employ more people, they do need to formalize their process for continually evaluating demand against the resources they have. They probably do not need a formal five-step S&OP process, however; instead, a simpler three-step process is used, as shown in Figure 20-17.

FIGURE 20-17

Small-business
S&OP.

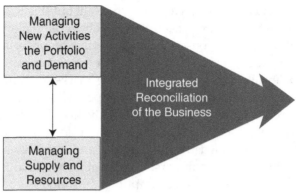

The first step is a combined managing new activities, portfolio, and demand review; this output is analyzed in a managing supply and resources step. The results of this evaluation then are discussed in the third step—integrated reconciliation of the business. Most of the players in the first two steps attend the third step, which is normally chaired by the managing director or owner of the business. The three-step process is also applicable in global, regional, and country environments. In businesses such as these, there are developed markets, which probably need the five-step model. There also are developing markets, which are likely to need only a three-step process. Managing the S&OP transition from developing to developed market is important.

Our recommendation is that smaller businesses should seek to engrain the culture of a formal S&OP process at an early stage of growth. New people joining the business as it grows have little choice but to act in a joined-up manner; new recruits with an inclination to functional behavior will recognize the danger of this silo optimization behavior if there is an existing process for integrating the business with S&OP. Another reason for implementing S&OP early is because implementing when the business has grown large enough to exhibit functional behavior will be more difficult. Developing a positive culture is one thing; making a cultural change is significantly more difficult.

A good analogy is that of training a climbing rose around an archway or up a trellis. Putting in the framework (i.e., arch, trellis, or S&OP framework) while the rose is small will enable the growth to be managed in the right direction with minimal training, pruning, and tying back of loose branches. Putting in the framework when the rose is mature is altogether more difficult. It involves lots of pruning, training of wayward branches going in opposite directions, many scratches and heartache, and time.

## SUMMARY

This chapter traces the evolution of S&OP from its inception in the late 1980s, when the primary objective was a medium- to long-term stable production plan, to the twenty-first century, when several successful businesses are using it as a dynamic business performance management process that enables and tracks their progress against their future portfolio and strategic intent. These businesses are maximizing the potential of S&OP by increasing profits and generating cash.

The foundational benefits of improved customer service and lower working capital are still important, but today, more advanced companies have built on a robust operational foundation and now use the latest view from S&OP to generate the quarterly forecast for corporate headquarters. These businesses no longer have annual budgeting as a separate exercise. They also demonstrate strong cross-functional behavior throughout the organization, and executive management is focused on long-term sustainability.

The evolution of S&OP, by examining it as a series of breakthroughs, has provided us with knowledge of what is important. Moreover, by understanding the big picture, we can learn the optimal way of successful implementation. S&OP evolving through time is shown (Figure 20-18).

Evolution from multiple sets of numbers to an S&OP process aligned with the future business agenda has taken place over 20 years. Unfortunately, some practitioners have tried to implement S&OP in the same direction as its evolution—left to right. They believe that their S&OP process eventually will align with the business and strategic agenda, but this usually takes far too long, and senior management loses patience. More often than not, the process becomes associated with one of the evolution steps and does not progress. A common problem is the objective of a single set of numbers when executive management wants to see a range—including highs and lows. Management then treats S&OP as a sideshow and sees S&OP the unifier as a false promise.

We believe that the powerful message in S&OP evolution is that you align the process with the strategic agenda and future portfolio from the outset. We have experience with many clients showing that you can approach any S&OP implementation,

**FIGURE 20-18**

Evolution of
S&OP from left to
right.

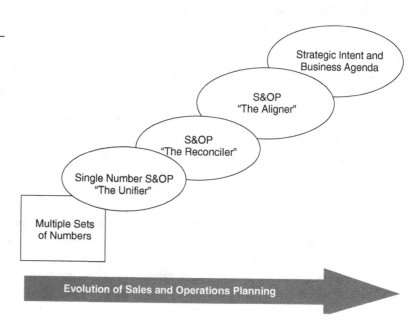

**FIGURE 20-19**

Breakthrough
S&OP—right to
left.

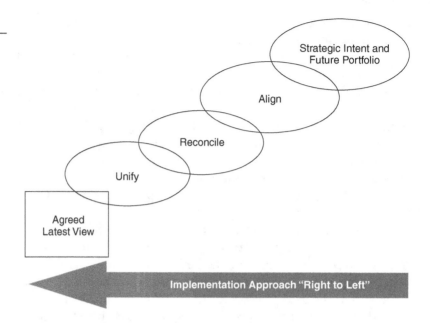

improvement program, or reimplementation and achieve dramatic results within six
months by using our right-to-left breakthrough approach (Figure 20-19).

Having aligned the process at the beginning with the strategic intent and future portfolio, we can reconcile the different views within the business to the future business agenda and sustainable success. The end result, achieved though cross-functional behavior, is
an agreed-on latest view over 18 to 24 months for both operational and business plans.

During the evolution, we were attempting to unify first, then reconcile, and finally,
align S&OP. We now know that the order must be reversed. We align the S&OP process
with the business agenda first, then reconcile different views, and finally, unification is
achieved through agreeing to the latest view of the future. The most rapid and sure way
to implement S&OP successfully is to implement from right to left.

To summarize, the key concepts of S&OP that are worthy of executive attention are

1. *Executive leadership.* The process should not be led by supply only. At the very
   least, the supply chain and finance should jointly lead the process, and this is
   appropriate in businesses following cost leadership as a strategy. In other cases,
   direction of the process should come from sales and marketing; when the
   strategic intent of the business is customer relationships, it should be sales led,
   and marketing should lead when a business follows product/service differentiation. Finance always should have a strong lead in the process because of its
   role in business planning, whatever the strategic intent. Ideally, executives
   should champion the process, but in large multinational companies, this is not
   always possible. Senior management does not normally buy the process; they
   buy the results. However, the process must be rigorous and disciplined so that
   the executives can have confidence in the integrity of the information.

2. *Importance of finance.* Volume and value must be integrated in the medium- to long-term views. Many businesses have separate processes for volume and financial forecasts.

3. *Roughly right versus precisely wrong.* Obsession with extrapolation of detailed forecasts is unhealthy and will turn off business management. Roughly right, not precisely wrong is the recommended approach. There must be an understanding of different ways of reducing and coping with uncertainty.

4. *New activities.* New activities must be integrated into demand/supply discussions from the outset.

5. *Knowledge versus data.* The key to success is to get a shared understanding of what the numbers in the latest view mean rather than just debating the numbers. Achieving this insight requires a focus on assumption changes, risks and opportunities to understand different views, and why there has been a change. Drowning in data but starved of knowledge is a very important concept.

6. *Cross-functional behavior.* The process must clearly join up functions. There must be recognition that different views add value and provide a richer understanding of trends and where the business is going. Cross-functional behavior and executive leadership are of paramount importance. Corrective action must be identified because flawless execution in all functions is demanded. A functional silo culture where the person who shouts loudest wins is detrimental to S&OP success; such organizations get stuck on a single set of numbers.

7. *Benefits in six months.* Successful implementation is achieved by setting an aggressive schedule, focusing on fast results (in six months), and a commitment to learning by doing. Right-to-left implementation means that the first step is alignment with the executive management agenda. The second step is recognition that different views add value, leading to a reconciliation process. An agreed-on latest view comes from reconciling these different views to the strategic intent and a valid operational plan.

8. *S&OP aligned with strategic intent.* S&OP as the umbrella for operational excellence is a flawed premise. Some think that operational excellence is a strategy, whereas it is a mandatory discipline for any business regardless of strategy. Operational excellence has a high impact in a business that is following cost leadership as a strategy, and a single set of numbers is more appropriate as an outcome of S&OP. However, many fast-moving consumer goods, pharmaceutical, and hi-tech companies often follow different strategies, such as customer relationships or product and service differentiation. These businesses have difficulty with the tenet of a single set of numbers because their rationale is high risk, innovation, and uncertainty, and they have a high incidence of new-product introduction. These businesses need an S&OP process that copes with these issues. Scenario planning with ranges and documented assumptions are main agenda items, and the ability to cope with ambiguity is critical.

9. *Customized S&OP.* Beware of prescription and one size fits all. Yes, the concept is applicable, but S&OP has to be tailored to your business needs. The dimensions of values and behaviors, processes, and resources must be understood. Our experience clearly shows that behavior change, data integrity, and simple integrated software are major issues to address in ensuring that your S&OP process delivers sustainable business results.

S&OP is applicable to all businesses—manufacturing or service, small or large, globally integrated or single site. Any company that desires transparency and the development of cross-functional strengths and medium- to long-term success can benefit.

## ABOUT THE AUTHORS

Dick Ling and Andy Coldrick have been collaborating on S&OP for 25 years. They are a very strong partnership and specialize in pushing the boundaries of S&OP. They help businesses all over the world to maximize S&OP's potential to generate more cash and increase profits. Dick created S&OP, and he and Andy are two of the leading thinkers and consultants on its evolution and advancement. They both now live in the United States and are working together even more closely. They assist clients with breakthrough success in implementing right-to-left S&OP.

They led the thinking on aligning the S&OP process with the strategic intent of the business and future portfolio. Before that they were the first to recognize that new-product activity and financial links into traditional S&OP were treated as afterthoughts but were not being truly integrated. They pioneered integration of these two pieces and also created the integrated reconciliation step to explode the single-number myth in vogue at the time. The importance of understanding change, assumption management, and scenario planning with a range of views all reinforced management's need for information that built knowledge and know-how rather than data just supplying more and more numbers. This led to the discovery that breakthrough S&OP necessitates strong cross-functional leadership and behavior. Using a range of views also presented the need to use a different process from the traditional demand/supply balancing model. During that time, they encountered companies in complex environments. Clients in just North America implement S&OP in one very large country—the United States. Europe is dealing with a more complex picture, having clusters of countries, and in many cases these clusters are combined into a region such as Europe, Africa, and the Middle East. In the 1990s, Europe and Asia were microcosms of the global picture. Dick and Andy pioneered the way to implement integrated S&OP in global, regional, and country environments based on experience in Europe. Their ability to work anywhere in the world with multinationals built an unrivaled experience in helping businesses maximize the benefits of global S&OP while understanding potential road blocks with different cultures, expectations, and systems.

Working with small developing countries within a regional context, together with parallel experience in helping small independent businesses, gave them the insights necessary for a simplified S&OP approach, albeit a three-step process.

They remain passionately committed to customizing S&OP for business environments. There are common principles and themes, but each successful S&OP user has its own uniqueness. Any successful business is striving to be different from its competition.

Dick and Andy express their appreciation to Bruce Bissell and David Whitewood for their insights as senior executives into the successful application of breakthrough S&OP.

The intellectual property rights of this chapter belong to the authors. Permission has been granted for the use in this text.

## BIBLIOGRAPHY

Ling, Richard C., and Walter E. Goddard. *Orchestrating Success: Improve Control of the Business with Sales and Operations Planning.* John Wiley and Sons: Hoboken, NJ, 1995.

Ling, Richard C., Andy Coldrick, Bruce Bissell, and David Whitewood, with Duncan Alexander. *Breakthrough Sales and Operations Planning: How Successful Is Your Business?* Ling-Coldrick: Mebane, NC 2010. Available at www.lingcoldrick.com.

## PART 4

# Looking Backward and Forward

## CHAPTER 21

# Historical Context

*"Those who cannot remember the past are condemned to repeat it."*
—GEORGE SANTAYANA
*"The farther back you can look, the farther forward you are likely to see."*
—SIR WINSTON S. CHURCHILL

## PRE-MRP INVENTORY CONTROL

With the invention and articulation of material requirements planning (MRP), the underlying assumptions around inventory needed to be challenged. Specifically, the following mainstays of pre-MRP inventory management became subject to reappraisal as to their validity, relevance, and applicability to manufacturing inventories:

1. The concept of stock replenishment
2. All techniques built around reorder points
3. The square-root approach to the economic order quantity
4. The analysis and categorization of inventory by function
5. The conventional notion of aggregate inventory management
6. The ABC inventory classification

### Stock Replenishment

Stock replenishment is a concept forcibly grafted onto a manufacturing inventory. Typical replenishment concepts are in conflict with basic management objectives of low inventory and high return on investment. The term *replenishment* means restoration to a state of (original) fullness. But manufacturing inventories should, if possible, be the very opposite of the mentioned *fullness*. Typical stock replenishment systems are based on the principle of having inventory items in stock at all times so as to make them available at the (poorly predictable) time of need. Stock replenishment is intended to compensate for the inability to determine the precise quantity and time of need in the short-term future.

In manufacturing, though, the idea is to have the inventory item available at the time of need (and, if possible, not before or after that time) rather than to carry it just so that it would be available when, and if, needed. To the extent that short-term need for individual manufacturing inventory items can be pinpointed in terms of both quantity and timing—and this is indeed possible through the use of modern computer-assisted methods—traditional stock replenishment techniques in a manufacturing environment prove undesirable and wasteful. One example of this is the use of kanbans in many lean implementations. These kanbans can contain inventory that is not necessary for some time but consume cash and space. The management of these kanbans through cards and other manual methods adds yet even another level of unnecessary complexity and waste when the desire is to eliminate waste.

## Reorder-Point Techniques

Reorder-point techniques, in their various forms, represent implementation of the stock replenishment concept. These techniques, including the statistical order point, min/max, ordering "up to," and maintenance of $N$ months' supply, represent variations on a common theme. Whether explicitly or implicitly, all of them forecast demand during replenishment lead time, and all attempt to provide for some safety stock to compensate for fluctuations in demand. Systems based on reorder-point techniques suffer from false assumptions about the demand environment, tend to misinterpret observed demand behavior, and lack the ability to determine the specific timing of future demand. These shortcomings, inherent in all systems of this type, manifest in a number of unsatisfactory performance characteristics, chief among them being an unnecessarily high overall inventory level, inventory imbalance, and stock-outs or shortages caused by the system itself.

## Economic Order Quantity

According to American Production and Inventory Control Society (APICS), *economic order quantity* is defined as:

> **economic order quantity (EOQ):** A type of fixed order quantity model that determines the amount of an item to be purchased or manufactured at one time. The intent is to minimize the combined costs of acquiring and carrying inventory.[1]

---

[1] *APICS Dictionary*, 12th ed. New York: Blackstone, 2008.

The basic formula is

$$\text{Quantity} = \sqrt{\frac{2AS}{iC}}$$

where $A$ is annual usage in units, $S$ is ordering costs, $i$ is annual inventory carrying cost rate as a decimal, and $C$ is unit cost.

The economic order quantity (EOQ) turns out to be a poor ordering quantity in the typical manufacturing demand environment. The EOQ equation is totally insensitive to the timing of actual, discrete demands (requirements) arising during the period that the EOQ is intended to cover following its arrival in stock. Once future requirements for an inventory item are precisely determined and positioned along a time axis, it can be seen that the square-root approach in the EOQ calculation does nothing to balance the lot size against either the timing or the quantity of actual requirements. For example, demand for an item over a 10-week period may be determined in advance to be, by week, 20-0-20-0-0-0-0-0-20-0. The EOQ for this item may turn out to be 50, more than needed to cover the first three weeks' requirements but not enough to cover the next requirement in the ninth week. The "remnant"' of 10 pieces will be carried for eight weeks without any purpose. Note that the EOQ still would be 50 if the 10-week demand were 20-0-40-0-0-0-0-0-0-0 or 20-0-0-0-0-0-0-0-0-0-40. In the first instance, the EOQ would fail to cover the first three weeks' requirements, and in the second, the excess 30 pieces would be carried for nine weeks without being able to satisfy the requirements of the tenth week. In any of these cases, the EOQ is determined solely on the basis of setup cost, unit cost, carrying cost, and annual usage. The derivation of the EOQ formula rests squarely on a basic assumption of uniform demand in small increments of the replenishment quantity, that is, gradual inventory depletion at a steady rate, which then allows the carrying cost to be calculated for an "average" inventory of one-half the order quantity. This basic assumption is grossly unrealistic vis-à-vis a manufacturing inventory and therefore fatal to the validity of the technique.

The stock replenishment, order-point, and order-quantity techniques dominated the pre-MRP world in both actual practice and the inventory control literature. The reason for this is historical. The field had been conditioned in favor of the philosophy expressed by these techniques because the pioneering theoretical work in inventory control generally was confined to the areas of order point and order quantity. This work had been stimulated by the fact that problems of order point and order quantity lend themselves to the application of mathematical statistical methods that have been known and readily available for quite some time. The inventory control problem was perceived as being essentially mathematical rather than one of massive data handling and data manipulation, the means for which simply did not exist before the computer and MRP.

## Analysis and Categorization of Inventory by Function

The fact that the chronic problems of manufacturing inventory management now can be solved, however, is due not to better mathematics but to better data processing and computer power. Inventory analysis and categorization by function, designed to account for a given total inventory in terms of the respective functions of its constituent inventory groupings, that is,

- Order sizing
- Fluctuation
- Stabilization
- Anticipation
- Transportation

is subject to revision not of concept but of the method of determining the value of the respective inventory categories. Of the functions listed, it is the first two that appear in a new light. Now the first question to be asked is "where" to put inventory rather than to answer the question of "how much."

Order sizing creates what is called *cycle stock* or *lot-size inventory*. But this category cannot be reckoned to approximate one-half the quantities being ordered, which has been the traditional approach. With the discarding of the EOQ in favor of discrete lot-sizing techniques (discussed in Chapter 8), the lack of validity of such an approximation becomes clearly apparent.

Actual order quantities for a given inventory item will be seen to equal the precalculated requirements for one or more planning periods without remainder, causing the quantity to vary freely from one order to the next. The number of periods covered by an order quantity will be dictated in part by the relative continuity of demand for the item in question. In cases of pronounced discontinuity, the order quantity will tend to equal requirements for one period. The same usually will be true for all assembled items (subassemblies) because of the typically minor assembly setup considerations.

The second category, called *safety stock*, serves primarily to compensate for or to absorb fluctuation in demand. In stock replenishment systems, the safety-stock quantity is (a known) part of the order-point quantity calculated for each inventory item, and the sum of the item safety stocks times the product cost represents a fair estimate of this category of inventory.

In effective MRP systems, safety stock on the item level tends to disappear. It is normally no longer calculated for each inventory item separately, but where used at all, it is incorporated at the end-item (master production schedule) level only. This has been developed further with the concept of actively synchronized replenishment, which locates inventory such that it buffers and compresses the overall manufacturing process.

## Conventional Notion of Aggregate Inventory Management

Aggregate inventory management is a concept and a set of techniques used for manipulating and controlling inventory in toto. As the term implies, the overall inventory investment level can become subject to direct management control through certain policy variables to the extent that the policies in question apply to the inventory across the board or if different policies apply to a limited number of inventory groups. Under the conventional approach to aggregate inventory management, the two inventory categories most susceptible to control by means of varying a policy are lot-size inventory and safety stock. When lot sizes are being determined through some form of the EOQ formula, it is possible to exert across-the-board control over them by manipulating the carrying-cost variable in this formula.

Carrying cost, a controversial value, is in all cases semiarbitrary (in practice, the values in use vary between 8 and 45 percent per annum from company to company) and therefore can be thought of as reflecting management policy. Increasing the carrying cost used in the EOQ computation will result in smaller lot sizes, and vice versa. Thus the inventory carrying cost in use at any given time reflects the premium that management is putting on the conservation of cash.

The idea is entirely sound, and its application is simple and direct. This brings the focus to the real driving force behind establishment of lot size, which is the setup time. The other factors in the EOQ equation (i.e., carrying cost, annual usage, and inventory carrying cost) are out of the direct control of the operations manager. The only item that can be affected directly by operations is the setup time to result in a reduced EOQ lot size. This is the basis for the single minute exchange of dies (SMED) that is so essential in lean implementations.

What now must be discarded, however, are the traditional methods of quantifying the results that can be expected as a consequence of a given individual policy change. In an EOQ environment, the theoretical relationship between incremental change in carrying cost and lot-quantity change (and, by extension, lot-size inventory change) is clean and straightforward. The EOQ varies inversely with the square root of the carrying cost. The value of EOQ squared doubles as a result of halving carrying costs and halves as a result of doubling this cost. Table 21-1 shows some typical lot-size reductions and the carrying-cost increases required to effect them. For example, a 10 percent reduction in lot size requires a 23 percent increase in the carrying cost.

**TABLE 21-1**

Carrying Cost and Lot Size Relationship

| Desired Lot-Size Reduction, % | Relative Values | | |
| --- | --- | --- | --- |
| | Carrying cost | EOQ | EOQ$^2$ |
| 10 | 123 | 90 | 81 |
| 20 | 156 | 80 | 64 |
| 30 | 200 | 70 | 50 |

Once a stock replenishment system, with its EOQ orientation, is replaced by an MRP system using discrete lot sizing, an exact mathematical relationship between incremental change in carrying cost and lot-size inventory no longer exists owing to the interrelationship of the parts.

Inventory carrying cost continues to figure in the determination of lot sizes where warranted by the economics of ordering, but other factors (see Chapter 8) exert a more direct influence on the lot size, which usually varies from order to order. Changes in lot sizes and in lot-size inventory therefore cannot be predetermined quantitatively on the basis of a change in carrying cost alone.

The other policy variable used in conventional aggregate inventory management is the *service level*, which enters into the traditional calculation of safety stock. In a stock replenishment system environment, the quantity of safety stock for end items is computed individually for each item in the inventory, and its principal determinant is the standard deviation of past demands per period from their arithmetic mean. A normal distribution of these demands being assumed, a desired service level, that is, incidence of item availability, determines the number of deviations represented by the safety stock. Historically, it was believed that the higher the service level desired, the higher was the safety stock, and vice versa. The investment in this inventory category therefore can be controlled by manipulating the service-level value. The principle has been invalidated by more recent developments such as actively synchronized replenishment (ASR), which reduces inventory while at the same time increasing customer service levels. In any case, this statistical safety-stock technique tends to become irrelevant because safety stock is normally not planned at the component item level in an MRP system.

## ABC Inventory Classification

ABC inventory classification[2] is a popular inventory control technique that is an adaptation of Pareto's law. In a study of the distribution of wealth and income in Italy, Vilfredo Pareto observed in 1897 that a very large percentage of the total national income was concentrated in a small percentage of the population. Believing that this reflected a universal principle, he formulated the axiom that the significant items in a given group normally constitute a small portion of the total items in the group and that the majority of items in the total will, in the aggregate, be of minor significance. Pareto expressed this empirical relationship mathematically, but the rough pattern is 80 percent of the distribution being accounted for by 20 percent of the group membership.

The 80-20 pattern holds in most inventories, where it can be shown that approximately 20 percent of the items account for 80 percent of total cost (unit cost times usage quantity). In the typical ABC classification, these are designated as A items, and the remaining 80 percent of the items become *B*'s and *C*'s, representing the middle 30 percent that account for 15 percent of cost and the bottom 50 percent that account for 5 percent of cost, respectively.

---

[2] Introduced by H. Ford Dickie in "ABC Inventory Analysis Shoots for Dollars," *Factory Management and Maintenance*, July 1951.

The idea behind ABC is to apply the bulk of the (limited) planning and control resources to the *A* items, "where the money is," at the expense of the other classes that have demonstrably much less effect on the overall inventory investment. The ABC concept is to be implemented by controlling *A* items "more tightly" than *B* items, and so on. Today, the principle of graduated control stringency may be somewhat difficult to comprehend, but in the precomputer days, the degree of control was equated with the frequency of reviews of a given inventory item record. Controlling *tightly* meant reviewing frequently. The frequency of review, in turn, tended to determine order quantity. *A* items would be reviewed frequently and ordered in small quantities to keep inventory investment down. A primitive but fairly typical ABC implementation is represented by policies shown in Table 21-2.

**TABLE 21-2**

Sample Ordering Rules Under ABC Classification

| Inventory Class | Review Frequency | Order Quantity |
| :---: | :---: | :---: |
| A | Monthly | 1 month's supply |
| B | Quarterly | 3 months' supply |
| C | Annually | 12 months' supply |

The rationale of ABC classification is the impracticality of giving an equally high degree of attention to the record of every inventory item owing to limited information-processing capacity. With modern computers and software, this limitation disappears and the ABC concept becomes significantly less relevant. Every practitioner knows that a single part, even one that is low in cost, can stop an entire shipment. Equal treatment of all inventory items, as far as planning is concerned, now becomes feasible. In a modern, well-implemented MRP system, every item, irrespective of its cost and volume, receives the same degree of care, the same stringent treatment.

Possible policy exceptions are certain extremely low-cost items, especially purchased ones, that may have safety stocks and be ordered in large quantities. These exceptions are made, however, not because of some inability of the computer system to plan and maintain the status of such items but rather because of the impracticality of accurate physical control. It simply does not pay to do exact counts of lock washers and cotter pins. The cost of counting can be several times larger than the total cost of the parts. Physical inventory control continues to be a problem in inventory management, and the ABC concept, when applied in this area (to inspection, storage, frequency of cycle checks, etc.), remains valid.

The techniques and concepts covered in the preceding discussion evolved during a time when, owing to very limited information-processing capacity and software availability, the precise pattern of future item demand could not be ascertained and reascertained; neither could the status of every inventory item be updated and reevaluated with sufficient frequency. The approaches embodied in these techniques and concepts reflect the former deficiencies of information in attempting to compensate for them by other means. They lost their relevancy and usefulness, however, once it became feasible, through computer-assisted MRP, to establish and maintain the formerly unavailable information.

## THE STORY OF MRP

In rereading the original Orlicky text, as well as the revision by George Plossl, the first-hand narrative of the history of operations management and system development from these two men had to be preserved to facilitate understanding of the concepts presented in this update. For example, below is an excerpt from the original Preface written by Joseph Orlicky in 1974—almost 40 years before this third edition. The issues he describes then are even more relevant today.

> Someone had to write this book.
>
> Since around 1960, when a few of us pioneered the development and installation of computer-based MRP systems, time-phased material requirements planning has come a long way—as a technique, as an approach, as an area of new knowledge. From the original handful, the number of MRP systems used in American industry gradually grew to about 150 in 1971, when the growth curve began a steep rise as a result of the "MRP Crusade," a national program of publicity and education sponsored by the American Production and Inventory Control Society (APICS).
>
> As this book goes to print, there are some 700 manufacturing companies or plants that either have implemented, or are committed to implementing, MRP systems. Material requirements planning has become a new way of life in production and inventory management, displacing older methods in general and statistical inventory control in particular. I, for one, have no doubt whatever that it will be the way of life in the future.
>
> Thus far, however, the subject of material requirements planning has been neglected in hard-cover literature and academic curricula, in favor of techniques that people in industry now consider of low relevance or obsolete. I suppose one of the reasons for this situation is the subject's position outside the scope of quantitative analysis and the view of it as being "vocational" rather than "scientific." The subject of production and inventory management is, of course, vocational in the sense that the knowledge is intended to be applied for solving real-life business problems. Like engineering or surgery, production and inventory management is oriented toward practice. Unlike many other approaches and techniques, material requirements planning "works," which is its best recommendation.
>
> In the field of production and inventory management, literature does not lead, it follows. The techniques of modern material requirements plan-

ning have been developed not by theoreticians and researchers but by practitioners. Thus the knowledge remained, for a long time, the property of scattered MRP system users who normally have little time or inclination to write for the public. The lag of literature became painfully evident to me when, in early 1973, I undertook to prepare a study guide for a publicly administered examination on material requirements planning. I chaired the committee of APICS in charge of developing this examination, under the guidance of Educational Testing Service of Princeton, and in preparing the study guide I wanted to list all written material pertaining to the subject.

I found that the entire MRP literature consisted of twenty-six items good, bad, and indifferent—all of which were either articles, excerpts, special reports, or trade-press "testimonials." Tutorial material on MRP basics was lacking entirely. The job of "getting it all together" remained to be done.

Someone had to write a book on "MRP from A to Z," and I concluded I may have to be the one. So I wrote this book.

In the text, I am quoting from, paraphrasing, and otherwise making free use of my own past writings on the subject. A section of Chapter 1 includes some of the material originally published in the *Proceedings of the 13th International Conference of APICS*, 1970, under the title, "Requirements Planning Systems: Cinderella's Bright Prospects for the Future." A good portion of this material, including slightly disguised illustrations, surfaced, without attribution, in a 1973 publication. Just so the unwary reader won't reach the wrong conclusion as to who is cribbing from whom, I felt obliged to reaffirm my authorship in a footnote on page 21.

Chapter 5 is largely based on my paper, "Net Change Material Requirements Planning," published in the *IBM Systems Journal*, vol. 12, no. 1, 1973.

Some of the material in Chapter 9 is adapted from a section called "The Problem of Data" in my book *The Successful Computer System*, McGraw-Hill, 1969.

Chapter 11 parallels, in part, the chapter called "Master Production Schedule Planning" in the *Communications Oriented Production Information and Control System (COPICS)*, IBM Corporation, 1972, which I helped to write.

Chapter 10 includes most of the material in the article, "Structuring the Bill of Material for MRP," published in *Production and Inventory Management*, vol. 13, no. 4, 1972, which I wrote in collaboration with George W. Plossl and Oliver W. Wight.

Finally, Chapter 12 is based on the keynote address I delivered at the 15th International Conference of APICS on October 12, 1972, in Toronto.

November, 1974
Joseph Orlicky
Stamford, Connecticut

George Plossl in his update of Orlicky's classic work begins with more of this historical timeline. Understanding the history best sets the stage to understand the direction for the future.

> In 1966, Joe Orlicky, Oliver Wight, and I met in an American Production and Inventory Control Society (APICS) conference. We found that we had all been working on material requirements planning (MRP) programs, Joe at J. I. Case Company and IBM, Oliver and I at The Stanley Works. We continued to meet and compare notes on MRP and other topics. In the early 1970s we organized the APICS MRP crusade, using the resources of the Society and the knowledge and experiences of a few "Crusaders" to spread the word on MRP among APICS members and others interested. All but a few APICS chapters participated.
>
> This crusade showed clearly the need for more professional literature and teaching aids on this powerful but new technique. Oliver's and my book, *Production and Inventory Control: Principles and Techniques*, published in 1967, was the first to include coverage of MRP—all of 16 pages' worth! Users of MRP were too busy trying to make it work to take time to write about it.

The purpose of this third revision of this book is to contribute to business literature a comprehensive, state-of-the-art treatment of a subject of relatively recent origin yet of first importance to the field of manufacturing operations management—effectively planning material in an increasingly complex and demand-driven world. This approach, its underlying philosophy, and the methods involved represent a sharp break with past theory and practice. The subject, broadly viewed, marks the coming of age of the field of production and inventory control and a new way of life in the management of a manufacturing business.

The commercial availability of computers in the mid-1950s ushered in a new era of business information processing with a profound impact of the new technology on the conduct of operations. George Plossl in the second edition discusses the impact of the computer. The hardware and software became available commercially in the early 1960s and were capable of handling data in volumes and at speeds previously scarcely imaginable; these lifted data-processing constraints that had handicapped inventory management. They made obsolete the older methods and techniques devised to live with these limitations.

In the early days, the most significant benefits were not achieved by the pioneering manufacturing firms that chose to improve, refine, and speed up existing procedures with computers but by those who undertook fundamental changes to their planning and control systems. However, abandoning familiar techniques, even those which had proven unsatisfactory, and substituting new, radically different approaches such as MRP-based systems required new knowledge not possessed by many people at that time. The first edition of this book provided an authoritative source of information on MRP to help develop the needed knowledge.

Since the early 1960s, when a few pioneered the development and installation of computer-based MRP systems, time-phased MRP has come a long way—both as a useful technique and as a source of new knowledge. From the original handful, the number of MRP systems in use in American industry grew gradually to about 150 in 1971, when the growth curve began a steep rise. By 1975, over 700 systems had been implemented, and the 1980s saw thousands more come into use. Today, it is a rare company in Western and Pacific Rim industrial countries that does not operate a material requirements plan. We have deliberately avoided saying that most use them successfully. Too few do! The important reasons are covered in this book.

In the early days, the subject of MRP was neglected in academic curricula in favor of intellectually challenging statistical and mathematical techniques. People in industry thought these irrelevant or obsolete. Academicians considered the study of MRP "vocational" rather than "scientific." Production and inventory management are vocational, of course, in the sense that knowledge is applied to solving real business problems. As with engineering or medicine, production and inventory management are oriented toward practice but must be based on a sound body of theoretical knowledge.

In the past also, communications were poor between practitioners in industry and those who taught production planning and inventory control. Over almost two decades, the first edition of this book helped to improve such communications because it combined both theoretical knowledge and practical experience.

## EVOLUTION OF THE ART

Manufacturing planning and control theory and practice have been evolving in the United States at an accelerating rate through the last half of the twentieth century. The first rigorous attempts to improve manufacturing processes and workers' productivity were made around the beginning of this century by pioneer industrial engineer Frederick W. Taylor. The standard-setting techniques developed by him and later applied widely by Henry Gantt, Frank and Lilian Gilbreth, and Harrington Emerson are still the basis for planning labor requirements and paying incentives to workers to reward them for producing more.

Ford Harris published in 1915 the first formula to calculate an EOQ to minimize the total of ordering-related and inventory carrying costs. In 1934, R. H. Wilson showed how statistics could be used to plan inventory cushions to reduce the impact of forecast errors and material shortages and to improve customer deliveries with minimum inventories.

During World War II, teams of British scientists applied mathematical techniques and scientific methods to complex problems involving the choosing among alternate uses of scarce resources. For over two decades after the war, attempts were made in Europe and America to apply tools of operations research, as it was called, mainly linear programming and queuing theory, to industrial logistics. These produced good results in some situations but had only limited application. Too often operations researchers resembled people with a familiar tool looking for something to fix, like a person with a screwdriver tightening all the screws around and then filing slots in nail heads to tighten those too. Real problems went unattacked and unsolved.

Business computer hardware and software became generally available in the early 1960s, making recordkeeping and use of complex planning techniques practicable for the myriad items found even in small manufacturers. This removed obstacles to the development of many planning techniques that were impractical to apply manually. Prominent among these were

1. George E. Kimball's "base stock system," aimed at eliminating wide variations in upstream demand caused by independent ordering of components of assembled products. This 1950's system foreshadowed Japanese kanban "pull" techniques by communicating end-product actual demands to each work center producing components and to outside suppliers. Lack of computers to handle the masses of data, long setups, large component-order quantities, and buffer stocks at many process steps prevented early wide use and any reaping of the potential benefits of this technique.

2. A forecasting technique called *exponential smoothing* was publicized by Robert G. Brown in 1959. This weighted-averaging technique found wide application in product forecasting because of small computer data storage requirements and flexibility in reacting to demand changes. As with many other mathematical techniques, variations were developed extending it to unusual demand patterns well beyond the point of diminishing returns. These techniques were developed in the late 1950s by IBM, with its IMPACT forecasting software providing the capability to apply order-point sophistication at all levels of stocking.

3. MRP driven by a master production schedule (MPS) was first applied successfully in 1961 by J. A. Orlicky on J. I. Case Company farm machinery. The rigorous logic and masses of data to be handled made this an ideal computer application. The enormous potential benefits over existing ordering techniques generated great interest worldwide.

4. Detailed capacity requirements planning (CRP) was known since Taylor and colleagues had showed how to develop work standards. While good computer software was available early in the 1960s, failure to develop adequate capacity planning resulted from poor-quality, incomplete processing data and work standards in most companies. Rough-cut (infinite) capacity planning techniques were used only for testing the validity of master production schedules. This neglect of capacity requirements planning contributed greatly to the early failures of MRP to realize its full potential.

5. Input/output (I/O) capacity control. Tight control of work input and output was impossible without sound capacity plans. This delayed and blunted attacks on long cycle times and made priority planning and control much less effective.

6. Operation simulation. O. W. Wright, J. D. Harty, and George Plossl developed at Stanley Tools in 1962 one of the first detailed computer simulations of plant

operations. Their Simulated Work Input and Flow Times (SWIFT) program showed us clearly the harmful effects of high work-in-process levels and erratic work input rates on schedule performance and costs.

Operations researchers were intrigued by the problem of determining optimal work sequences in complex manufacturing environments. Sophisticated and expensive computer programs to plan such optimal schedules began to appear in the early 1970s; Werner Kraus of IBM explained his program for Model 1401 computers to George Plossl over dinner in Stuttgart in mid-1972. As bigger, faster computers became available, IBM people in Great Britain converted KRAUS into Capacity Loading And Sequence Scheduling (CLASS) for the Model 1401 computers and later into Capacity Planning and Operation Sequence Scheduling (CAPOSS) for the 360 series and later computers. Other computer software suppliers and independent programmers in Europe and America developed similar optimizing programs, but successful applications were the rare exception.

One of these was R. L. Lankford's 1971 homegrown program at Otis Engineering in Texas; this worked very well for him in predicting schedule deteriorations and altering processing to avoid them. The problem with most such "finite loading" was not fatal flaws in the programs but lack of support of sound planning, execution, and control.

Increasing interest in manufacturing planning and control and a growing need for education led to the founding of the American Production and Inventory Control Society (APICS) in 1957. Chapters were quickly formed in the United States, Canada, Mexico, Europe, and South Africa, providing a forum to disseminate information among practitioners in industry, consultants, academics, and others interested. The body of knowledge was codified, its language formalized, and examinations developed to test individuals' professional competence by the APICS Certification Program Council, which George Plossl led in the early 1970s.

In the 1990s, the Internet once again revolutionized business. Nowhere has this impact probably been greater, at least potentially, than in the area of manufacturing logistics, that is, inventory management and production planning. Until the advent of the computer, these functions constituted a chronic, truly intractable problem for the management of virtually every plant engaged in the manufacture of discrete items passing through multiple stages of conversion from raw material to product. The Internet brought new business models, including customer expectations of faster response with higher variety.

Known and available solutions to the problem in question were imperfect, only partial, and generally unsatisfactory from a management point of view. The first computer applications, around 1960, in the area of manufacturing inventory management represented the beginning of a break with tradition. The availability of computers capable of handling information in volumes and at speeds previously scarcely imaginable constitutes a lifting of the former heavy information-processing constraint and the sudden obsolescence of many older methods and techniques devised in light of this constraint. Traditional inventory management approaches, in the precomputer days, obviously could not go beyond the limits imposed by the information-processing tools available at the time. Because of this, almost all those approaches and techniques suffered from imperfection. They simply represented the best that could be done under the circumstances. They acted as a crutch and incorporated summary, shortcut, approximation methods often based on tenuous or quite unrealistic assumptions, sometimes force fitting concepts to reality so as to permit the use of a technique.

The breakthrough in this area lies in the simple fact that once a computer becomes available, the use of such methods and systems is no longer obligatory. It becomes feasible to sort out, revise, or discard previously used techniques and to institute new ones that heretofore it would have been impractical or impossible to implement. It is now a matter of record that among manufacturing companies that pioneered inventory management computer applications in the 1960s, the most significant results were achieved not by those who chose to improve, refine, and speed up existing procedures but by those who undertook a fundamental overhaul of their planning and control systems. Many things have changed, but this fact remains the same today. The result was abandonment of techniques proven unsatisfactory and a substitution of new, radically different approaches that the availability of computers made possible. In the area of manufacturing inventory management, the most successful innovations are embodied in what has become known as MRP systems.

When implemented, such systems not only demonstrated their operational superiority but also afforded an opportunity for the student of inventory management to gain new insights into the manufacturing inventory problem. The new, computer-aided methods of planning and controlling manufacturing inventories made the true interrelationships and behavior of items constituting these inventories highly visible, thus illuminating the tenuousness of many previous assumptions and revealing the causes of inadequacies (always admitted) of many traditional methods. It became evident that the basic tenet of the old inventory control theory, namely, that inventory investment can be reduced only on pain of a lower service level (and vice versa) no longer holds true. The successful users of the new systems reduced their inventories and improved delivery service at the same time. A revolutionary change occurred, and a new premise was established. Orthodox approaches and techniques became open to question, and existing inventory control literature—indeed, an entire school of thought—was marked for reexamination.

## EVOLUTION OF MRP AND PLANNING SYSTEMS

Even in these simpler, more predictable times, MRP was really successful, as measured by significant bottom-line results, including dramatic inventory reduction in only a small percentage of companies that implemented the tool. The early adopters showed significant results, but as MRP came into more widespread use, the same results were not achieved. This significant failure rate of MRP was a major point of discussion in the APICS meetings at the time. One big reason was that MRP was intended to do only that— plan material. APICS professionals at the time knew that capacity was a critical consideration. However, the computer power at the time was limited, and even if the capacity algorithms were available, it was just not possible to calculate both at the time. Remember that the first MRP packaged systems were written in only 8 kB of memory! This evolution continued with the improvement of technology that enabled the planning of capacity sequentially to the material plan, and a closed-loop MRP process was possible in 1972. However, computers quickly became more powerful, and closed-loop MRP developed to answer the problems of the day. *Closed-loop MRP* is defined as:

> A system built around material requirements planning that includes the additional planning processes of production planning (sales and operations planning), master production scheduling, and capacity requirements planning. Once this planning phase is complete and the plans have been accepted as realistic and attainable, the execution processes come into play. These processes include the manufacturing control processes of input-output (capacity) measurement, detailed scheduling and dispatching, as well as anticipated delay reports from both the plant and suppliers, supplier scheduling, and so on. The term closed loop implies not only that each of these processes is included in the overall system but also that feedback is provided by the execution processes so that the planning can be kept valid at all times.[3]

Closed-loop MRP was the next evolution and allowed the planning of both material and capacity. Still, the development and implementation of an MRP system were far from a guarantee of success. The tool was far more sophisticated. The availability of APICS education provided the necessary people who understood how the tools worked, but still the implementation was not a guarantee of success. Technology became more powerful, and the client-server age was on us. In the 1980s, MRP II (manufacturing resource planning) was developed to provide further integration to the core business system by incorporating the financial analysis and accounting functions. *MRP II* is defined as:

> A method for the effective planning of all resources of a manufacturing company. Ideally, it addresses operational planning in units, financial planning in dollars, and has a simulation capability to answer what-if questions. It is made up of a variety of processes, each linked together: business planning, production planning (sales and operations planning), master production scheduling, material requirements planning, capacity requirements planning, and the execution support systems for capacity and material. Output from these systems is integrated with financial reports such as the business plan, purchase commitment report, shipping budget, and inventory projections in dollars. Manufacturing resource planning is a direct outgrowth and extension of closed-loop MRP.[4]

---

[3] *APICS Dictionary*, 12th ed. New York: Blackstone, 2008, p. 21
[4] *APICS Dictionary*, 12th ed. New York: Blackstone, 2008, p. 78.

Commercially available MRP II systems became more available through a variety of vendors. No longer was it necessary for companies to develop their own systems. Software companies catering to the needs of different industries and platforms provided a wide variety of software products off the shelf. At the same time, the APICS education and certification program provided industry with professionals capable of using these systems. Still, these systems that were so advanced at the time still were no guarantee of bottom-line success. Simply getting the software implemented and running was not sufficient to ensure bottom-line results.

In the 1990s, as technology began to move to Internet architecture, enterprise resource planning (ERP) was the next evolution. ERP brought all the resources of an enterprise under the control of a centralized integrated system and database. *ERP* is defined as:

> Framework for organizing, defining, and standardizing the business processes necessary to effectively plan and control an organization so the organization can use its internal knowledge to seek external advantage.[5]

Companies continued to invest in technology, pursuing the Holy Grail of integrated planning, and yet significant bottom-line results were not achieved. The underlying driver was that if only a company could deliver on time, there was more than sufficient market to be addressed. In the mid-1990s, advanced planning and scheduling (APS) systems leveraged the visibility of the company's resources in ERP and promised to keep all scarce resources busy all the time. *The APICS Dictionary* defines an *APS* as:

> Techniques that deal with analysis and planning of logistics and manufacturing during short-, intermediate-, and long-term time periods. APS describes any computer program that uses advanced mathematical algorithms or logic to perform optimization or simulation on finite capacity scheduling, sourcing, capital planning, resource planning, forecasting, demand management, and others. These techniques simultaneously consider a range of constraints and business rules to provide real-time planning and scheduling, decision support, available-to-promise, and capable-to-promise capabilities. APS often generates and evaluates multiple scenarios. Management then selects one scenario to use as the "official plan." The five main components of APS systems are (1) demand planning, (2) production planning, (3) production scheduling, (4) distribution planning, and (5) transportation planning.[6]

Once again, the implementation of these complex systems was rarely a significant bottom-line success. This is not to say that the software didn't get implemented or didn't run. The reality was that the improved bottom-line results promised in the business case were the exception rather than the rule. The merits of a system or approach must be measured by the results it achieves.

---

[5] *APICS Dictionary*, 12th ed. New York: Blackstone, 2008, p. 45.

[6] *APICS Dictionary*, 12th ed. New York: Blackstone, 2008, p. 4.

George Plossl in the second edition of this book (1994) claimed the following problems handicapping planning systems:

1. *Weak elements.* Overselling MRP as a "system," not just the priority planning element, diverted attention from the implementation of master scheduling to drive it and capacity requirements planning to plan the resources needed to support it.

2. *Missing elements.* Almost universally, capacity control was a missing link. Believing that capacity planning was weak, companies failed to utilize input/output control for the major benefits this made possible, even with crude capacity planning.

3. *Oversophistication.* MRP program designers, obsessed with the potential power of computers and software, attempted to build into MRP capabilities to cope with every eventuality in manufacturing and to include every known technique, however little use these would be. Part-Period Balancing with Look-ahead/Look-back lot sizing (see Chapter 8) is a classic example.

4. *Invalid data.* MRP to many users has meant "More Ridiculous Priorities" because of data errors. This subject is discussed in depth in Chapter 10 and mentioned in several others.

5. *Lack of integration.* Data must flow from files in design engineering, process engineering, order entry, purchasing, plant floor, and many other activities into planning and control files and back out. Transposing these data manually between files delayed the flow, disrupted the timing, and destroyed accuracy. Planning, however well done, was late, unresponsive, and invalid. People could not be held accountable for executing such plans.

These problems continue to plague most planning systems today. It was long known that the steady flow of work on assembly lines was far better for many reasons than batch production in functional work centers with similar processing operations. In the early 1970s, group technology (GT) was introduced to improve batch production. This technique grouped together machines and equipment used to make families of parts having similar physical characteristics (i.e., shape, weight, material, and dimensions) or processing operations (i.e., machining, welding, and automatic insertion). Advantages included reduced processing cycle times, lower work-in-process, less material handling, and tighter supervision. These tangible costs carried more weight in management thinking than the intangible benefits of faster processing, greater flexibility, and tighter control; group technology was not widely applied in America or Europe.

In the early 1980s, wide interest developed in both America and Europe in Japanese manufacturing practices because of their great advances in quality and productivity. Japan's increasing share of U.S. markets for autos, motorcycles, electronic equipment, cameras, machine tools, and many other products generated great concern among U.S.

competitors. Many study missions to Japan, technical societies, business schools, and politicians attempted to explain how the Japanese had made such great gains.

Among many factors, attention centered on just-in-time (JIT) and kanban "pull systems" and quality circles. One effect was to revive interest in group technology, newly named *manufacturing cells*, in which one or more machines and operators produce a family of similar parts or products very quickly and flexibly in small lots. Cells, renamed *flexible manufacturing systems* (FMS), became highly automated as electronic controls were applied to storage systems, material handling equipment, and production machines. Details of these are beyond the scope of this book.

The evolution of these systems made obvious the need for integration of procurement, preproduction, production, and postproduction activities. Design and process engineering, tooling, purchasing, production, quality assurance, and of course, planning and control personnel had to become tightly knit teams to achieve the benefits enjoyed by the best Japanese firms. Even traditional cost-accounting practices had to be changed. All these are now evolving—too fast for some and too slow for others.

## PLANNING, EXECUTION, AND CONTROL

These three terms are very common and are used frequently by many people at all organization levels in manufacturing firms. Their meanings are often different among these people, however. Here are the generally accepted definitions:

*Planning*—assigning numbers to future events to create plans

*Execution*—converting plans to reality

*Control*—tracking execution, comparing execution to plans, measuring deviations, sorting the significant from the trivial, and instituting corrective actions in either the plans or the execution

These definitions and statements of purpose make eminently clear why planning and execution will require different techniques and that control activities provide interfaces between the two. The planning, execution, and control (often misnamed "information") system must support both purposes, recognizing their differences. Five principles apply to these systems:

1. *There is one system framework common to all types of manufacturing.* As startling as this is and as unlikely as it seems, it derives inevitably from the common logic, the similarity of needs of the major parties, the nature of the resources employed, and the identical types of data involved. As will become evident when the elements of the system are defined, the importance of each element along with the techniques employed can differ among various types of businesses.

2. *There is no single best way to control a manufacturing business.* The second principle may seem at first glance to contradict the first, but it is simply saying that

the same vehicle can be used in myriad ways to achieve the different objectives of a number of users with varied interests and to react to changes. This is not to say that a concept may not fit a wide variety of businesses but rather that the application will be unique company by company.

The tools of planning, execution, and control include many techniques and tools usable in many ways. Selecting and applying the right tools is important for success, but a sound strategy for directing their use is vital for manufacturing enterprises to remain competitive.

3. *Do not commit any flexible resource to a specific use until the latest possible moment.* All resources used in manufacturing have flexibility, some more than others. People can handle a wide variety of tasks and acquire new skills. Materials can be made into many different items. Machinery can produce a myriad of parts and products. Money can buy an almost unlimited variety of goods and services. Fuels and energy have many uses. But once a resource has been consumed in the production or procurement of some specific material, component, product, or facility, its flexibility has been lost. This will be highlighted in depth later in this book.

Plans cover future occurrences, subject to "the slings and arrows of outrageous fortune." They undoubtedly will change, and the amount and effects of changes will be more severe as planning horizons lengthen. The later specific allocations of resources are made, the fewer the changes are necessary, the less severe their effects, and the less waste of resources. The strategy just stated derives from the first principle in this list and explains the sources of benefits of faster flows of materials and information.

Each organizational group within a company has its own functions and objectives, but its plans and actions must be communicated to and coordinated with those of other groups through integrated systems. Mislabeled "information systems," these really process only data. Information is the fraction of such data with specific use by or meaning to people doing their work. Data-processing systems must be designed to produce needed information for users in proper formats. Behaving as if soft data are precise leads to huge and costly synchronization problems.

Planning and execution data are very different. For example, plans show that an order for 50 pieces of a component is to be released on week 26, six weeks hence, and is due to be completed by week 30. The actual dates the order is released and due probably will be earlier or later, and even the quantity can be changed. Planning data, called *soft data*, look precise but often are changed.

If a decision has been made to start an order for one of this component's parents in week 23, this is a firm need date for the component. If work has started on 24 pieces, this too is firm. Execution data need to be precise; they are hard data.

4. *Resources must be adequate and willing.* Planning can be done by computer-based systems with minimum human involvement, but successful execution depends on people taking the proper actions when needed. Effective planning systems are necessary but not sufficient; they make tight control possible, but people make it happen.

5. *Garbage in, garbage out.* The output of any computer-based system is dependent on the accuracy of the inputs. If the data that are input are incorrect, then the outputs will be less than worthless. Accuracy is required in inventory records, bills of materials, customer orders, shop orders, and every other necessary input. The effectiveness of the planning system can only be as good as these inputs. Failure of the system to work is not where the software doesn't run, it is when the information gained from the system is misleading.

Almost 40 years later we are at another time of reexamination and transition. Shortly after the turn of the millennium, the world of manufacturing turned upside down. Production became more efficient in the United States. Eastern Europe was incorporated into the European Union, putting low-cost production very close to a lucrative market. China became the manufacturing powerhouse for the world. Manufacturing capacity now far exceeds current market requirements. Exacerbating this situation are customers who have become increasingly fickle. Product life cycles have plummeted. The Internet now allows global sourcing with a few clicks of a mouse. Manufacturing companies worldwide are faced with more volatility than ever before. No longer can a company achieve a sustainable competitive advantage with the old rules. These fundamental shifts taking place in current global manufacturing environments are forcing companies to reexamine the rules and tools that manage their businesses. The world has changed, and further technology barriers have been removed. Companies will succeed not because they improve, refine, and speed up the enforcement of obsolete rules and logic but because they are able to fundamentally adapt their operating rules and systems to the new global circumstances. A new approach to planning is required. This is the subject of this last part of this update.

## CHAPTER 22

# Blueprint for the Future: Demand-Driven MRP Logic

**D**emand-driven manufacturing is a manufacturing strategy of dramatic lead-time compression and the alignment of efforts to respond to market demands. This includes careful synchronization of planning, scheduling, and execution with consumption. It was coined by PeopleSoft in 2002 and embraced later by several research companies. It is important to point out that demand-driven manufacturing is not synonymous with make-to-order manufacturing. Demand driven manufacturing requires a fundamental shift from the centrality of inventory to the centrality of demand. To be successful, a company must be able to sense and adapt to market changes.

The traditional push approach has proven to be grossly inadequate in a highly volatile and variable manufacturing landscape dominated by more complex planning scenarios than ever. Seeing the benefits of being demand-driven, many companies have attempted to build walls around or disable the push-based aspects of traditional material requirements planning (MRP) in an attempt to use it in a more demand-driven fashion. At the same time, the limited set of materials planning and inventory control tools in pull-based philosophies such as lean and drum-buffer-rope (DBR) are also proving to be grossly inadequate, even counterproductive, to the implementation of demand-driven manufacturing. A new type of MRP is required to deal effectively with today's circumstances and fully capitalize on and implement pull-based philosophies.

The traditional MRP rules that were conceived codified and commercialized in the '50s, '60s and '70s under the old "Push and Promote" mode of operation are now breaking down. This includes the general industry love affair with better forecasting algorithms. Working to forecast has long been compared to driving a car by looking on the rear view mirror. Today, however, the road is a twisty mountain road in dense fog and the penalties for error are significant. Paying large sums of money for more sophisticated forecast algorithms simply means now you have a more expensive rear view mirror. Any appreciable gains by these "smarter" algorithms are being more than offset by the rise of volatility.

## THE MRP CONFLICT

Figure 22-1 is a conflict diagram depicting many companies' current dilemma around MRP. The failure to resolve this dilemma consistently and satisfactorily results in poor organizational and supply-chain performance, conflicting modes of operation between planning and manufacturing, and countless numbers of workarounds.

This diagram is read from left to right. Today, few people in manufacturing would disagree with an objective to be "agile." What is agility? The twelfth edition of the *APICS Dictionary* defines *agility* as:

> The ability to successfully manufacture and market a broad range of low-cost, high-quality products and services with short lead times and varying volumes that provide enhanced value to customers through customization. Agility merges the four distinctive competencies of cost, quality, dependability, and flexibility.

A problem with this definition is not whether it is a desirable state to achieve; the problem is that it is too difficult to achieve given no shortage of challenging circumstances relative to the manufacturing environment. One of those challenges is the conflict defined earlier. In many manufacturers, the inability to resolve this conflict means that agility is completely unrealistic, and the company attempts to find a compromise position.

There are two critical needs coming into contention behind the compromises. From a manufacturing perspective, a company must have a realistic way to respond, pace, and produce to actual demand. This way must include both capacity and materials, as discussed previously. Within increasingly shorter horizons that are inherently more variable and volatile, MRP tools simply do not create the correct demand signals, nor do they facilitate materials availability. Additionally, many pull-based or demand-driven scheduling techniques (e.g., lean and drum-buffer-rope) are effectively blocked by this lack of material synchronization. MRP systems appear to be overly complex and lack clear visibility for the quick and effective decision making that agility requires. In most cases,

**FIGURE 22-1**

The MRP conflict.

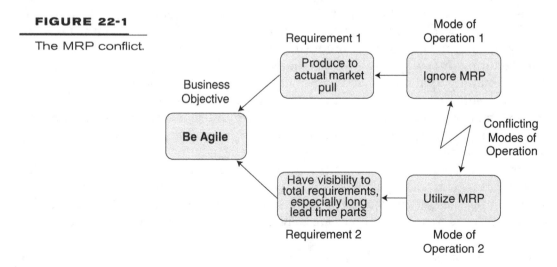

owing to the shortcomings listed earlier, this leads many manufacturing personnel within companies to think that they should ignore MRP. In fact, a frequent milestone for a lean or DBR implementation is that the computer planning system has been eliminated or effectively marginalized.

On the other hand, from a planning and purchasing perspective, companies must have a way to see, plan, synchronize, and manage the availability of all materials, components, and end items effectively. This is especially true for critical and/or long-lead-time manufactured and purchased parts. With today's increasingly complex planning scenarios (discussed earlier), it leads planning personnel to insist on using MRP. A common reaction to an expressed desire to ignore or turn off MRP leads to a fairly common reaction by planning personnel: "You think it's bad now? Wait 'till we turn it off—we will be flying completely blind." The person who made the suggestion immediately loses credibility with the planning side of the organization, which view the suggestion as one based on oversimplification driven by a lack of fundamental understanding of materials planning.

However, both the opposing perspectives are correct. This is why the conflict is so pervasive and chronic. The more complex the manufacturing environment, the more acute this conflict tends to be. The inability to reconcile the dilemma in those environments effectively leads to the ineffective MRP compromises categorized earlier and also will relegate lean, DBR, and Six-Sigma implementations to lip service. This completely squanders their potential and the time, effort, and money already put into them.

The business requirements that are driving both sides of the dilemma must be achieved without the conventional inaccuracy, inconsistency, and massive additional efforts and waste associated with the current set of compromises. If companies want to be agile, there is no other way.

## DEMAND-DRIVEN MRP INTRODUCTION

MRP, as noted earlier in this book, has some very valuable core attributes in today's more complex planning and supply scenarios (e.g., bill of material visibility, netting capability, and maintenance of sales order/work order connection between demand allocations and open supply). In other words, critical aspects of it are still relevant—perhaps even more relevant than in the past four decades. The key is to keep those attributes but eliminate MRP's critical shortcomings while integrating the pull-based replenishment tactics and visibility behind today's demand-driven concepts into one system in a dynamic and highly visible format. The solution is called *demand-driven MRP* (DDMRP).

DDMRP builds from the still relevant foundations of Orlicky's MRP. It takes advantage of advances in technology over the past 60 years, as well as incorporating innovative new logic with regard to the lead-time compression required to achieve and sustain a competitive advantage in the demand-driven world. In addition to these innovations, DDMRP leverages the complete toolbox, including core MRP and distribution requirements planning (DRP) logic, theory of constraints (TOC), and lean principles. Figure 22-2 illustrates this fusion. A list of new terms is available in Appendix C at the end of this book.

**FIGURE 22-2**

Demand-driven MRP.

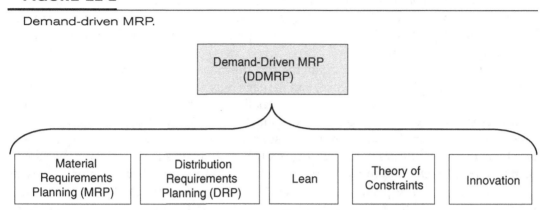

Demand-driven MRP is a dynamic and effective demand-driven solution to answer the challenges of today's manufacturing landscape. Through innovative approaches in inventory and product structure analysis, new demand-driven planning rules and integrated execution tactics, DDMRP is designed to tie material availability and supply directly to actual consumption throughout the bills of material (BOMs). When used holistically across a supply chain, it removes the cascading and compounding disruptions that most supply chains face. Additionally, this approach is a prerequisite to using and sustaining pull-based scheduling and execution methods such as lean and DBR effectively in more complex manufacturing environments. DDMRP has a unique way to incorporate required elements of strategic planning in the sales and operations plan with little exposure to the variability and volatility experienced with traditional forecasting techniques.

Many readers might look at the suggested solution and see a staggering amount of suggested change. The irony is that much of the solution is already known and accepted. These elements have not been packaged together with the key innovations that will enable demand-driven manufacturing for the global manufacturing and supply landscape of the twenty-first century. The following chapters will describe the package and introduce these new innovations. Demand-driven MRP is focused around critical parts called *strategically replenished parts*. These strategically replenished parts often will "drive" the system and thus requirements for all parts (stocked and nonstocked).

## THE FIVE PRIMARY COMPONENTS
## OF DEMAND-DRIVEN MRP

Each of these five components will be explored in depth in the five chapters that follow. All are necessary to remove the undesirable MRP conflict symptoms and compromises and open the door to agility. Ignoring any of these components will reduce the value of the solution dramatically in most environments.

## 1. Strategic Inventory Positioning

The first question of effective inventory management is not "How much inventory should we have?" Nor is it "When should we make or buy something?" The most fundamental question to ask in today's manufacturing environments is, "Given our system and environment, where should we position inventory (within BOMs and the facilities) to have the best protection?" Think of inventory as a breakwall to protect boats in a marina from the roughness of incoming waves. Out on the open ocean, the breakwalls have to be 50 to 100 feet tall, but in a small lake, the breakwalls are only a couple feet tall. In a glassy smooth pond, no breakwall is necessary. A company will need to carefully analyze its environment and then position and build the necessary inventory breakwalls.

## 2. Buffer Profiles and Level Determination

Once the strategically replenished positions are determined, the target levels of those buffers have to be set initially based on several factors. Different materials and parts behave differently, but many also behave nearly the same. Demand-driven MRP groups parts and materials chosen for strategic replenishment and that behave similarly into buffer profiles. Buffer profiles take into account important factors, including lead time (relative to the environment), variability (demand or supply), whether the part is made or bought or distributed, and whether significant order multiples are involved. These buffer profiles are made up of *zones* that produce a unique buffer picture for each part as their respective individual part traits are applied to the group traits.

## 3. Dynamic Buffers

Over the course of time, group and individual traits can and will change as new suppliers and materials are used, new markets are opened and/or old markets deteriorate, and manufacturing capacities and methods change. Dynamic buffer levels allow the company to adapt buffers to group and individual part trait changes over time through the use of several types of adjustments. Thus, as more or less variability is encountered or as a company's strategy changes, these buffers adapt and/or are adjusted to fit the environment.

## 4. Demand-Driven Planning

As discussed earlier in this book, the world of push and promote is dead. The holdovers of that era, both rules and tools, must be stripped away, greatly changed or enhanced, or completely reconstructed. Instead of making things too complex or too simple, it is time to define a planning suite of rules that meet at least two requirements. First is to take advantage of the sheer computational power of today's hardware and software. Second is to take advantage of the new demand-driven approaches. When these two elements are combined, then there is the best of both worlds: relevant approaches and tools for the way the world works today *and* a system that promotes better and quicker decisions and actions at the planning and execution levels.

## 5. Highly Visible and Collaborative Execution

Simply launching purchase orders (POs), manufacturing orders (MOs), and transfer orders (TOs) from any planning system does not end the materials and order management challenge. These POs, MOs, and TOs have to be managed effectively to synchronize with the changes that often occur within the execution horizon. The execution horizon is the time from which a PO, MO, or TO is opened until the time it is closed in the system of record. Demand-driven MRP is an integrated system of execution for all part categories in order to speed the communication of relevant information and priorities throughout an organization and supply chain (Figure 22-3).

These five components work together to dampen, if not eliminate, the unnecessary nervousness of traditional MRP systems and the resulting bullwhip effect in complex and challenging environments. In using this approach, planners will no longer have to try to respond to every single message for every single part that is off by even one day. This approach provides real information about those parts that are truly at risk of negatively affecting the planned availability of inventory. Demand-driven MRP sorts the significant few items that require attention from the magnificent many parts that are currently being managed. Under the demand-driven MRP approach, fewer planners can make better decisions more quickly. This means that companies will be better able to leverage their working and human capital as well as the significant investments they have made in information technology.

*Authors' Note:* There is a challenge associated with writing this book. A large portion of the solution involves high visibility. A large portion of that visibility is accomplished through easy-to-interpret color signals. This book is printed in monochromatic format. Printing monochromatically does not bring that visibility to life very well. The reader will have to use some amount of imagination to get the proper sense of visibility.

**FIGURE 22-3**

The five components of demand-driven MRP.

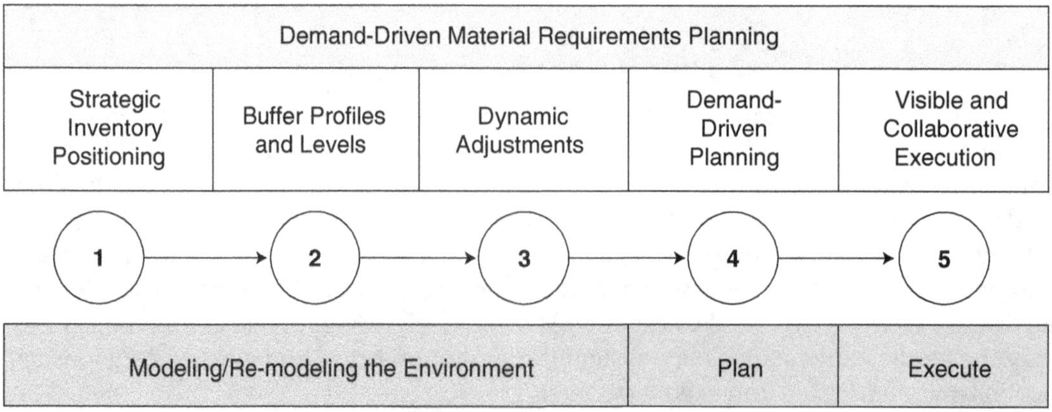

| Demand-Driven Material Requirements Planning | | | | |
|---|---|---|---|---|
| Strategic Inventory Positioning | Buffer Profiles and Levels | Dynamic Adjustments | Demand-Driven Planning | Visible and Collaborative Execution |
| 1 | 2 | 3 | 4 | 5 |
| Modeling/Re-modeling the Environment | | | Plan | Execute |

## CHAPTER 23

# Strategic Inventory Positioning

The first step in demand-driven material requirements planning (DDMRP) is to thoroughly consider *where* inventory should be placed. The six positioning factors from Chapter 4 are used to determine the initial positioning strategy. Below and in Figure 23-1 is a summary of those factors.

1. Customer tolerance time
2. Market potential lead time
3. Variable rate of demand
4. Variable rate of supply
5. Inventory leverage and flexibility
6. The protection of key operational areas.

Under DDMRP, these six factors are applied systematically across the entire bill of material (BOM), routing structure, manufacturing facilities, and supply chain to determine the best positions for purchased, manufactured, and finished items (including service parts). The bigger the manufacturing or supply-chain system these factors are applied to, the more significant the results from better synchronization can be. Later in this book we will examine the impact of the solution on an integrated supply chain.

## ASR LEAD TIME: A NEW TYPE OF LEAD TIME

The positioning example from Chapter 4 began by using manufacturing and cumulative lead times (MLT and CLT) as factors for determining the right position. Within the example, however, a critical point can be realized; there is actually another type of lead time

**FIGURE 23-1**

Critical factors for properly positioning inventory.

| Strategic Inventory Positioning Factors | |
|---|---|
| Customer Tolerance Time | The amount of time potential customers are willing to wait for the delivery of a good or a service. |
| Market Potential Lead Time | The lead time that will allow an increase of price or the capture of additional business either through existing or new customer channels. |
| Demand Variability | The potential for swings and spikes in demand that could overwhelm resources (capacity, stock, cash, etc.). |
| Supply Variability | The potential for and severity of disruptions in sources of supply and/or specific suppliers. This can also be referred to as supply continuity variability. |
| Inventory Leverage and Flexibility | The places in the integrated BOM structure (the Matrix BOM) or the distribution network that leave a company with the most available options as well as the best lead time compression to meet the business needs. |
| Critical Operation Protection | The minimization of disruption passed to control points, pace-setters or drums. |

that needs to be recognized, calculated, and made visible. That lead time will be a critical factor in:

- Understanding how to best leverage inventory
- Setting inventory levels properly
- Compressing lead times
- Determining realistic due dates when needed

The truth is that MLT and CLT are only realistic under two extremes. In more complex manufacturing operations, these two extremes rarely exist. To make MLT a realistic planning input, all components at every level would need to be stocked and reliably managed so that they are always readily available. MRP assumes that all components are available at the time of order release. This assumption is seldom achieved or achieved only through significantly bloated inventory positions.

Alternatively, the real MLT is much longer as production control personnel attempt to clear the shortage. In reality, MLT represents a dramatic underestimation of real lead time. To make CLT a realistic planning input means that no component parts on the longest leg will be stocked. CLT assumes that components on the longest path will not be available within their respective lead times but rather require a complete make to order from the components below. Reality demonstrates that experienced planning personnel are not so naive. In addition, when the demands of the twenty-first-century competitive environment are considered, the reality is that the market will not tolerate it.

Stocking inventory for components anywhere other than on the longest leg can waste capital, space, attention, and possibly capacity. This statement assumes that these stocked parts are not on the longest leg of other parts. Figure 23-2 illustrates this with a simple example. In this example, the parent item is Part 101. The BOM and the component parts' discrete lead times (both manufactured and purchasing) are noted with a

**FIGURE 23-2**

Lead-time analysis for Part 101.

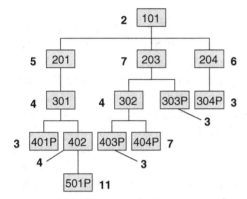

**Part 101 BOM with all part Lead Times**

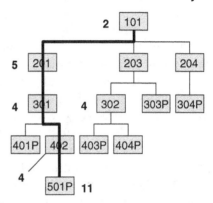

**Cumulative Lead Time = 26 days**

large number to the side of each part. The manufacturing lead time (MLT) of Part 101 is 2 days. The cumulative lead time (CLT) is 26 days and is indicated by the bold line terminating at the Part 501P.

Figure 23-3 identifies actual stocked positions represented by shaded boxes. This is a hypothetical example of an actual stock position, not an illustration of an ideal stock position. Stocked items are one subassembly (Part 203) and three purchased components indicated by the *P* suffix in their part numbers. Note that in this case only one immediate component of Part 101 is stocked. If the MLT of Part 101 (2 days) is used either as an input to build a stock position or to commit to a customer order, that commitment will be

**FIGURE 23-3**

Nonstocked components extending MLT.

Realistic Lead Time ≠ MLT

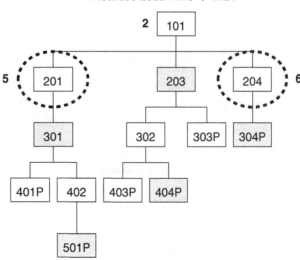

a gross *underestimation* of how long it takes to reliably produce a Part 101. Experienced planning personnel are very aware of this phenomenon and have developed alternative processes to work around the issue. One option is to stock every component. Another is an adjustment of manufacturing lead times in order to try to reflect the reality in responding to an order.

In this example, using CLT is not realistic either. The CLT path revealed in the figure has two stocked components on the path (Parts 301 and 501P). On average, parts probably will be available more often than they are not. These stocked positions essentially "decouple" the CLT path, making 26 days a significant *overestimation* for Part 101. This is demonstrated in Figure 23-4. Once again, planners tend to adjust the lead times to plan these items effectively. Every planner knows intuitively that in most cases the realistic reliable lead time is neither the MLT nor the CLT; it is something in between. This realization opens the door for a new type of calculated lead time and a different way to perceive, analyze, and control a BOM.

Figure 23-5 shows that the realistic lead time can be determined by the MLT of the immediate component part that is not stocked. In this case, it is defined by the path through Part 204 (the bold line). This path is broken at a stocked position (Part 304P). The true, realistic lead time is defined by the longest unprotected or unbuffered sequence in the BOM for a particular parent (in this case Part 101). This is called the *actively synchronized replenishment lead time* (ASR lead time), which is the core concept behind DDMRP. This will be described further in Chapter 26.

With the BOM decoupled by possibly several embedded stock positions, there will be several ASR lead times (ASRLT). In Figure 23-6, the BOM for Part 101 has three independent ASRLT stratifications or layers. Part 101 has an ASRLT of 8 days. Part 203 has an

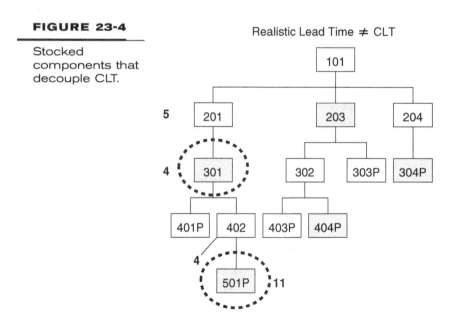

**FIGURE 23-4**

Stocked components that decouple CLT.

Realistic Lead Time ≠ CLT

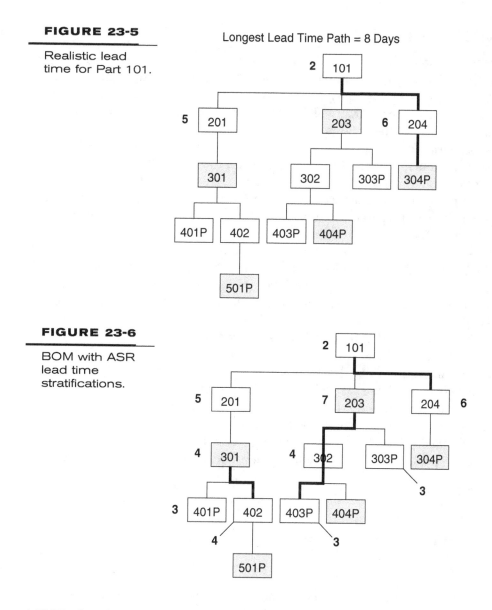

**FIGURE 23-5**

Realistic lead
time for Part 101.

Longest Lead Time Path = 8 Days

**FIGURE 23-6**

BOM with ASR
lead time
stratifications.

ASRLT of 14 days. Part 301 has an ASRLT of 8 days. There are three independent ASRLT groups within the Part 101 BOM.

ASRLT is a qualified CLT. By using the ASRLT approach, planners now can calculate and/or determine more realistic dates for the replenishment or delivery of a part. Of course, using this approach requires each discrete part number to have an MLT or production lead time (PLT), and those lead times should be as accurate as possible. ASRLT is a pivotal building block for where to position inventory, the size of those inventory positions, and critical date-driven alerts and priorities.

## ASRLT AND MATRIX BOMS

Leveraging the previous concept of ASRLT in combination with other tools can give unprecedented visibility across a company's various BOMs in order to identify where to stock and where not to stock. Many manufacturers make many types of end items. Despite each end item having a unique BOM, frequently there is a substantial number of shared components across these end-item BOMs.

In these scenarios (significant number of shared components across BOMs), using a matrix BOM in combination with ASRLT becomes a powerful inventory leverage and lead-time compression tool. A matrix BOM is "a chart made up from the bills of material for a number of products in the same or similar families. It is arranged in a matrix with components in columns and parents in rows (or vice versa) so that requirements for common components can be summarized conveniently."[1]

Figure 23-7 is a simple example in which a company makes four different end items: 101, 1H01, 20H1, and 20Z1. The shaded parts are parts that are currently stocked. In this case, all end items are stocked. No intermediate components are stocked. Four different purchased items are currently stocked. The direct material costs of all purchased and end items are also provided.

Figure 23-8 represents the matrix BOM associated with the BOMs in Figure 23-7. Shaded boxes in the column and row headers are parts that are currently stocked as per the shaded boxes in Figure 23-7. Clearly, this is a very simple example. For companies that have hundreds of end items, as well as deep BOMs with many shared components, the matrix BOM can get quite complex and very large.

A matrix BOM is a much broader picture than the where-used report. The where-used report is oriented to a particular component to see which parents it goes into and its usage per parent. The *APICS Dictionary* defines a *where-used list* as "a listing of every par-

## FIGURE 23-7

Bills of material.

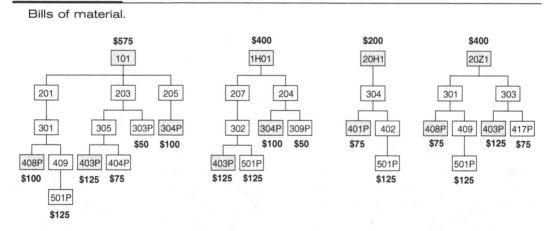

---

[1] *APICS Dictionary*, 12th ed. New York: Blackstone, 2008, p. 82.

## FIGURE 23-8

Matrix bill of material for Figure 23-7.

| | | 101 | 1H01 | 2H01 | 2Z01 | 201 | 203 | 204 | 205 | 207 | 301 | 302 | 303 | 304 | 305 | 402 | 409 |
|---|---|---|---|---|---|---|---|---|---|---|---|---|---|---|---|---|---|
| | | | | | | | | | | | | | | | | | *Parent Items* |
| Component Items | 201 | 1 | | | | | | | | | | | | | | | |
| | 203 | 1 | | | | | | | | | | | | | | | |
| | 204 | | 1 | | | | | | | | | | | | | | |
| | 205 | 1 | | | | | | | | | | | | | | | |
| | 207 | | 1 | | | | | | | | | | | | | | |
| | 301 | | | | 1 | 1 | | | | | | | | | | | |
| | 302 | | | | | | | | | 1 | | | | | | | |
| | 303 | | | | 1 | | | | | | | | | | | | |
| | 304 | | | 1 | | | | | | | | | | | | | |
| | 305 | | | | | | 1 | | | | | | | | | | |
| | 304P | | | | | | | 1 | 1 | | | | | | | | |
| | 307P | | | | | | 1 | | | | | | | | | | |
| | 309P | | | | | | | 1 | | | | | | | | | |
| | 401P | | | | | | | | | | | | | 1 | | | |
| | 402 | | | | | | | | | | | | | 1 | | | |
| | 403P | | | | | | | | | | | 1 | 1 | | 1 | | |
| | 404P | | | | | | | | | | | | | | 1 | | |
| | 408P | | | | | | | | | | 2 | | | | | | |
| | 409 | | | | | | | | | | 2 | | | | | | |
| | 417P | | | | | | | | | | | | 1 | | | | |
| | 501P | | | | | | | | | | | 1 | | | | 1 | 2 |

ent item that calls for a given component, and the respective quantity required, from a bill-of-material file. See: implosion."[2] A where-used list is a very specific slice of a matrix BOM.

Identifying shared components to stock in order to compress lead times and leverage inventory across parent items is a primary objective. Using a where-used report *might* identify a candidate for that objective. That candidate is 501P. It is not stocked, and it is common to all end items. Figure 23-9 shows 501P in relation to all parent-item BOMs. But this is not enough information to draw definitive and consistent conclusions about

## FIGURE 23-9

A shared component.

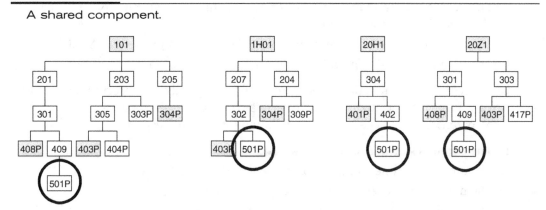

---

[2] Ibid, p. 93.

## FIGURE 23-10

ASRLT chains.

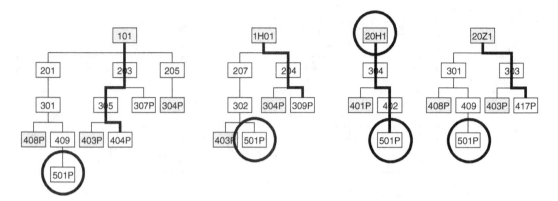

whether stocking a common component such as 501P is a good application of capital to achieve the desired compression and leverage. There is a missing critical element. For example, if 501P did not factor into the ASRLT chains of the parent items, then the company will not get the compression or leverage it seeks. Figure 23-10 is an example of ASRLT chains where 501P was only a factor with regard to one end item. In this case, the benefits of stocking of 501P may be relatively insignificant.

However, if 501P factored into more or all of the ASRLT chains, then the value of stocking 501P for compression and leverage purposes increases dramatically. The value equation would be determined by the following factors. If the parent is stocked, then the following items are needed:

1. The parent's starting ASRLT
2. The average on-hand inventory position, assuming the starting ASRLT (see Chapter 24 for further description)
3. The amount of time compression achieved (the reduction of total ASRLT)
4. The required investment in the component (average on-hand position)

Use only the direct material costs of items. By using only the direct material costs, the true cash impact associated with making this compression is identified.

If the parent is not stocked, then the following items are needed:

1. The parent's starting ASRLT
2. The amount of time compression achieved (the reduction of total ASRLT)
3. The market impact as a result of time compression
4. The required investment in the component (average on-hand position)

**FIGURE 23-11**

ASRLT chains terminating with 501P.

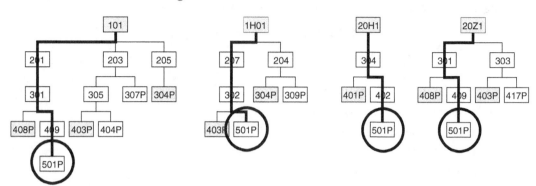

If the ASRLT chains for the end items all terminated with 501P such as described in Figure 23-11, it represents an obvious opportunity to compress all end-item lead times and leverage a lower-cost item against multiple end-item positions.

Figure 23-12 is the matrix BOM with parts identified that lie on an ASRLT path. Boxes that are shaded within the matrix represent those parts.

Below are the numbers associated with this compression and leverage. Average on-hand in relation to strategic buffers will be discussed in Chapter 24.

**FIGURE 23-12**

Matrix BOM with ASRLT components highlighted.

| | | Parent Items | | | | | | | | | | | | | | | |
|---|---|---|---|---|---|---|---|---|---|---|---|---|---|---|---|---|---|
| | | 101 | 1H01 | 2H01 | 2Z01 | 201 | 203 | 204 | 205 | 207 | 301 | 302 | 303 | 304 | 305 | 402 | 409 |
| Child Items | 201 | 1 | | | | | | | | | | | | | | | |
| | 203 | 1 | | | | | | | | | | | | | | | |
| | 204 | | 1 | | | | | | | | | | | | | | |
| | 205 | 1 | | | | | | | | | | | | | | | |
| | 207 | | 1 | | | | | | | | | | | | | | |
| | 301 | | | | 1 | 1 | | | | | | | | | | | |
| | 302 | | | | | | | | | 1 | | | | | | | |
| | 303 | | | | 1 | | | | | | | | | | | | |
| | 304 | | | 1 | | | | | | | | | | | | | |
| | 305 | | | | | | 1 | | | | | | | | | | |
| | 304P | | | | | | | 1 | 1 | | | | | | | | |
| | 307P | | | | | | 1 | | | | | | | | | | |
| | 309P | | | | | | | 1 | | | | | | | | | |
| | 401P | | | | | | | | | | | | | 1 | | | |
| | 402 | | | | | | | | | | | | | 1 | | | |
| | 403P | | | | | | | | | | | 1 | 1 | | 1 | | |
| | 404P | | | | | | | | | | | | | | 1 | | |
| | 408P | | | | | | | | | | 2 | | | | | | |
| | 409 | | | | | | | | | | 2 | | | | | | |
| | 417P | | | | | | | | | | | | 1 | | | | |
| | 501P | | | | | | | | | | | 1 | | | | 1 | 2 |

Part 101

$575 total direct material cost

ASRLT reduction = 6 days/19 days (31.5%)

Average on-hand position goes from 100 to 80

Total savings = $11,500

Part 1HO1

$400 total direct material cost

CLT reduction = 5 days/12 days (41.6%)

Average on-hand position goes from 176 to 130

Total savings = $18,400

Part 20H1

$200 total direct material cost

ASRLT reduction = 2 days/8 days (25%)

Average on-hand position goes from 220 to 180

Total savings = $8,000

Part 20Z1

$400 total direct material cost

ASRLT reduction = 4 days/12 days (33%)

Average on-hand position goes from 95 to 70

Total savings = $10,000

Total parent-part average on-hand inventory savings = $47,900

However, this end-item inventory savings does not come without an investment. In this case, in order to support the end-item activity, the average on-hand position of 501P is 220 units. This represents an average cash commitment of $27,500. This means that there is a net average cash savings of $20,400. Remember that leveraging these strategic stocking points is not just about lower working capital; there are other advantages to stocking 501P. Most companies would acknowledge a benefit associated with compressing lead times above and beyond the working-capital reduction. The manufacturing schedule is now less susceptible to supplier disruptions associated with 501P. If there is a critical resource that 501P feeds, then the benefits of this lower susceptibility is further amplified. In summary, the company gets compressed lead times with less disruption for less investment.

Identifying 501P as a stocked part will alter the ASRLT chains. The change in the ASRLT paths has been reflected in an updated matrix BOM in Figure 23-13.

This updated matrix BOM demonstrates how stocking a critical shared component part will shift the chains to different paths. By stocking 501P, three of the paths shift to entirely different legs of the BOM. This shift is illustrated in Figure 23-14. In this example, there is not another shared component that is common to more than one end item and that is a factor in the ASRLT chains of each end item.

**FIGURE 23-13**

Updated matrix BOM.

| | | 101 | 1H01 | 2H01 | 2Z01 | 201 | 203 | 204 | 205 | 207 | 301 | 302 | 303 | 304 | 305 | 402 | 409 |
|---|---|---|---|---|---|---|---|---|---|---|---|---|---|---|---|---|---|
| | | | | | | | | **Parent Items** | | | | | | | | | |
| | 201 | 1 | | | | | | | | | | | | | | | |
| | 203 | 1 | | | | | | | | | | | | | | | |
| | 204 | | 1 | | | | | | | | | | | | | | |
| | 205 | 1 | | | | | | | | | | | | | | | |
| | 207 | | 1 | | | | | | | | | | | | | | |
| | 301 | | | | 1 | 1 | | | | | | | | | | | |
| | 302 | | | | | | | | | 1 | | | | | | | |
| Child Items | 303 | | | | 1 | | | | | | | | | | | | |
| | 304 | | | 1 | | | | | | | | | | | | | |
| | 305 | | | | | | 1 | | | | | | | | | | |
| | 304P | | | | | | | 1 | 1 | | | | | | | | |
| | 307P | | | | | | 1 | | | | | | | | | | |
| | 309P | | | | | | | 1 | | | | | | | | | |
| | 401P | | | | | | | | | | | | | 1 | | | |
| | 402 | | | | | | | | | | | | | 1 | | | |
| | 403P | | | | | | | | | | | 1 | 1 | | 1 | | |
| | 404P | | | | | | | | | | | | | | | 1 | |
| | 408P | | | | | | | | | | 2 | | | | | | |
| | 409 | | | | | | | | | | 2 | | | | | | |
| | 417P | | | | | | | | | | | | 1 | | | | |
| | 501P | | | | | | | | | | | 1 | | | | 1 | 2 |

This is not to say that there is no merit in exploring compression possibilities for each individual parent item. There still can be tremendous value from a market and/or inventory perspective in continuing to test the impact of stocking components that lie on individual parent's ASRLTs. Figure 23-15 represents a decision flowchart for using ASRLT at either the individual part level or across all parts with a matrix BOM. This decision-making matrix can and should be iterated periodically (i.e., monthly, quarterly, and semi-annually depending on the level of change in the environment). It is designed to use ASRLT to compress lead times and/or inventory positions from a single parent BOM perspective or from a matrix BOM perspective. There is an illustrated linkage between the two perspectives.

**FIGURE 23-14**

Shifted ASRLTs for each end item.

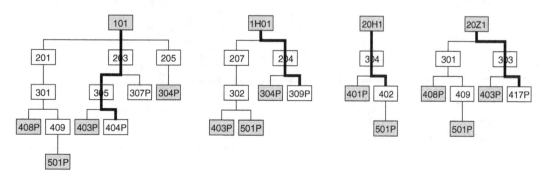

# FIGURE 23-15

ASRLT-enabled decision matrix.

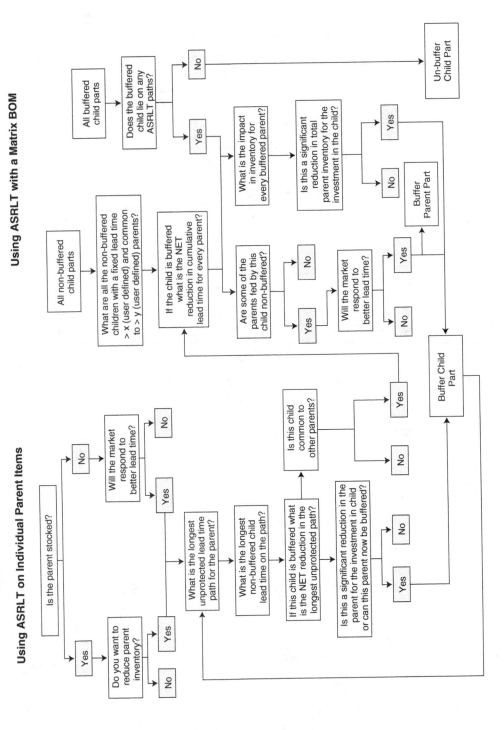

## Using ASRLT on Individual Parent Items

## Using ASRLT with a Matrix BOM

Enough iterations of this decision matrix will yield ASR equilibrium. ASR equilibrium is a point where there is a general indifference to parent-item lead time and inventory reduction for the investments involved with stocking components. Companies must be diligent about reapplying this decision matrix because the equation involved can be extremely dynamic with changes to the relevant inputs in the matrix BOM. Those inputs include

- Any change to the manufacturing or purchased lead times of parts
- Additions and/or deletions of component parts that are shared across a significant number of parent parts
- Significant additions or deletions of parent items

## CHAPTER 24

# Buffer Profiles and Level Determination

**P**roperly considering where to put inventory is necessary to eliminate the most common types of problems associated with inventory and materials management. However, it is far from sufficient to eliminate these problems. Once parts have been selected that merit a strategic position designation, the process of setting their respective buffer levels begins.

## INVENTORY: ASSET OR LIABILITY REVISITED

In order to better understand how to determine the buffer levels of strategic positions/parts, we must answer a question: Is inventory an asset or a liability? According to the balance sheet, it is an asset. The 1980s and 1990s saw many large companies play interesting paper games with inventory. Despite having no demand, many companies continued to build inventory, realize the accounting value-add from that inventory, and declare profits against it. In the process, the company was drained of cash and went deeply into debt but according to generally accepted accounting practice (GAAP) was profitable. Today, with the proliferation of methodologies such as lean and theory of constraints (TOC), in addition to the global economic meltdown starting in 2008, fewer companies can afford to play these games. Wall Street also has become aware of this ruse and the penalties of too much inventory.

    With regard to inventory and planning, assume that the word *asset* means that inventory is available in a quantity sufficient to capture a valid market opportunity and nothing more. Extrapolating this definition further, it can be concluded that there is liability when a company has more inventory than is necessary to meet market requirements (overages) and when it does not have enough (shortages).

    Figure 24-1 is a simple chart illustrating this principle. The Y axis determines whether the inventory position is an asset or a liability, as defined earlier. Asset and lia-

**FIGURE 24-1**

Inventory asset-
liability curve.

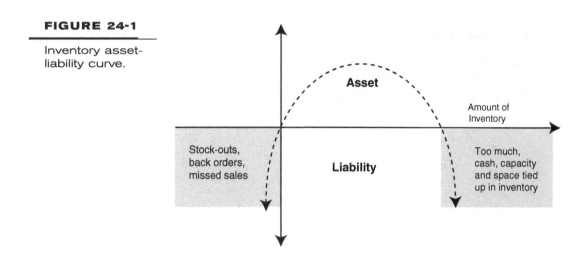

bility are delineated by the $X$ axis, which depicts quantity. Above the $X$ axis, the inventory position is an asset; below, it is a liability. Where the axes intersect, the quantity is zero. As described earlier, there are two cases that create a liability to a business. The left-hand case is obvious. It starts with having no inventory and quickly gets worse as back orders grow. The right-hand case is something companies are much more appreciative of today. As quantities grow beyond the market's desire, the organization wastes cash, capacity, and space.

The shape of the curve will shift depending on circumstances or inputs. Unfortunately, companies often flip-flop back and forth between these two undesirable positions at a part/stock-keeping-unit (SKU) level. At an aggregate inventory level, companies often find themselves in both positions at the same time (too much of the wrong inventory while also having too little of the right inventory). This graphic clearly depicts two points that represent the limits that a company must manage to stay within with regard to both individual part/SKU buffer levels and its aggregate inventory position. One of those limits is obvious; it is the quantity zero. If there is no demand, then zero inventory is an excellent inventory position. The next section will discuss how to determine the other limit (referred to as the *buffer level*) and then how to manage between these two limits.

## BUFFER PROFILES

Buffer levels and the zones that comprise them are determined by a combination of globally managed attributes and policies and critical individual traits of the respective material, part, or end item. The first step in applying these traits is to create *buffer profiles*. Obviously, different materials, parts, and end items behave differently. Conversely, many also behave very similarly. Buffer profiles are families or groups of parts for which it make sense to devise a set of rules, guidelines, and procedures that can be applied the same way to all members of a given buffer profile. Devising and revising rules, guidelines, and procedures for hundreds or thousands of parts individually quickly would prove overwhelming. These families should not be confused with the traditional notion of product families described in Chapter 11, which tend to be components or end items grouped by like characteristics in terms of physical configuration or markets. With buffer profiles, the familial connection is made based on several factors relating to behavioral or policy-related traits. The buffer level or top-side limit of each material, part, or end item will be the summation of *zones*. These zones are color-coded and sized by a combination of various inputs. These color designations will become critical to management of the buffers between the defined limits.

There are four key factors that will tend to form the various groups in most environments.

## Factor 1: Item Type

The first grouping will be made by determining whether an item is manufactured (M), purchased (P), or distributed (D). The reasons to group by these designations are

- Frequently, companies designate the control of these different item types to different people or groups. Intuition about behavior is frequently limited to those groups.
- From an organizational perspective, there tends to be a varying degree of direct control over these item types. The assumption is that companies have more direct control over something within their physical control. Sometimes the amount of control that extends to purchased and distributed items depends on the vertical integration of the enterprise.
- Relative lead-time horizons can be very different among these item types. Short lead times for purchased items could be up to a week. Short lead times for manufactured items could be one to two days. Depending on the positioning model, lead times for distributed items will be inbound transportation time plus any administration, pack, and unpack times.

## Factor 2: Variability

Variability can be segmented into three slices—high, medium, and low—with the two dimensions of demand and supply. In this case, however, the supply and demand variability is only with regard to the discrete part or SKU number.

> **demand variability:** The potential for spikes in demand against this particular part or SKU number. Once again, a variability designation can be calculated by a variety of equations or determined by rules of thumb with intuitive planning personnel. Mathematically, demand variability or uncertainty can be calculated through standard deviation, mean absolute deviation (MAD), or variance of forecast errors.[1]

---

[1] *APICS Dictionary*, 12th ed. New York: Blackstone, 2008 p. 101.

Heuristically, companies can use the following segmentation:

- *High demand variability.* This part is subject to frequent spikes.
- *Medium demand variability.* This part is subject to occasional spikes.
- *Low demand variability.* This part has little to no spike activity—its demand is relatively stable.

Supply variability is the potential for and severity of disruptions in sources of supply for this part or SKU number. This can be calculated by examining the variance of promise dates from actual receipt dates. The caution here is that many of these dates are often determined and managed initially through critical flaws in how material requirements planning (MRP) has been used. Finally, the number of alternate sources for a part or material could factor into the supply variability equation because the net effect of more sources might be more reliable supply. Supply variability can be considered as:

- *High supply variability.* This part or material has frequent supply disruptions.
- *Medium supply variability.* This part or material has occasional supply disruptions.
- *Low supply variability.* This part or material has reliable supply (either a highly reliable single source or multiple alternate sources that can react within the purchasing lead time).

Purchased parts tend to be influenced almost exclusively by supply variability. One exception is in pure make-to-order environments, where there are no sub-component, intermediate-component, or end-item buffers. A pure make-to-order environment would indicate that the inventory positioning factors dictated buffering only for some purchased items. This is an example of why companies cannot skip the inventory positioning step. It can dramatically alter which items will end up in which buffer profiles.

Manufactured parts can be subject to both supply and demand variability depending how the positioning model is formulated. Manufactured parts are less subject to demand variability if the manufactured item feeds another level of buffered component or end item. These parts are less subject to supply variability if they consume critical parts that are replenished strategically. This is due to the dampening nature of the buffer breakwalls. However, in many cases it is usually a blend of being fed by some buffered positions and feeding only some buffered positions. An example of this type of manufactured part is one that is used in subassemblies or the end item (some of which might be buffered) but is also a service part (which might go directly to the customer). This type of manufactured part probably would be subject to more demand variability than a part that fed only some buffered subassemblies or end items. It is imperative that companies carefully apply the positioning factors in Chapter 4.

Distributed parts or SKUs will tend to be affected by one variability type depending on their respective locations in the internal supply chain. Distributed parts/SKUs at central buffers can be largely immune from variability if the downstream positions that they feed are sized and managed properly and supply is reliable. Part/SKU buffers at down-

**FIGURE 24-2**

Different variability factors for combinations of buffers.

Supply Variability                                                    Demand Variability

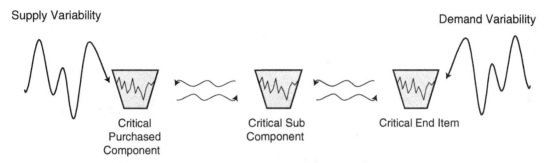

Critical                Critical Sub            Critical End Item
Purchased               Component
Component

stream locations will be affected almost exclusively by demand variability because they are protected by the central buffer on the supply side. See Chapter 4 for more detail on inventory positioning in distribution networks.

Figure 24-2 illustrates how buffers at different stages within a manufacturing process can experience different levels of variability depending on their relationships with each other. Arrowed lines that move from left to right represent supply variability. Coming out of a buffer, they are smoother and imply more consistent availability. Arrowed lines that move from right to left represent demand variability. Coming out of a buffered positioned, they are smoother and convey more consistent order quantities and/or intervals.

## Factor 3: Lead Time

Lead time can be segmented simply into three categories: short, medium, and long. These designations are relative to the company's specific environment and part type. Typically, there is a large distribution spread in the size of lead times associated with purchased parts. This spread could be anywhere from almost zero lead time for on-site supplier-managed inventory to lead times measured in months or years. Purchased parts that are reliably at very short lead times are not candidates for strategic replenishment designation. Little benefit can be gained from the additional management of these parts. Figure 24-3 details the distribution of lead times for purchased parts identified for strategic replenishment in a sample environment. In this case, the shortest lead time is 3 days, whereas the longest is 56 days.

There are differing circumstances that dictate what should be deemed short, medium, and long. Eventually, it will come down to a comfort level for the planners in that environment. Later in this chapter the lead-time-designation influence on buffer levels and zones is discussed.

Manufactured parts have three distinct types of calculated lead times that can be evaluated in order to determine what is short, medium, and long. As discussed in

**FIGURE 24-3**

Purchased-part lead time.

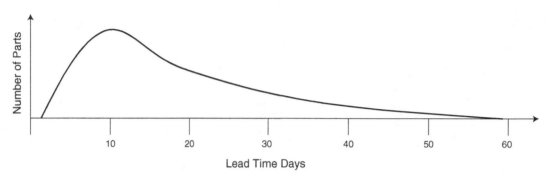

Chapter 23, two of these lead times, manufacturing lead time (MLT) and cumulative lead time (CLT), are most often unrealistic, either an underestimation or an overestimation, respectively. To this extent, actively synchronized replenishment lead time (ASRLT) is best used to determine what is short, medium, and long. Figure 24-4 provides an example of what the difference in distribution may look like among MLT, CLT, and ASRLT. Clearly, this affects what planning personnel view as short, medium, and long.

Figure 24-5 represents an example of short, medium, and long designation against the distribution of manufacturing parts chosen for replenishment.

Short = 1 to 10 days.

Medium = 11 to 25 days.

Long = 26+ days.

## Factor 4: Significant Minimum Order Quantity (MOQ)

Ordering policies, including order minimums, maximums, and multiples, complicate planning and supply scenarios but are a fact of life for planners. Many of these ordering policies are based on valid data and sound assumption; some are not. It is a given that there will be parts/SKUs that do require minimum order quantities. Minimum order quantities (MOQs) can affect buffer levels and zones. Significant MOQs definitely will affect buffer levels and zones. The qualifying characteristics of what makes an MOQ *significant* will be examined later in this chapter. Additionally, frequently, times with MOQs also will be designated long-lead-time parts/SKUs.

Based on these four factors, there are 54 potential buffer profiles. Depending on the manufacturing environment, there could be even more derivations and permutations than this. If there is a certain global attribute that makes sense by which parts should be grouped that is not related to variability, lead time, part type, and order quantity, then another type of buffer profile should be explored. Figure 24-6 details the 54 different buffer profile combinations. Each buffer profile has been designated with a code based

**FIGURE 24-4**

(A) Part lead-time distribution using MLT. (B) Part lead-time distribution using CLT. (C) Part lead-time distribution using ASRLT.

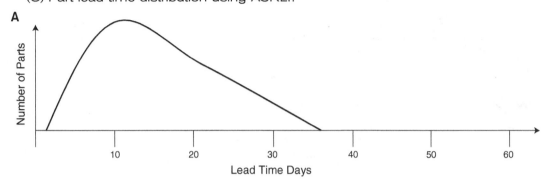

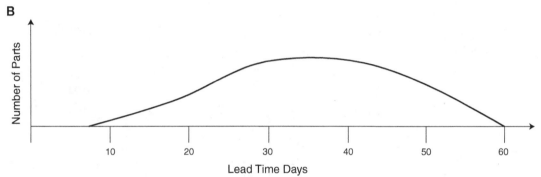

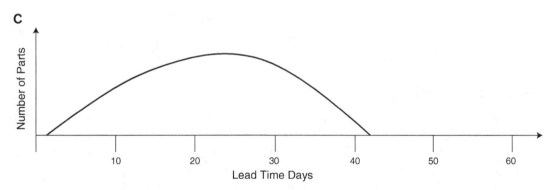

on its combination of attributes. Within that naming code, the first letter signifies part type, either *M* for make, *B* for buy, or *D* for distributed. Next is the variability category, 1 for low, 2 for medium, or 3 for high. Next will be the lead-time category, 0 for short, 1 for medium, or 2 for long. Finally, the buffer profile name will be appended with an MOQ if a significant MOQ exists. For example, a distributed part with medium variability and short lead time is in the buffer profile D20. A purchased part with high variability, long lead time, and a significant MOQ will be in the buffer profile B32MOQ.

**FIGURE 24-5**

Part lead-time distribution using ASRLT to determine lead-time factor.

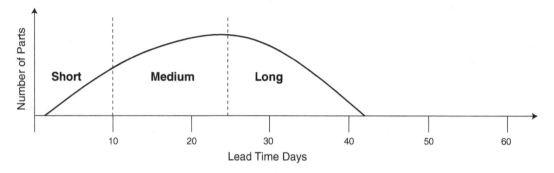

**FIGURE 24-6**

Buffer profile combinations.

| | | Make = M | Buy = B | Distributed = D | | |
|---|---|---|---|---|---|---|
| | Low = 1 | M10 | B10 | D10 | Short = 0 | |
| | | M11 | B11 | D11 | Medium = 1 | |
| | | M12 | B12 | D12 | Long = 2 | |
| Variability Categories | Medium = 2 | M20 | B20 | D20 | Short = 0 | Lead Time Categories |
| | | M21 | B21 | D21 | Medium = 1 | |
| | | M22 | B22 | D22 | Long = 2 | |
| | High = 3 | M30 | B30 | D30 | Short = 0 | |
| | | M31 | B31 | D31 | Medium = 1 | |
| | | M32 | B32 | D32 | Long = 2 | |
| | | M10MOQ | B10MOQ | D10MOQ | | |
| | | M11MOQ | B11MOQ | D11MOQ | | |
| | | M12MOQ | B12MOQ | D12MOQ | | |
| | | M20MOQ | B20MOQ | D20MOQ | | |
| MOQ Application | | M21MOQ | B21MOQ | D21MOQ | | MOQ Application |
| | | M22MOQ | B22MOQ | D22MOQ | | |
| | | M30MOQ | B30MOQ | D30MOQ | | |
| | | M31MOQ | B31MOQ | D31MOQ | | |
| | | M32MOQ | B32MOQ | D32MOQ | | |

## BUFFER ZONES

There are three color-coded zones that comprise the total buffer. The color coding is universally intuitive: green, yellow, red. These zones will determine both planning and execution priorities. Green represents an inventory position that requires no action. Yellow represents a part that has entered its rebuild or replenishment zone. Red represents a part that is in jeopardy and may require special attention. Figure 24-7 shows the zone stratification within a stock buffer. Please remember that in this book the colors will appear as different intensity of grayscale due to the limitations of the black and white printing process.

Figure 24-8 displays the meaning of each colored zone. In addition to the three intuitive zones, there are two additional color-coded zones. These zones do not affect the actual buffer sizing calculation; they are determined after that calculation is made. The first additional zone describes an overstocked position or over top of green (OTOG). This zone's color is light blue. Some authors refer to this as the "white zone." Finally, there is

**FIGURE 24-7**

Moving to zone
stratifications in a
stock buffer.

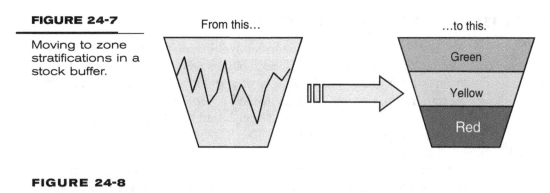

**FIGURE 24-8**

The meaning of each buffer zone.

| Stock Out | ALERT! | Rebuild | OK | OTOG/Too Much |
|-----------|--------|---------|----|----|

a color-coded zone that displays an out-of-stock situation with demand. This zone's color is dark red. Some authors refer to this as the "black zone."

Figure 24-9 shows the inventory asset-liability curve with the overlaid color-coded zones.

This color-coding system will be used for both planning and execution priority management. Visibility is an integral part of the power of the DDMRP solution. From a planning perspective, the color coding will determine if additional supply is needed based on the available stock position [on hand + open supply – demand (including qualified spikes)]. The nature of the available stock equation and its relation to planning activity and buffer status will be explored in Chapters 25 and 26. From an execution perspective, the color coding determines actions (primarily expediting or resource schedule manipulation) based on different types of alerts. This will be explained in depth in Chapter 27.

**FIGURE 24-9**

Asset/liability
curve with buffer
zones.

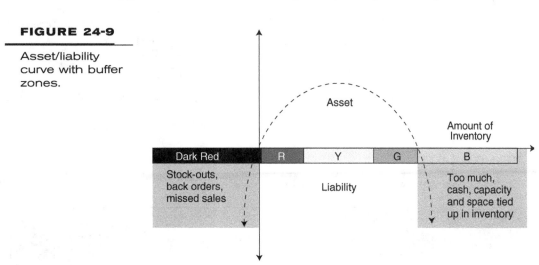

# CALCULATING BUFFER LEVELS

As mentioned earlier, the summation of the green, yellow, and red zones will yield the top level of the buffer, or top of green (TOG). Each of these zones is determined by applying rules determined for a global attribute against certain individual part traits. Figure 24-10 shows the minimum amount of data needed to determine zone sizes and thus buffer levels for purchased, manufactured, and distributed parts. Note that there is a special input for distributed parts called *location*. This allows parts/SKUs at different forward locations to have different buffer profiles and individual inputs applied.

Each zone within the buffer (G, Y, or R) is sized by an expression of average daily usage (ADU) over a percentage of lead time (expressed in days). The green zone also will have the option of being expressed as the minimum order quantity (if present) when that minimum order quantity is significant. Determining the significance of the MOQ will be discussed later. The yellow zone for all buffer profiles is usually set at 100 percent of usage over lead time. The red zone has two subzones. The summation of these two subzones will define the total red zone quantity. One subzone is called *red zone base*. The other subzone is called *red zone safety*. These subzones are affected by different buffer profile inputs. This is diagrammed in Figure 24-11.

Buffer level or TOG = red zone base + red zone safety + yellow zone + green zone

**FIGURE 24-10**

Determining
buffer levels and
zones.

| Determining Buffer Levels and Zones | |
|---|---|
| Group Trait Inputs | Individual Part/SKU Inputs |
| • Lead Time Category<br>• Make, Buy, or Distributed<br>• Variability Category<br>• Significant MOQ Factor | • Average Daily Usage<br>• Discrete Lead Time<br>  – ASR Lead Time<br>  – Purchasing Lead Time<br>  – Transportation Lead Time<br>• Ordering Policy<br>  – Minimum<br>  – Maximum<br>  – Multiple<br>• Location (Distributed Parts) |

**FIGURE 24-11**

Buffer zones with red zone base and red zone safety.

| Red Zone Base | Red Zone Safety | Yellow | Green |
|---|---|---|---|

## Calculating Average Daily Usage (ADU)

Any average is only as relevant as the period over which the equation was applied. Taking too short a range produces just as unrealistic a number as taking too long a range. If a significant event occurs that has altered demand profiles dramatically within a relevant range, then planners should be cautious. If the significant event is an anomaly and will revert back to normal behavior, then the abnormal usage should be excluded. However, if the dramatic shift is indicative of what the future might look like for this part, then that data should be included or at the least factored in. Calculating some ADUs will require the planner to consult with others in the organization to formulate a valid approach.

For new-product introductions, there will be no history to calculate ADU. For new-product introductions, there may be only a business case that predicts a certain rate and volume of market acceptance. In this case, an ADU can be formulated based on that plan. However, the plan may represent a significant over- or underestimation. At this point, planners can use the formulated ADU, but additional information is required (described in Chapter 27) to protect against the over- or underestimations. The second case of no history to calculate from occurs when the system of record does not record usage. This often happens in legacy systems. If a company has a modern planning system and these data are not available, then the lack of availability would be due to poor implementation. The fortunate news is that most of the time parts chosen for strategic buffering represent the minority of total parts, a very important minority, but a minority nonetheless. The Pareto principle is valid here. Planners may need to consult with other relevant personnel to calculate an approximation of ADU that seems realistic. Once the baseline has been established, then the ADU can be tracked accurately from that point forward.

## Lead-Time Category Buffer Impact

The lead-time category will have a direct impact on the size of the green and red zone bases. The basic assumption with this category is that the longer the lead time of the part/SKU, the higher would be the inventory. The cost associated with covering demand is also higher owing to the higher probability of variance within that longer lead time. This reinforces the need for the lead-time compression tactics described in Chapter 23.

Figure 24-12 describes the recommended ranges of impact for the green and red zone bases for each lead-time category. This is expressed as a range to allow for flexibility owing to the unique circumstances involved with every company. Planning personnel will need to set the value based on their company's environment and their comfort level.

This figure illustrates the attempt to build leaner but more frequently replenished buffers for the longer-lead-time parts. By minimizing the green zone, the yellow zone

**FIGURE 24-12**

Recommended impact ranges.

|  | Green Zone Impact | Red Zone Base Impact |
|---|---|---|
| Long Lead Time | 20-40% Usage over LT | 20-40% Usage over LT |
| Medium Lead Time | 41-60% Usage over LT | 41-60% Usage over LT |
| Short Lead Time | 61-100% Usage over LT | 61-100% Usage over LT |

(rebuild zone) will be penetrated more frequently. This means that there will be a steady stream of open supply orders. This extends a larger percentage of the buffer position to inbound supply. Having more open supply orders also means that the red zone base can be minimized. The presence of a significant MOQ often will block or challenge this green zone strategy owing to the inherent lumpiness caused by the MOQ.

Two steps are necessary to determine if an MOQ is significant. First, the part's green zone is calculated based on the lead-time category it falls in without consideration of MOQ. Second, a comparison must be made of the calculated green zone quantity and that MOQ. If the order minimum is greater than the size of the calculated green zone, then the MOQ is significant. The green zone then is sized to the MOQ.

For example, Part XYZ has an ADU of 6 and a lead time of 20 days. It is coded a long-lead-time part, in which its green zone will be 30 percent of usage over lead time (36 units). If this part has an MOQ of 48 units, this would qualify as a significant MOQ (48 > 36). This designation means that the buffer profile of this part will be changed to an MOQ profile. Under MOQ profiles, the green zone is set to the MOQ (48 units). In this example, the MOQ equals 8 days of average daily usage [48 units (MOQ)/6 units (ADU) = 8 days]. Eight days, incidentally, means that the green zone will end up being sized at 40 percent of usage over lead time.

## Variability Category Buffer Impact

The variability category will size the red zone safety portion of the total red zone. Early in development of this solution, the thinking was that red zone safety might best be described as a factor of ADU (e.g., 2 $\times$ ADU, 3 $\times$ ADU, etc.). This quickly creates a problem for the variability factor to be managed effectively at a global level. Within a buffer profile there still can be relatively wide discrepancies between the discrete lead times of parts, especially long-lead-time parts. The same issue applies to expressing red zone safety as a percentage of usage over lead time. The most effective heuristic is that red zone safety should be an expression of a percentage of the red zone base. By doing this, there is a smoothing effect that happens between parts that have lead-time disparities within the same buffer profile.

Figure 24-13 shows how red zone safety is sized. A range is provided in order to allow customization to individual environments.

**FIGURE 24-13**

Red zone safety definition.

| | Red Zone Safety Impact |
|---|---|
| High Variability | 60-100% Red Zone Base |
| Medium Variability | 41-60% Red Zone Base |
| Low Variability | 20-40% Red Zone Base |

## Buffer Level Examples

To help illustrate the application of buffer level determination, three examples are presented: one purchased, one manufactured, and one distributed.

***First Example: Purchased Part 403P***   Part 403P will follow the buffer profile B11MOQ. The purchasing lead time for this part is 21 days. The ADU is 17. The MOQ is 300. The MOQ represents 84 percent of usage over lead time. This is a medium-lead-time category part, so the MOQ is significant. The MOQ of 300 will be the green zone. The yellow zone will be an entire lead time's usage (17 units/day × 21 days = 357 units). The red zone will be calculated by combining the red zone base and safety factors. For red zone base 403P is in the medium-lead-time category (41 to 60 percent usage over lead time). In this particular case, we will use 50 percent (once again, this is determined by the planners for the specific environment). Red zone base will be 179 (357 × 0.5 = 178.5). For red zone safety, 403P is in the low variability category (20 to 40 percent of red zone base). In this particular case we will use 30 percent. Red zone safety will be 54 (179 × 0.3 = 107.1). The top limit of the buffer (top of green) will be 890 units, top of yellow will be 590, and top of red is 233. Figure 24-14 presents the buffer summary.

***Second Example: Manufactured Part 707***   Part 707 will follow the buffer profile M21. The ASRLT for this item is 13 days. The ADU is 6. There are no ordering policies for the item. Part 707 is a medium-lead-time part. This will define both the green zone and red zone base at a range of 41 to 60 percent usage over lead time. In this environment, the planners chose to set this at 45 percent. The green zone for Part 707 will be 35 (78 × 0.45 = 35.1). Red zone base also will be 35. Yellow zone will be set to usage over one full lead time: 78. For red zone safety, Part 707 is in the medium variability category (41 to 60 percent of red zone base). Planners have set this at 50 percent. Red zone safety will be 18 (35 × 0.5). The top of green will be 166 units, top of yellow is 131, and top of red is 53. Figure 24-15 presents the buffer summary.

**FIGURE 24-14**

Part 403P buffer summary.

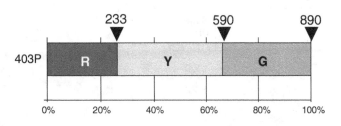

**FIGURE 24-15**

Part 707 buffer summary.

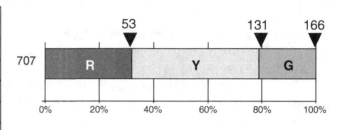

| Part: 707 Lead Time: 13 Days | Buffer Profile: M21 |
|---|---|
| Green Zone | 35 |
| Yellow Zone | 78 |
| Red Zone Base | 35 |
| Red Zone Safety | 18 |

***Third Example: Distributed Part 501D*** Part 501D will follow the buffer profile D22MOQ. The location is a distribution center in Newark, NJ. The lead time of this item at this location is 28 days (the source is Southeast Asia). The ADU at this location is 40. There is an MOQ of 1,000. Part 501D at this location is a long-lead-time part (20 to 40 percent usage over lead times for both green zone and red zone base). In this case, planners have determined that 30 percent of usage over lead time will be applied to green zone and red zone base. The MOQ far exceeds what the calculated green zone would be (MOQ = 1,000, calculated green zone = 336). The MOQ (1,000) will be treated as the green zone. The yellow zone will be 100 percent usage over lead time: 1,120. The red zone base is calculated at 336. Part 707 is in the medium variability category (41 to 60 percent of red zone base). In this case, the planners have chosen 60 percent usage over lead time as the variability factor. Red zone safety will be 202 (336 × 0.6). The top of green will be 2,658, top of yellow is 1,658, and top of red is 538. Figure 24-16 presents the buffer summary.

## BUFFER LEVEL SUMMARY

Figure 24-17 is an example of a group of four parts (r457, f756, h654, and r672) that are in a specific buffer profile. Quantity on hand is represented by the *Y* axis. The part numbers

**FIGURE 24-16**

Part 501D buffer summary.

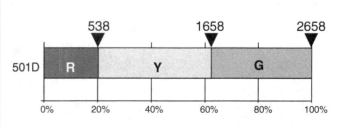

| Part: 501D | Buffer Profile: D22MOQ |
|---|---|
| Green Zone | 1000 |
| Yellow Zone | 1120 |
| Red Zone Base | 336 |
| Red Zone Safety | 202 |

**FIGURE 24-17**

Buffer profile.

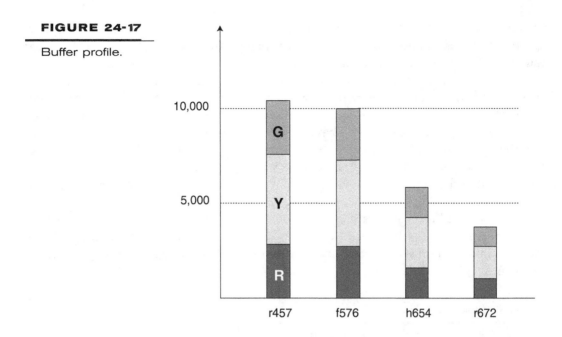

populate the X axis. Since they are in the same buffer profile family, they have the same global attributes but have different buffer levels because they have different individual attributes (ADU and discrete lead times).

Figure 24-18 is an example of one distributed part's different buffer profiles and levels across four forward distribution points.

**FIGURE 24-18**

Distributed part buffer profile.

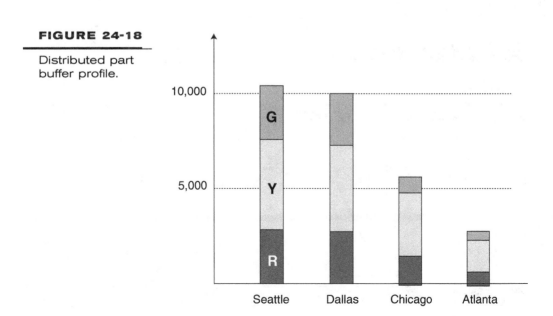

When first starting to use these DDMRP techniques, companies commonly find themselves out of balance from an inventory perspective. When parts are chosen for strategic buffering and those parts are given appropriate buffer profiles, companies often observe a bimodal distribution that looks like a saddle, as illustrated in Figure 24-19. Examining the on-hand inventory positions of these strategically replenished parts will show that there is often a large bulge to the left representing parts that are in the red or stocked out. Some of these parts will have no open supply orders. At the same time, there is another lump to the right that represents parts that are over top of green or over-stocked. It is not uncommon to see some parts with an on-hand inventory position above 5,000 percent of the top of green level. Surprisingly, it is also not uncommon to see these parts with additional open supply orders.

When the buffer profiles are correct, the on-hand (not available stock) inventory position averages in the lower half of the yellow zone, as illustrated in Figure 24-20. Getting to this position is a constant challenge and should be a target by which planning effectiveness is judged.

**FIGURE 24-19**

Current on-hand inventory misalignment.

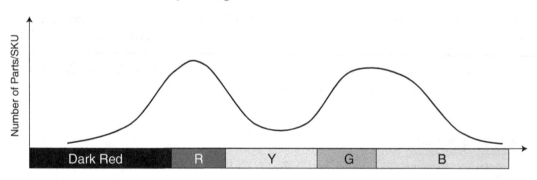

**FIGURE 24-20**

Desired on-hand inventory alignment.

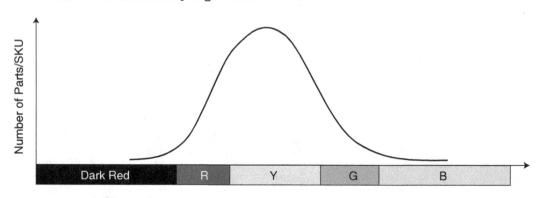

Depending on how severe the current "saddle" situation is, transitioning a company to DDMRP can represent a challenge. In order to bring inventory into balance, inventory positions will have to be built and others will need to drain off. Unfortunately, the buildup is an immediate requirement, whereas the drain-off tends to take longer. One way to mitigate this is to examine any open supply against parts/SKUs that are well over the top of green and make attempts at cancellation or delay to free up cash to shift to parts/SKUs whose positions must be built up.

Figure 24-21 is an example from an actual company that illustrates the movement from a saddle inventory distribution to the one-hump ideal. This figure covers a 12-month period for parts chosen for strategic buffering. Notice that the real impact happens within the first four months.

The realignment involves both inventory reductions and additions. The net effect, however, is a significant decrease in total inventory over the course of a year while providing better market protection. Figure 24-22 shows the dollars required to build positions, dollars that drain off, and the net effect by period. Each period is one month, with the exception of period 7, which represents months 7 to 12.

## FIGURE 24-21

Sample company's inventory realignment.

Projected Inventory Reductions Over 1 Year

|  | Mo 1 | Mo 2 | Mo 3 | Mo 4 | Mo 5 | Mo 6 | Mo 7-12 | Totals |
|---|---|---|---|---|---|---|---|---|
| Inventory Beginning of Period | $ 9,564,443 | 7,738,294 | 7,869,529 | 7,569,990 | 7,392,764 | 7,278,870 | 7,194,836 | $ 9,564,44 |
| Projected Inventory Reductions (consumption) | (2,415,391) | (796,305) | $ (313,565) | $ (187,875) | $ (113,894) | $ (84,033) | $ (794,430) | $ (4,705,49 |
| Projected Inventory Increases - (purchases) | $ 589,242 | $ 927,540 | $ 14,025 | $ 10,649 | $ - | $ - | $ - | $ 1,541,45 |
| Net Inventory Reduction by Period | $ (1,826,149) | $ 131,235 | $ (299,540) | $ (177,226) | $ (113,894) | $ (84,033) | $ (794,430) | $ (3,164,03 |
| Inventory End of Month | $ 7,738,294 | $ 7,869,529 | $ 7,569,990 | $ 7,392,764 | $ 7,278,870 | $ 7,194,836 | $ 6,400,406 | $ 6,400,40 |
|  |  |  |  |  |  |  |  |  |
| % Cumlative Inventory Reduction | 19.1% | 17.7% | 20.9% | 22.7% | 23.9% | 24.8% | 33.1% | 33.1 |

## FIGURE 24-22

Sample company's inventory increases, decreases, and net inventory effect by period.

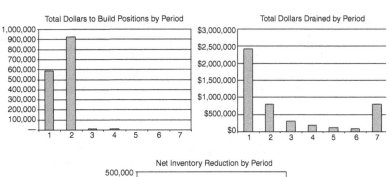

## SUMMARY

Grouping items chosen for strategic replenishment into buffer profiles allows for global management of items that behave similarly. Despite providing 54 potential combinations of buffers, typically even some of the most complex environments use fewer than 20 different profiles at any one time. These profiles will establish how individual item attributes will determine the top of a buffer by influencing the size of the zones that comprise those buffers. The additional benefit of the zonal approach is that the zones can be color-coded. This allows for highly visible planning and execution prioritization that will be discussed in later chapters.

## CHAPTER 25

# Dynamic Buffers

**O**ver the course of time, the conditions that placed parts/stock-keeping units (SKUs) into certain buffer profiles, as well as their individual traits, can and will change. New suppliers and materials are used, new markets are opened and grow while others deteriorate, and manufacturing capacity and methods change. These changes will affect the inputs into the buffer equation. Letting buffer levels adjust themselves to these changes allows a company to adapt its working capital to a dynamic environment. There are three types of adjustments to be considered: recalculated adjustments, planned adjustments, and manual adjustments.

## RECALCULATED ADJUSTMENTS

The first category of adjustment is most often automated (the manner of automation will be greatly determined by the planning system's capabilities). There are two types of recalculated buffer adjustments: average daily usage (ADU)–based adjustments and zone occurrence–based adjustments. While both approaches are discussed, the recommendation is for ADU-based adjustments.

### ADU-Based Recalculation

As more or less variability is encountered through time, buffers should adapt and change to fit the environment. An easy way to do this is to recalculate the ADU based on a rolling horizon. The length and frequency of the horizon are user-defined. In most cases, the usage frequency is daily (thus ADU). The length of the rolling time horizon, however, is specific to the environment. Some companies may choose a 3-month rolling horizon, whereas others feel compelled to use 12 months. Too short of a rolling horizon may yield buffer changes that are overreactive. Too long of a horizon may yield changes that are underreactive. Regardless of what the length of the horizon is that a company settles on for each part/SKU, circumstances may occur that will make it either overreactive or underreactive. This is the purpose for alerts or early-warning indicators. More on those types of alerts is provided in the manual adjustments section.

Figure 25-1 illustrates how a buffer can adjust based on changes to ADU. The initial buffer size (based on its buffer profile and individual part traits) can be seen at the far left of the figure. The black sawtooth line represents the available stock (on-hand + open supply – qualified demand) position, whereas the gray smooth line represents the ADU. If this time frame represented a 24-month time period, note that the average daily usage would rise dramatically, begin to stabilize, and eventually reach maturation. The total change over that time is an ADU of 6 to 48. These changes then flex the target buffer via a recalculated ADU. In this case, it may be a three-month rolling recalculation.

## Occurrence-Based Recalculation

Another way to adjust buffers is by measuring the number of defined occurrences that happen within a prescribed interval with regard to a particular part/SKU. Companies usually use this kind of adjustment in conjunction with a fixed-interval/reorder inventory model. The basic logic is that based on the lead time and demand profile of the part, there should be an average order interval. If the buffer is sized improperly, then situations will occur with unacceptable frequency. For example, a number of red zone occurrences or stock-outs within that interval could trigger a buffer increase. Alternatively, a sustained green zone available stock position (meaning no additional supply order generation) over the defined interval could trigger a decrease in the buffer.

The difficulty associated with this method is simply defining and maintaining all the relevant parameters, which include

- Number of occurrences
- Size of the interval
- Size of the adjustments based on the number of occurrences

**FIGURE 25-1**

Dynamic adjustment of a part.

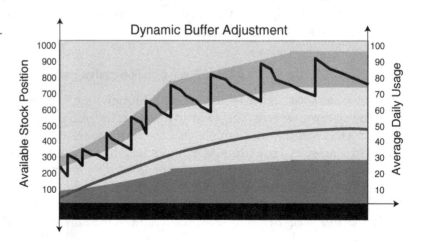

The answers to all of these questions are also specific to the environment and individual planner experience and intuition.

# PLANNED ADJUSTMENTS

Buffers also can be manipulated through something called *planned adjustments*. Planned adjustments are based on certain strategic, historical, and business intelligence factors. In demand-driven material requirements planning (DDMRP), these planned adjustments represent the necessary elements of planning and risk mitigation required to help resolve the conflict between necessary elements of plan predictability and the use of demand-driven operational methods.

These planned adjustments are manipulations to the buffer equation that affect inventory positions by raising or lowering buffer levels and their corresponding zones at certain points in time. This manipulation occurs through adjusting ADU to a historically proven or planned position based on an approved business case. Planned adjustments are used for common situations such as seasonality and product ramp-up and ramp-down. Product ramp-up and ramp-down are caused by product introduction, product deletion, and product transitions.

## Seasonality

Many companies have challenges with seasonality. Figure 25-2 shows a product that has a substantial bulge in demand once per year. The ADU (represented by the dark smoothed line) "flexes" the buffer levels to create that bulge. During peak seasonality, ADU is more than double that of the low period.

Planned adjustments also must work in concert with the actively synchronized replenishment lead time (ASRLT) of the part/stock-keeping unit (SKU). This is particularly important with regard to long-lead-time parts/SKUs. Figure 25-3 is an example of a sea-

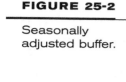

**FIGURE 25-2**

Seasonally adjusted buffer.

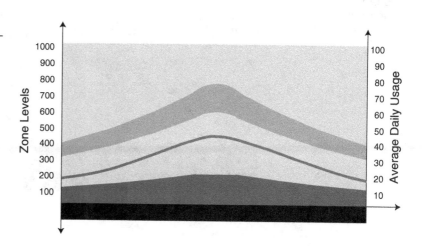

**FIGURE 25-3**

Planning with
seasonal
adjustment up
and ASRLT
factored in.

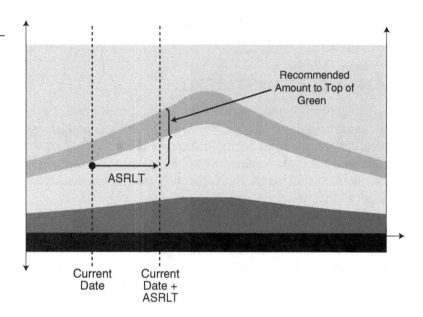

Recommended
Amount to Top of
Green

ASRLT

Current
Date

Current
Date +
ASRLT

sonal part. Note that at the current date, the part's available stock position is in the rebuild
zone without factoring for anything else. In this case, however, the recommended resupply
amount will not be determined by the current date's top of green level. Instead, it will be
determined by the top of the green level on a date equal to the current day plus the ASRLT.

Figure 25-4 is another example of the effect of using ASRLT in combination with
planned adjustments. Once again, the part's available stock position resides in the top of
the rebuild zone without considering other factors. When taking the current date plus

**FIGURE 25-4**

Planning with
seasonal
adjustment down
and ASRLT
factored in.

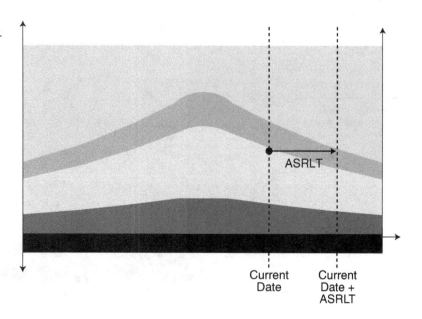

ASRLT

Current
Date

Current
Date +
ASRLT

ASRLT, however, the actual available stock position is at the top of green if not slightly over top of green.

Complicating the seasonality and ASRLT picture could be available capacity over the course of a time period. Ice cream sales in the North America are an example. Obviously, the summer is the peak time. Ice cream manufacturing capacity does not have a counter-cyclic product to balance when there is less demand for ice cream. This means that companies are reluctant to carry excess capacity above and beyond what the high-season requirements are. Figure 25-5 is a seasonal example where all end items are buffered, and the aggregate buffered inventory positions exceed plant capacity during the peak period. In this example, the plant's capacity is 60,000 gallons a day, on average. During the peak season, buffer requirements exceed the total plant's average daily requirements.

This often leads companies to believe that they must level-load the facilities over the course of the year. In doing this, companies are trying to spread capacity requirements out over the entire year to meet seasonal demand peaks. By level-loading, they are setting an artificial batch determined by a perceived capacity limitation within a certain interval. The obvious problem with this is that it requires companies to commit to larger batches of specific inventory well in advance of demand. Inevitably, during peak seasons, there are shortages of high-moving items and massive surpluses of other slow-moving items. Figure 25-6 is an example of a level-loaded plant.

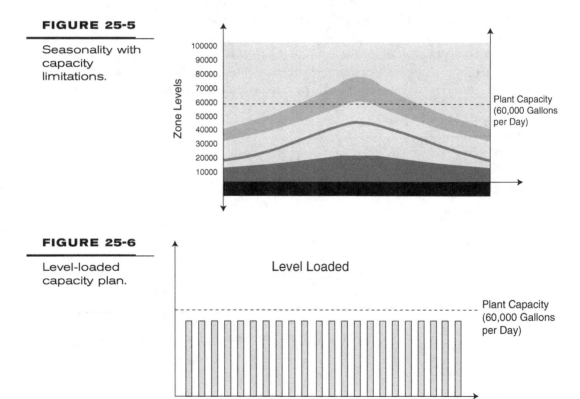

**FIGURE 25-5**

Seasonality with capacity limitations.

**FIGURE 25-6**

Level-loaded capacity plan.

Trying to use level-loading in conjunction with seasonally adjusted buffered items poses a problem. Seasonal buffer profiles will mirror their bulge effect with regard to capacity requirements. The biggest capacity contention occurs during the ramp-up period going into a high-demand period. Generally, during periods in which capacity is under more contention, batch sizes should be increased (assuming that there is a time cost associated with setup). During times when capacity is under less contention, setups should be less of an issue, and batches can track to demand or replenishment requirements more closely.

Figures 25-7, 25-8, and 25-9 are examples of how the preceding aggregate buffer profile might translate to batch requirements to build up the seasonal buffer positions and then be responsive to requirements during the high season. Figure 25-7 is meant to show an average batch size that grows during the ramp-up period. This is due to the fact that during the buffer ramp-up period, load exceeds average plant capacity of 60,000 gallons per day. This means that capacity has become a bottleneck during this time period. A *bottleneck*, by definition, is "a facility, function, department, or resource whose capacity is less than the demand placed upon it. For example, a bottleneck machine or work center exists where jobs are processed at a slower rate than they are demanded."[1] When capacity becomes a bottleneck, saving setup is an appropriate tactic (assuming that there is a capacity cost to a setup). During this time, the plant will be running larger batches and possibly using overtime, additional shifts, or even temporary labor. In so doing, its output probably will exceed its average calculated capacity.

Notice how the average batch size grows during ramp-up but quickly reduces when buffer positions have been established. Now the factory has "sprint" or surge capacity in order to react to how the buffers are performing while it is still peak season. The factory is essentially storing capacity in strategic or safe positions, as defined by the buffers, and

**FIGURE 25-7**

Variable batch size to coincide with buffer strategy.

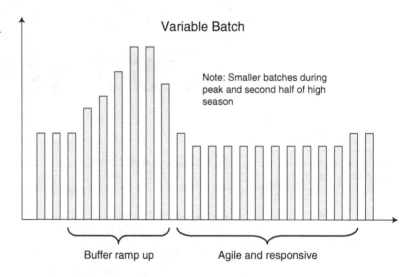

Variable Batch

Note: Smaller batches during peak and second half of high season

Buffer ramp up          Agile and responsive

---

[1] *APICS Dictionary*, 12th ed. New York: Blackstone, 2008, p. 14.

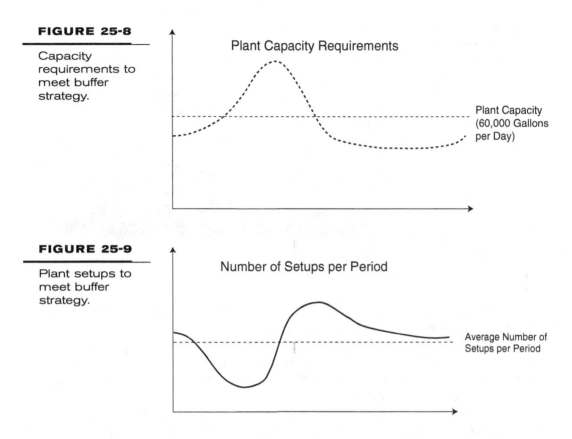

**FIGURE 25-8**

Capacity requirements to meet buffer strategy.

Plant Capacity Requirements

Plant Capacity (60,000 Gallons per Day)

**FIGURE 25-9**

Plant setups to meet buffer strategy.

Number of Setups per Period

Average Number of Setups per Period

then transforming into quick-response mode in order to react to how the market affects those buffer positions. Setups might go up, but shortages will go down.

Clearly, this is not an efficient mode of operation, as viewed from the traditional cost-efficiency model. The asset base is being used at different levels over the course of the year to reduce shortages, maximize sales potential, minimize old or unnecessary inventory, and limit expedite-related expenses. There is nothing inefficient about the combination of these effects.

## Ramp-Up and Ramp-Down

Planned adjustments also can be used for part/SKU introduction, deletion, and transition. The example of part/SKU introduction is examined first. In the part/SKU ramp-up example (Figure 25-10), there is a part that is being ramped up based on a sales and marketing plan. The figure covers a 12-month period. The sales and marketing plan calls for the SKU to reach market maturation within a 9-month period in its initial distribution region with an average daily usage of over 40. Instead of immediately spending the cash and capacity to bring the buffer to full size, a planned adjustment will be applied. This planned adjustment will ramp up the planned ADU over a period of time, thus creating

**FIGURE 25-10**

**FIGURE 25-10**

Part ramp-up
example.

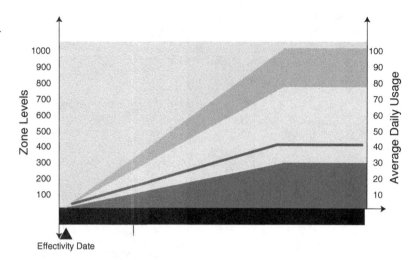

a buffer profile and zone strategy that also grows over time. The buffer can be adjusted based on the real performance against that plan.

In the part ramp-down example (Figure 25-11), a part that is being discontinued is displayed. The far right of the figure represents the date at which the part is planned to be inactive (the "effectivity" date). The ADU is ramped down to create a gradually diminishing buffer level and zone definition.

Figure 25-12 incorporates both ramp-down and ramp-up adjustments. In this case, the new part (on the right) is obsolescing the old part (on the left). The old part is allowed to drain off as the new part is being ramped up to a full ADU-buffered position that coincides with the old part's "effectivity" date. In this case, if there is a spike that will deplete the old part's buffer at a faster than planned rate, the new part has stock to cover that demand.

**FIGURE 25-11**

Part ramp-down
example.

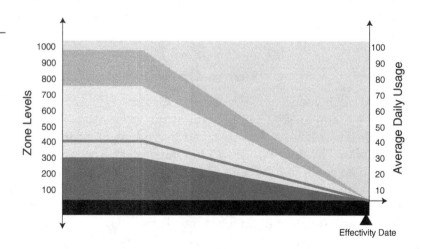

**FIGURE 25-12**

Part transition example.

Transition Date

Note in Figure 25-12 that the ramp-up curve is steeper than the ramp-down curve. A company with the available capacity to make that ramp-up curve occur does not commit resources earlier than necessary. The key is that this example has planned the ramp-up curve to be at 100 percent of ADU at the time the old part is no longer active while at the same time having an amount of new inventory in advance of that date. This strategy will minimize or eliminate obsolete inventory while allowing for a seamless transition to a new part from the market's perspective. This will reduce or eliminate the risks of missed sales owing to shortages that tend to occur through poorly managed transitions.

Figure 25-13 is what a planned adjustment management screen might look like. Note that the planned adjustment on the "Summer up" line creates a bulge in the months of July and August. In this case, the percentage number within a monthly column represents the ADU factor on the last day of the month. Thus a part/SKU moving from 110 percent on May 31 to 130 percent on June 30 can have the increase in ADU spread incrementally over the 30 days of June.

In this case a new product, "SR Phase In" coincides with an old product "DC Phase Out." Thus buffers that are part of the SR phase-in are designed to be at 100 percent ADU

**FIGURE 25-13**

Planned adjustment screen.

Today's Date: January 1st

| Planned Adjustment | January | February | March | April | May | June | July | August | September | October | November | December |
|---|---|---|---|---|---|---|---|---|---|---|---|---|
| Summer Up | 80% | 80% | 80% | 90% | 110% | 130% | 140% | 140% | 1100% | 90% | 70% | 80% |
| SR Phase In | 0% | 0% | 0% | 33% | 66% | 100% | 100% | 100% | 100% | 100% | 100% | 100% |
| DC Phase Out | 100% | 80% | 60% | 40% | 20% | 0% | 0% | 0% | 0% | 0% | 0% | 0% |
| Raptor Launch | 0% | 0% | 0% | 0% | 0% | 20% | 50% | 80% | 100% | 100% | 100% | 100% |

when the buffers associated with DC phase-out go to 0 percent ADU. Figure 25-14 represents graphic depictions of the four planned adjustments. SR phase-in and DC phase-out are displayed together because they are involved in a transitional relationship.

**FIGURE 25-14**

Planned
adjustment
depictions.

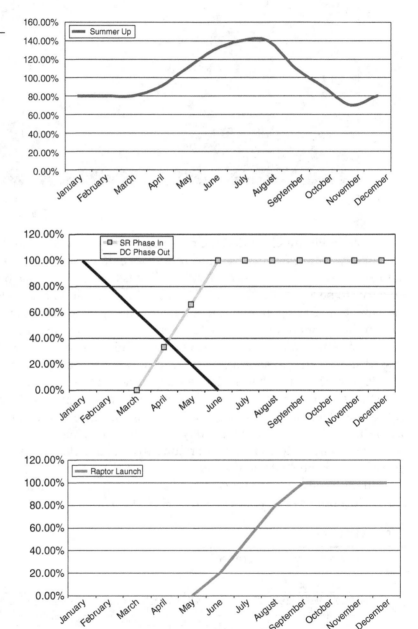

# MANUAL ADJUSTMENTS

Manual adjustments usually are prompted by alerts that are designed to provide visibility to unplanned changes where the rolling ADU calculation may not react fast enough. These unplanned changes could include events or trends that are known to one part of the organization but are not communicated to planning personnel.

One type of alert that could prompt manual adjustments is called an *ADU alert*. An ADU alert is designed to warn planners of a serious trajectory change in ADU over a shorter time frame than the rolling horizon. Both factors of what constitutes serious and shorter horizon are completely specific to the environment and buffer profile.

The severity will be determined by the *ADU alert threshold*. An ADU alert threshold is a defined level of change in ADU that triggers the alert within the ADU alert horizon. The *ADU alert horizon* is a defined shorter rolling range within the broader rolling horizon used to calculate ADU. This can be thought of in terms of a statistical process control run chart that can identify a special cause of variation.

The ADU alert threshold should vary based on the variability codes of a buffer profile. For example, low variability profiles could be set at 25 percent standard deviation over the ADU alert horizon. Medium variability profiles might be set at 50 percent and high variability profiles at 75 percent. These percentages and assignments are only examples. Planning personnel will need to consider their environment and make the appropriate setting.

For example, take a part that is subject to a three-month (12-week) roll in order to calculate ADU. ADU has been hovering around six per day for the last several months. In this example, the ADU alert horizon has been set up based on a two-week rolling horizon. The alert occurs if ADU is above 12 (double the ADU) over that time. The alert is intended only to make a planner aware that something critical may have changed. It will require follow-up. That follow-up may require discussions with several people representing different functions within the organizations. For example, if the ADU alert on this part is triggered and the planner follows up with appropriate personnel and learns that it appears to be an anomaly, no corrective action is required. Conversely, the planner may discover that marketing has just launched a new Web-based strategy that includes a direct, easy-to-use online catalogue for service parts ordering. A new ADU may need to be calculated based on the success of this new program.

## CHAPTER 26

# Demand-Driven Planning

The volume of material requirements planning (MRP) reschedule messages can seem impossible to work before more changes occur and the process begins again. Frequently, critical actions are missed or incomplete pictures are painted. Most purchasing, materials, and fulfillment organizations have limited capacity and trust when it comes to sorting through the current demand signals and planned orders generated by MRP. A significant understanding of MRP logic is required to even begin to understand the implications of a reschedule message. At times, it is easier to just leave it alone rather than risk really disrupting the operation. However, ignoring some of these messages will generate the need for expensive corrective actions at a later time (e.g., expedites, premium freight, overtime, etc.). Clearly, all action messages are not created equal.

Generating, coordinating, and prioritizing actionable material signals becomes much simpler when the environment is modeled in a demand-driven way. The current inventory status is evaluated for potential negative impacts. Flags are set for alert that meet specific spike criteria against open supply orders and demand allocations that include future sales orders. Planners then have the ability to see quickly where the signals are really coming from and react appropriately before they get into trouble. This better matches the current intuition of the planners, and now they have real visibility to establish correct and comprehensive priorities.

## PART PLANNING DESIGNATIONS

There are five different part planning designations in demand-driven MRP (DDMRP). All these designations still will have actively synchronized replenishment lead time (ASRLT) applied to them if appropriate. The five designations focus attention on the parts that are the most critical or strategic and bring designation-specific tools to bear when appropriate. These designations are:

- *Replenished parts.* Replenished items are strategically chosen parts managed by a color-coded buffer system for planning and execution. Buffers are calculated by a combination of globally managed traits relative to the buffer profile into which the part falls and a few critical individual part attributes. Additionally, these positions are designed to be dynamic or recalculated within defined intervals. OTOG = over top of green; TOY = top of yellow; TOR = top of red; and OUT = stocked out. See Chapters 24 and 25 for a discussion on setting the initial buffer and how the buffer adapts to changes.
- *Replenished override (RO) parts.* Replenished override items are strategically chosen parts managed by a color-coded buffer system for planning and execution but whose buffer and zone levels are defined and static (as opposed to calculated and dynamic for the replenished parts). RO parts occur when there are defined limits or dictated levels of inventory within the planning environment. An example might be spaces in a vending machine; there is a finite amount of space that must be divided sensibly and in multiples. RO designation can be extremely valuable in this instance. Without dynamic calculation of the buffer equation, the color-coding system becomes that much more important for planners to prioritize planning- and execution-related activity. Figure 26-1 shows a conceptual rendering of replenished and replenished override buffers.
- *Min-max (MM) parts.* The MM designation is for less strategic and/or readily available stocked parts/stock-keeping units (SKUs). There is still a role for traditionally defined MM tactics in DDMRP. These tactics are limited, however, to the types of parts that are managed by those tactics. The *APICS Dictionary* defines MM as:

  > **min-max system:** A type of order point replenishment system where the "min" (minimum) is the order point and the "max" (maximum) is the "order up to" inventory level. The order quantity is variable and is the result of the max minus available and on-order inventory. An order is recommended when the sum of the available and on-order inventory is at or below the min.[1]

  Min-max buffers can be dynamically altered or adjusted in the same way as replenished parts if max and min are calculated as a factor of average daily usage (ADU). OMAX = over maximum; MAX = order up to inventory level; MIN = order point; OUT = stocked out. Figure 26-2 shows a conceptual rendering of a Min-max buffer.
- *Nonbuffered (NB) parts.* NB parts/SKUs are not stocked. They are transferred, made, or purchased to order or actual demand. In most environments, most parts will fall under this designation.
- *Lead-time-managed (LTM) parts.* LTM parts are nonbuffered parts that require special attention. These are parts that do not come in sufficient quantity to justify

---

[1] *APICS Dictionary*, 12th ed. New York: Blackstone, 2008, p. 83.

## FIGURE 26-1

Replenished and
replenished
override part
buffer schema.

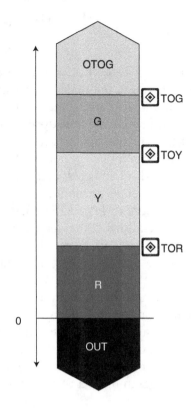

## FIGURE 26-2

Min-max buffer
schema.

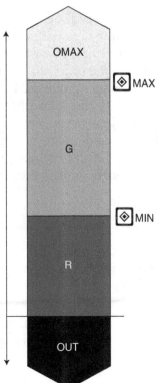

stocking via a replenishment designation, but when there is a requirement for them, historically it has been difficult to maintain control and visibility. Ask most seasoned materials managers in major manufacturers and they can immediately recite a list of these types of components and the suppliers that make them. These components can be very difficult to manage, especially if they have long lead times and/or are sourced from a remote supplier. Without an effective way to manage these parts, there is a risk of major synchronization problems, costly expediting, and/or poor service-level performance.

Figure 26-3 is a conceptual picture of how LTM parts work. In this case, the part has a 60-day lead time. The last third of that lead time becomes the *LTM alert zone*. In this case, the LTM alert zone is 21 days. The LTM alert zone will have three distinct color-coded subzones: green, yellow, and red. In the example, each subzone will be seven days in length. Additionally, there will be a zone dedicated to an order that is late. A notification is given to the planner/buyer each time the part enters a subzone. Green zone entry = 21 days in advance of due date; yellow zone entry = 14 days in advance of due date; red zone entry = 7 days in advance of due date; late = an order that is past due. These notifications are intended to prompt planning and purchasing personnel to follow up and document the status of these orders beginning at a reasonable time frame in advance of the order being due.

In traditional planning systems there is very little done about the management of these types of parts. They are managed by due date with no formal system of visibility and proactive management to reflect real priorities. The assumption remains that all the parts will be available by the release time of the order that needs them. The problem is identified only when the part is late. Orders using that part then are possibly released short those parts, causing possible rework on the shop floor and increasing work-in-process. Alternatively, some companies will begin to pull parts ahead of time to identify this kind of shortage. This process results in a storehouse of partially filled kits and a manual system to track the missing parts. Robbing from one pre-pulled kit to fill another kit makes the situation even worse. Additional information on LTM parts will be avail-

**FIGURE 26-3**

Lead-time-managed buffer schema.

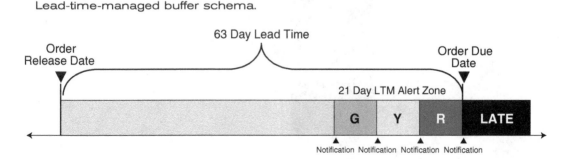

**FIGURE 26-4**

DDMRP part designations.

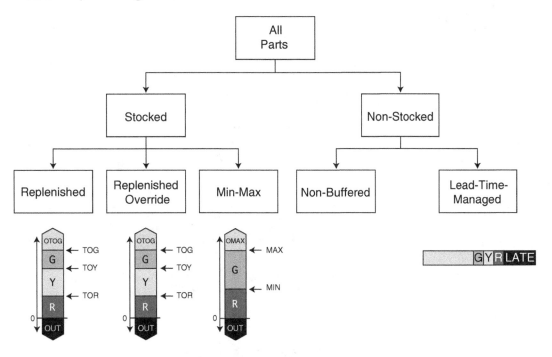

able in Chapter 27. Figure 26-4 is a quick reference chart on the five different part designation types.

Figure 26-5 is how an indented bill of material (BOM) with both replenished and min-max items might look. Parts with three colored zones are replenished (e.g., FPA). Parts with two colored zones are min-max (e.g., PPD). Gray boxes represent nonbuffered parts (e.g., SAC). The bracketed numbers represent lead times. The first number in the bracket represents the particular part's ASRLT, whereas the second number represents the part's manufacturing lead time.

## THE DDMRP PROCESS

The DDMRP process includes innovations for both stocked and nonstocked parts. In both cases, new planning rules have been implemented to promote unprecedented levels of visibility for planning personnel.

## SUPPLY GENERATION FOR STOCKED ITEMS

Stocked items (i.e., replenished, replenished override, and min-max) are resupplied as actual demand forces the "available stock" equation of parts into their respective rebuild zones.

**FIGURE 26-5**

Indented BOM
with different part
designations and
ASRLT factored
in.

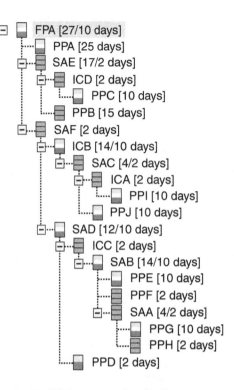

FPA [27/10 days]
PPA [25 days]
SAE [17/2 days]
ICD [2 days]
PPC [10 days]
PPB [15 days]
SAF [2 days]
ICB [14/10 days]
SAC [4/2 days]
ICA [2 days]
PPI [10 days]
PPJ [10 days]
SAD [12/10 days]
ICC [2 days]
SAB [14/10 days]
PPE [10 days]
PPF [2 days]
SAA [4/2 days]
PPG [10 days]
PPH [2 days]
PPD [2 days]

## Available Stock Equation

It is important to note that the buffer level driving demand generation is based on an *available stock equation*. Available stock is calculated by taking on-hand balance + on-order stock (also referred to as *open supply*) – unfulfilled qualified actual demand. Definitions of each from the *APICS Dictionary* are as follows:

**on-hand balance:** The quantity shown in the inventory records as being physically in stock.

**on-order stock:** The total of all outstanding replenishment orders. The on-order balance increases when a new order is released, and it decreases when material is received against an order or when an order is canceled

**actual demand:** Actual demand is composed of customer orders (and often allocations of items, ingredients, or raw materials to production or distribution).[2]

Using actual demand removes forecasted orders from the equation. It is important to note, however, that the preceding available stock equation uses "qualified" actual demand. For stocked parts, actual demand is qualified by the combination of time and

---

[2] Ibid, p. 3.

quantity factors. Only a portion of actual demand at any one time will be involved in the available stock equation rather than the total demand. Typically, for stocked positions,

$$\text{Actual demand} = \text{sales/customer orders due today} +$$
$$\text{any past-due sales/customer orders} + \text{qualified spikes}$$

## Qualified Order Spikes

Demand forecast consumption rules are some of the most complex areas to understand in even the most rudimentary MRP system. Complexity stems from how to handle the actual overage or underconsumption against the forecasted quantities. When MRP was planned in weekly time buckets, this was a bit easier because the week time bucket smoothed some of this volatility. However, now with MRP daily or in real time, the forecast error can be almost impossible to identify and respond to in a timely fashion.

A qualified spike is a quantity of cumulative demand (usually in a daily bucket) that represents a threat to the integrity of the buffer and is within a critical timing window. In order to define a threatening quantity, the *order spike threshold* must be determined. An order spike threshold (OST) is managed globally through the buffer profiles and is expressed as a percentage of the red zone. Revisiting the three example parts from Chapter 24 (403P, 707, and 501D), assume that the buffer profile to which each part is assigned has an OST set to 50 percent of the red zone. In Figures 26-6, 26-7, and 26-8 you will see each part's OST calculation and position in relation to the entire buffer.

An OST is set for all stocked items. For replenished and replenished override parts, it is typically set at 50 percent of the red zone. For min-max parts, it is a percentage of the minimum. In each case, a daily cumulative amount of actual demand equal to or above the OST will be qualified as a spike. This is then added into the available stock equation. In order to qualify as a spike, the cumulative amount must trip the threshold and occur within the critical timing window. The critical timing window is the *order spike horizon*. The order spike horizon is set to a realistic reaction time. In most cases, the order spike horizon is set at a minimum of one ASRLT for each buffered part.

## FIGURE 26-6

Order spike threshold setting for Part 403P.

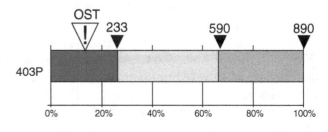

| Part: 403P Lead Time: 21 Days | Buffer Profile: B11MOQ |
|---|---|
| Red Zone Base | 179 |
| Red Zone Safety | 54 |
| OST % | 50% |
| OST Level | 117 |

**FIGURE 26-7**

Order spike threshold setting for Part 707.

| Part: 707<br>Lead Time:<br>13 Days | Buffer<br>Profile:<br>M21 |
|---|---|
| Red Zone Base | 35 |
| Red Zone Safety | 18 |
| OST % | 50% |
| OST Level | 27 |

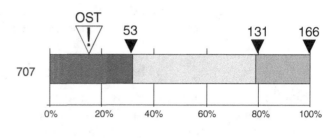

**FIGURE 26-8**

Order spike threshold setting for Part 501D.

| Part: 501D | Buffer<br>Profile:<br>D22MOQ |
|---|---|
| Red Zone Base | 336 |
| Red Zone Safety | 202 |
| OST % | 50% |
| OST Level | 269 |

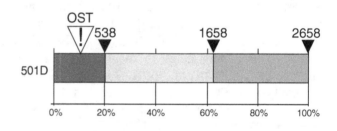

Figure 26-9 provides an example part called XYZ. From left to right, we have depicted

- The size of the red zone (dark-shaded box labeled *R*)
- The quantity of the OST (50 percent) depicted by the dotted line
- Bars that represent sales order quantity (The sales order numbers are attached to the bars at the top of the graphic, for example, SO #1234.)
- The order spike horizon (lightly shaded area)
- A sales order quantity axis that goes from top to bottom (The bigger the sales order, the longer is the representative bar.)

According to Figure 26-9, there are a total of nine sales orders within the order spike horizon (1234, 1235, 1236, 1237, 1238, 1239, 1240, 1241, and 1242). There are two sales orders outside the horizon (1243 and 1244). One sales order (SP #1242) has tripped the threshold; thus it is shaded darker than the other sales orders. SO #1244 is for a larger quantity than SO #1242 but is not shaded dark because it is outside the spike horizon. This sales order is technically not a spike yet. The order spike horizon is set to a time range in which the environment can reasonably react to spikes. It is a waste to react too early. This sales order could be changed in ways that could affect whether it is a spike (quantity or date) before it passes into the spike horizon.

**FIGURE 26-9**

Example of qualifying order spikes for Part XYZ.

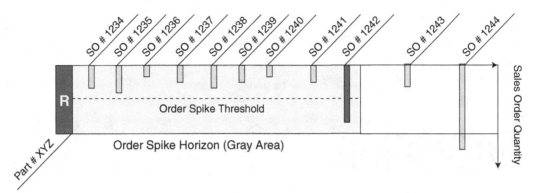

Figure 26-10 is an example of what a demand source screen might look like with regard to Part XYZ. Note that the sales orders that are within the horizon are in shaded lines and one (SO #1242) has been bolded and appended with a "SPIKE!" identifier. This tells the planner that SO #1242 has been placed in the available stock equation. All other sales orders are not included in the available stock equation because they are not due today or past due and they are not qualified as a spike.

Figure 26-11 illustrates, from a different perspective, what is considered actual demand in relation to the available stock equation. Note that the bars represent actual demand and the quantity of that actual demand by day. Order spikes can be the summation of several sales orders on a day or simply one sales order. Within the order spike horizon, three days have been tagged with a spike quantity (marked with an $X$ above the bar). These bars go above the dotted line, marking the order spike threshold. The only actual demand that is considered in the available stock equation are the three spikes identified out of a total of 20 different days with actual demand.

**FIGURE 26-10**

Order source screen for Part XYZ with threshold and qualified spike.

| Part # XYZ   Order Spike Horizon = 14 days   ASRLT = 14 days   Today's Date: 06/21 | | |
| --- | --- | --- |
| Demand | Quantity | Due Date |
| SO# 1234 | 50 | 06/23 |
| SO# 1235 | 70 | 06/25 |
| SO# 1236 | 20 | 06/27 |
| SO# 1237 | 35 | 06/28 |
| SO# 1238 | 50 | 06/29 |
| SO# 1239 | 35 | 06/30 |
| SO# 1240 | 20 | 07/01 |
| SO# 1241 | 35 | 07/03 |
| SO# 1242 – SPIKE! | 120 | 07/05 |
| SO# 1243 (outside horizon) | 50 | 07/08 |
| SO# 1244 (outside horizon) | 160 | 07/11 |

**FIGURE 26-11**

Order spike qualification.

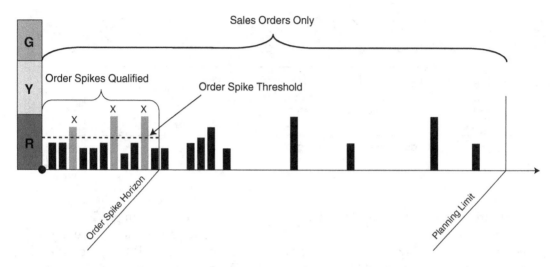

Supply against a part will be generated by its available stock position relative to its buffer and zone levels. For example, consider four parts (r457, f576, h654, and r672). These four parts are assigned to a specific planner (GSC).

Figure 26-12 shows the difference in relative buffer position between a calculated available stock level and actual on-hand. The black arrows indicate the on-hand position, and the no-fill arrows represent the available stock position. This type of visibility gives

**FIGURE 26-12**

Available stock versus on-hand position.

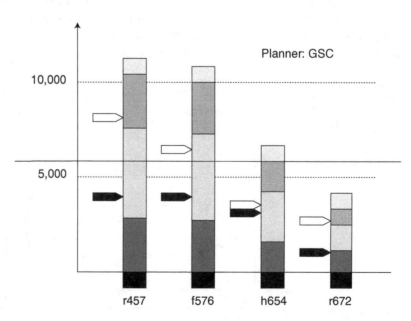

clear signals from both planning and execution perspectives. The actual on-hand inventory position relative to the buffer zones will provide execution priority, as discussed in Chapter 27.

Figure 26-13 is what a planning screen would look like in relation to Figure 26-12. The sequence from top to bottom is determined by the priority of the planning action. The planning action priority is determined by a general reference (color) and a discrete reference (percentage of buffer penetration). In this case, Part h654 is the highest priority from a planning perspective because it is in its rebuild zone and has the deepest available stock penetration against its buffer.

This relative priority distinction is a crucial differentiator between the traditional list of planner alerts and action messages associated with traditional MRP and the highly focused and visible approach of DDMRP. Under this type of approach, planners can quickly judge the relative priority across orders without massive amounts of additional analysis and data queries.

In Figure 26-13, two orders require planning action—one requires execution action and one requires no action (planning or execution). In the case of both Parts h654 and f576, there is a recommended planning action to launch a supply order in such a quantity to bring the available stock position to the top of green. There is no planning action for Parts r457 and r672 because their calculated available stock positions place them in the green. There will need to be some additional investigation on Part r672 in order to determine whether an expedite of open supply is warranted given the status of its on-hand position. This is an execution-related matter and will be covered in Chapter 27.

Sometimes a recommended quantity might exceed the top of green level if the part has an order multiple. Figure 26-14 is an example of an order multiple pushing the available stock position over top of green. Part 408 follows the buffer profile M21. The buffer levels and available stock position are provided. Note that this part has a minimum order quantity, an order multiple, but no order maximum. This means that at least 20 units always must be ordered, and any amount above 20 must be ordered in increments of 20. The minimum order quantity is not significant because it represents only 20 percent of the calculated green zone.

**FIGURE 26-13**

Planning screen example.

| Part | Available Stock | Open Supply | On-hand | Demand | Recommended Supply Qty | Action | Planner Code |
|------|-----------------|-------------|---------|--------|------------------------|--------|--------------|
| h654 | 4038 (66%) | 530 | 3721 (60%) | 213 | 2162 | Place New Order | GSC |
| f576 | 6872 (69%) | 3358 | 4054 (41%) | 540 | 3128 | Place New Order | GSC |
| r457 | 8265 (83%) | 5453 | 4012 (39%) | 1200 | 0 | No Action | GSC |
| r672 | 3852 (90%) | 2743 | 1332 (30%) | 223 | 0 | Expedite Open Supply (Execution) | GSC |

**FIGURE 26-14**

Planning and
order multiple for
Part 408.

| Part: 408 Lead Time: 10 Days | Buffer Profile: M21 |
|---|---|
| Top of Green | 500 |
| Top of Yellow | 400 |
| Top of Red | 200 |
| Available Stock | 350 |
| Order Min | 20 |
| Order Multiple | 20 |
| Order Max | None |

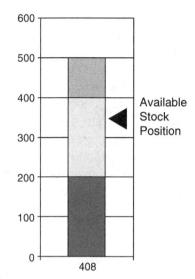

In this case, the available stock position is at 350, placing it in the rebuild zone. In order to get to the top of green, a recommended quantity of 150 would be required. In this case, however, there is an order multiple that does not allow ordering 150. Instead, to reach the top of green, 160 (eight multiples) must be ordered.

Figure 26-15 shows the available stock position after the quantity of 160 has been ordered. Note that the available stock position is now above the top of green. Rules can be constructed to affect the recommended supply quantity in relation to order multiples to minimize over top of green situations. An example would be to limit over top of green situations to occurrences where the over top of green with an additional order multiple is less than the under top of green without the additional order multiple.

## Highly Visible Priority

As introduced in the preceding example (Figure 26-13), all replenished buffer parts are managed using highly visible zone indicators, including the percentage of depletion of the buffer (frequently called *buffer penetration*). This is a far simpler and faster approach than to have to sift through the planning-message queue to determine real and relative priority. The buffer status in DDMRP relates to the available stock position. The recommended order quantity will be the quantity to bring the available stock position to the top of the green (which is the top of the buffer).

Figure 26-16 shows what a planning screen could look like under DDMRP tactics. This view is filtered by critical, high, and medium priority (all of which require planning action). Note that there is both the general color reference and discrete reference to percentage remaining of the buffer. The highest-priority item is PPJ. The PPJ available stock position places it in a net negative position by a quantity of over the total buffer top of green (TOG).

**FIGURE 26-15**

Order multiple pushing available stock over top of green for Part 408.

| Part: 408 Lead Time: 10 Days | Buffer Profile: M21 |
|---|---|
| Top of Green | 500 |
| Top of Yellow | 400 |
| Top of Red | 200 |
| Available Stock | 510 |
| Order Min | 20 |
| Order Multiple | 20 |
| Order Max | None |

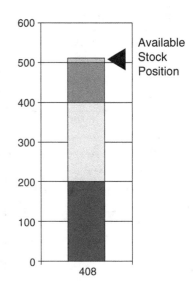

**FIGURE 26-16**

Example DDMRP planning screen.

| Priority | Percent | Part | Profile | ASRLT | On Hand | Supply Orders | Demand Allocations | Available Stock | TOG | Today's AF | Due Date | Reorder Quantity | Vendor | Location |
|---|---|---|---|---|---|---|---|---|---|---|---|---|---|---|
| Critical | −101.7 | PPJ | B10 | 10 | 0 | 120 | 242 | −122 | 120 | 100% | 11/05 | 242 | Sony | Plant |
| Critical | −84.2 | PPG | B10 | 10 | 50 | 70 | 221 | −101 | 120 | 100% | 11/05 | 221 | Philips | Plant |
| Critical | −83.3 | PPI | B12 | 10 | 0 | 132 | 242 | −110 | 132 | 100% | 11/05 | 242 | Sony | Plant |
| Critical | −75.4 | PPE | B11 | 10 | 30 | 96 | 221 | −95 | 126 | 100% | 11/05 | 221 | Philips | Plant |
| Critical | 0.0 | FPA | B10 | 10 | 0 | 0 | 0 | 0 | 40 | 100% | 11/05 | 40 | | Region 2 |
| Critical | 0.0 | FPA | B10 | 10 | 0 | 0 | 0 | 0 | 40 | 100% | 11/05 | 40 | | Region 1 |
| Critical | 0.0 | FPA | B10 | 10 | 0 | 0 | 0 | 0 | 40 | 100% | 11/05 | 40 | | Region 3 |
| High | 14.3 | SAD | M10 | 12 | 0 | 189 | 162 | 27 | 189 | 100% | 11/09 | 162 | | Plant |
| High | 14.5 | SAB | M10 | 14 | 0 | 221 | 189 | 32 | 221 | 100% | 11/11 | 189 | | Plant |
| High | 33.1 | ICB | M11 | 14 | 0 | 242 | 162 | 80 | 242 | 100% | 11/11 | 162 | | Plant |
| Medium | 35.6 | 425-1001 | B20 | 15 | 60 | 40 | 0 | 100 | 281 | 100% | 11/12 | 181 | | Region 4 |
| Medium | 45.9 | 425-1001 | B11 | 10 | 80 | 20 | 0 | 100 | 218 | 83% | 11/05 | 118 | | Region 3 |
| Medium | 46.0 | PPA | B10 | 25 | 0 | 300 | 162 | 138 | 300 | 100% | 11/26 | 162 | Siemens | Plant |

# SUPPLY GENERATION FOR NONSTOCKED ITEMS

The supply-generation formula for nonstocked items is straightforward. All actual demand orders get a corresponding supply order. The supply order's release date, however, is a factor of the actual demand-order due date and the ASRLT of the item. The release date is determined by backscheduling the length of the ASRLT from the actual demand due date using traditional backscheduling logic.

Figure 26-17 is an example of a nonbuffered part's (Part 123) actual demand orders and their corresponding release dates. The upper level of the figure represents the sales-order quantities and due dates. The left-pointing arrows below that level represent the backschedule for the due dates. The lower level depicts where those release dates fall on

**FIGURE 26-17**

Order release date versus demand date.

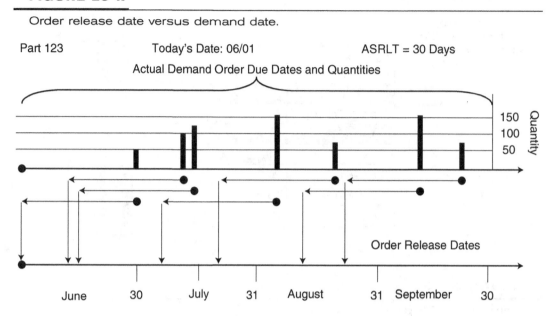

Part 123              Today's Date: 06/01              ASRLT = 30 Days

a time line. Note that the current day is June 1. It is important to note that this initial date is irrespective of quantity/capacity. It is simply driven by the ASRLT in the part's bill of material (BOM). Once the order with due date and quantity is sent to capacity requirements planning (CRP), a different release and/or promise date may be assigned to deal with capacity limitations. Figure 26-18 is what the planner would see in relation to Part 123's planning screen.

This example is a relatively simple one illustrating the materials scheduling effect with a nonbuffered parent item only. A more complex example with a nonbuffered part has a BOM comprised completely of nonbuffered parts. In Figure 26-19, the BOM is displayed in a horizontal rather than the traditional vertical fashion. This view provides more of a project management type of feel. The parent item (FPZ) is at the far right of the

**FIGURE 26-18**

Planning screen
for Part 123.

| Part: 123 | | ASRLT = 30 Days | |
|---|---|---|---|
| Demand | Quantity | Due Date | Release Date |
| SO# 1234 | 50 | 6/30 | 6/02 |
| SO# 1235 | 100 | 7/13 | 6/14 |
| SO# 1236 | 125 | 7/15 | 6/16 |
| SO# 1237 | 155 | 8/06 | 7/07 |
| SO# 1238 | 75 | 8/20 | 7/21 |
| SO# 1240 | 155 | 9/12 | 8/13 |
| SO# 1241 | 75 | 9/23 | 8/24 |

**FIGURE 26-19**

Bill of material and ASRLT for Part FPZ.

Bill of Material for Part FPZ (Displayed Horizontally)

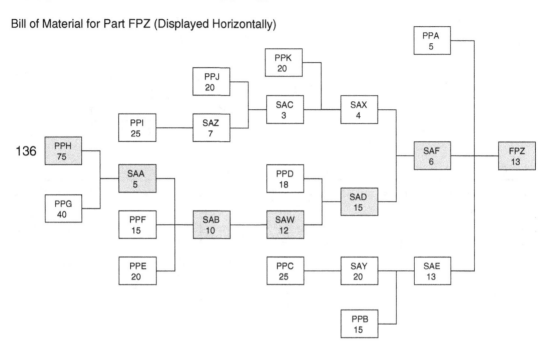

figure. Each box represents an item or material that is either a subassembly (e.g., SAF) or a purchased part (e.g., PPI). The number inside each box represents each part's manufacturing or purchasing lead time in days (e.g., 20 = 20 days). The shaded path represents the ASRLT of FPZ. In this case, the path is 136 days. Since there are no buffered components, the ASRLT and the cumulative lead time (CLT) are one and the same.

A staggered order-release schedule is created for all orders associated with FPZ. Figure 26-20 shows those staggered releases along a 140-day timeline. The part numbers with boxes represent orders on that lie on the ASRLT chain.

Figure 26-21 is what a planning screen would look like in relation to the planning of Part FPZ. The shaded lines represent items that lie on the ASRLT chain. Note that all purchased parts are coded *LTM*. This means that special attention will be paid to these parts as time gets closer to their due dates. As explained earlier, the assumption is that this visibility and preemptive following up (referred to as *pre-expediting* by many) will reduce disruptions from suppliers by either knowing about delays well in advance or by keeping the issue in front of suppliers so that they stay on track with the promise date.

The objective of planners in situations such as the preceding example is to limit variability (internal or external) from affecting the ASRLT chain. Variability passed to that chain will have an impact on the promise date of the end item. Figure 26-22 provides an

## FIGURE 26-20

Staggered schedule with critical ASRLT orders identified.

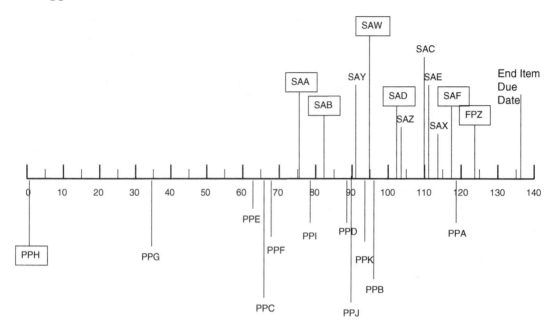

## FIGURE 26-21

Example planning screen for Part FPZ.

| Part | Order Number | Profile | Start Date | ASRLT | Request Date | Promise Date | Late | Quantity | Vendor | Customer | Source |
|---|---|---|---|---|---|---|---|---|---|---|---|
| PPH | PO-000591 | LTM | 2/15 | 75 | 2/28 | 2/28 | | 1 | SpaceTech | | Driven by WO-000589, SAA |
| PPG | PO-000590 | LTM | 1/19 | 40 | 2/28 | 2/28 | | 1 | Siemens | | Driven by WO-000589, SAA |
| PPE | PO-000592 | LTM | 2/13 | 20 | 3/05 | 3/05 | | 1 | Sony | | Driven by WO-000587, SAB |
| PPF | PO-000593 | LTM | 2/18 | 15 | 3/05 | 3/05 | | 1 | SpaceTech | | Driven by WO-000587, SAB |
| PPC | PO-000594 | LTM | 2/18 | 25 | 3/15 | 3/15 | | 1 | Boeing | | Driven by WO-000582, SAY |
| SAA | WO-000589 | NB | 2/28 | 80 | 3/05 | 3/05 | | 1 | | | Driven by WO-000587, SAB |
| PPI | PO-000596 | LTM | 3/03 | 25 | 3/28 | 3/28 | | 1 | Raytheon | | Driven by WO-000588, SAZ |
| SAB | WO-000587 | NB | 3/05 | 90 | 3/15 | 3/15 | | 1 | | | Driven by WO-000585, SAW |
| PPD | PO-000595 | LTM | 3/09 | 18 | 3/27 | 3/27 | | 1 | MicroTech | | Driven by WO-000583, SAD |
| PPJ | PO-000598 | LTM | 3/15 | 20 | 4/04 | 4/04 | | 1 | SpaceTech | | Driven by WO-000586, SAC |
| SAY | WO-000582 | NB | 3/15 | 45 | 4/04 | 4/04 | | 1 | | | Driven by WO-000580, SAE |
| SAW | WO-000582 | NB | 3/15 | 102 | 3/27 | 3/27 | | 1 | | | Driven by WO-000583, SAD |
| PPK | PO-000599 | LTM | 3/18 | 20 | 4/07 | 4/07 | | 1 | Siemens | | Driven by WO-000584, SAX |
| PPB | PO-000597 | LTM | 3/20 | 15 | 4/04 | 4/04 | | 1 | Sony | | Driven by WO-000580, SAE |
| SAD | WO-000583 | NB | 3/27 | 117 | 4/11 | 4/11 | | 1 | | | Driven by WO-000581, SAF |
| SAZ | WO-000588 | NB | 3/28 | 32 | 4/04 | 4/04 | | 1 | | | Driven by WO-000586, SAC |
| SAE | WO-000580 | NB | 4/04 | 58 | 4/17 | 4/17 | | 1 | | | Driven by WO-000579, FPA |
| SAC | WO-000586 | NB | 4/04 | 35 | 4/07 | 4/07 | | 1 | | | Driven by WO-000584, SAX |
| SAX | WO-000584 | NB | 4/07 | 39 | 4/11 | 4/11 | | 1 | | | Driven by WO-000581, SAF |
| SAF | WO-000581 | NB | 4/11 | 123 | 4/17 | 4/17 | | 1 | | | Driven by WO-000579, FPA |
| PPA | PO-000600 | LTM | 4/12 | 5 | 4/17 | 4/17 | | 1 | SpaceTech | | Driven by WO-000579, FPA |
| FPZ | WO-000579 | NB | 4/17 | 136 | 5/01 | 5/01 | | 1 | | NASA | Driven by Atlantis, line 1, NASA |

**FIGURE 26-22**

Slippage in Part FPZ schedule.

| Part | Order Number | Profile | Start Date | ASRLT | Request Date | Promise Date | Late | Quantity | Vendor | Customer | Source |
|------|------|---------|------------|-------|------|------|------|----------|--------|----------|--------|
| PPH | PO-000591 | LTM | 2/15 | 75 | 2/28 | 2/28 | | 1 | SpaceTech | | Driven by WO-000589, SAA |
| PPG | PO-000590 | LTM | 1/19 | 40 | 2/28 | 2/28 | 2 Days | 1 | Siemens | | Driven by WO-000589, SAA |
| PPE | PO-000592 | LTM | 2/13 | 20 | 3/05 | 3/05 | | 1 | Sony | | Driven by WO-000587, SAB |
| PPF | PO-000593 | LTM | 2/18 | 15 | 3/05 | 3/05 | | 1 | SpaceTech | | Driven by WO-000587, SAB |
| PPC | PO-000594 | LTM | 2/18 | 25 | 3/15 | 3/15 | | 1 | Boeing | | Driven by WO-000582, SAY |
| SAA | WO-000589 | NB | 2/28 | 80 | 3/05 | 3/05 | 2 Days | 1 | | | Driven by WO-000587, SAB |
| PPI | PO-000596 | LTM | 3/03 | 25 | 3/28 | 3/28 | | 1 | Raytheon | | Driven by WO-000588, SAZ |
| SAB | WO-000587 | NB | 3/05 | 90 | 3/15 | 3/15 | 2 Days | 1 | | | Driven by WO-000585, SAW |
| PPD | PO-000595 | LTM | 3/09 | 18 | 3/27 | 3/27 | | 1 | MicroTech | | Driven by WO-000583, SAD |
| PPJ | PO-000598 | LTM | 3/15 | 20 | 4/04 | 4/04 | | 1 | SpaceTech | | Driven by WO-000586, SAC |
| SAY | WO-000582 | NB | 3/15 | 45 | 4/04 | 4/04 | | 1 | | | Driven by WO-000580, SAE |
| SAW | WO-000582 | NB | 3/15 | 102 | 3/27 | 3/27 | 2 Days | 1 | | | Driven by WO-000583, SAD |
| PPK | PO-000599 | LTM | 3/18 | 20 | 4/07 | 4/07 | | 1 | Siemens | | Driven by WO-000584, SAX |
| PPB | PO-000597 | LTM | 3/20 | 15 | 4/04 | 4/04 | | 1 | Sony | | Driven by WO-000580, SAE |
| SAD | WO-000583 | NB | 3/27 | 117 | 4/11 | 4/11 | 2 Days | 1 | | | Driven by WO-000581, SAF |
| SAZ | WO-000588 | NB | 3/28 | 32 | 4/04 | 4/04 | | 1 | | | Driven by WO-000586, SAC |
| SAE | WO-000580 | NB | 4/04 | 58 | 4/17 | 4/17 | | 1 | | | Driven by WO-000579, FPA |
| SAC | WO-000586 | NB | 4/04 | 35 | 4/07 | 4/07 | | 1 | | | Driven by WO-000584, SAX |
| SAX | WO-000584 | NB | 4/07 | 39 | 4/11 | 4/11 | | 1 | | | Driven by WO-000581, SAF |
| SAF | WO-000581 | NB | 4/11 | 123 | 4/17 | 4/17 | 2 Days | 1 | | | Driven by WO-000579, FPA |
| PPA | PO-000600 | LTM | 4/12 | 5 | 4/17 | 4/17 | | 1 | SpaceTech | | Driven by WO-000579, FPA |
| FPZ | WO-000579 | NB | 4/17 | 136 | 5/01 | 5/01 | 2 Days | 1 | | NASA | Driven by Atlantis, line 1, NASA |

example of slippage occurring within the ASRLT chain. The promise date of PPG has been moved back two days. It is two days late. This will result in a delay to SAA. SAA lies on the ASRLT chain. The two-day delay of PPG transfers variability from a noncritical path to a critical path and is pushing the promise date of the end item by two days.

Coding feeding paths into that chain with an LTM designation is one way to gain better visibility and protection for the ASRLT chain. In many cases, the LTM designation itself will not prevent slides of those feeding legs from affecting the ASRLT chain, but the LTM designation notifies the planner about the slide sooner. There is value in this knowledge in order to allow the development of another tactic that will absorb the variation in BOM legs that affect the ASRLT chain.

Figure 26-23 displays the total length of all legs in the BOM by a large number at the left termination of the leg. The ASRLT is represented by the shaded boxes and remains 136 days. The longest non-ASRLT path is 101 days (terminating in PPG). This is a difference of 35 days. This leg can be started a sensible period of time ahead of the late start date to account for potential delays without changing or redefining the ASRLT chain. A time buffer can be employed that will absorb potential disruption from a supplier without adding any lead time to the total assembly schedule and better protect the ASRLT. All activity/legs feeding into the ASRLT chain that are not part of that chain can be buffered with time. Figure 26-24 is FPZ's BOM with these time buffers inserted. The shaded rectangular boxes represent the time buffers. The time buffers are an insertion into the lead time of non-ASRLT legs. The length of each of those legs has been lengthened, and the release or start dates for items on those paths are adjusted accordingly.

**FIGURE 26-23**

FPZ BOM with lengths of all legs indentified.

Bill of Material for Part FPZ (Displayed Horizontally)

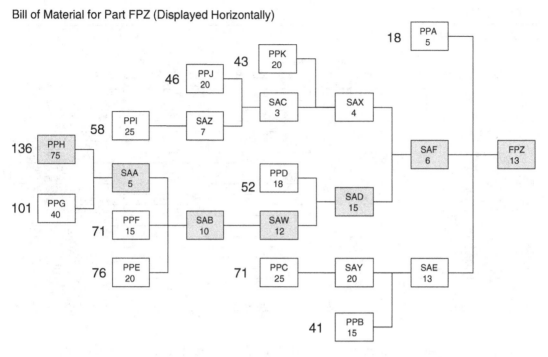

**FIGURE 26-24**

FPZ BOM with time buffers on feeding orders and/or paths.

Bill of Material for Part FPZ (Displayed Horizontally)

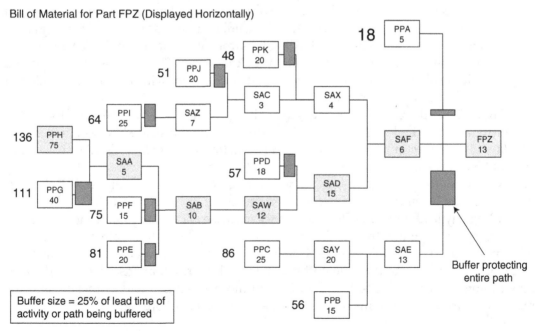

Buffer size = 25% of lead time of activity or path being buffered

The size of the time buffer varies according to the length on the path that is being buffered. In Figure 26-24, time buffers have been sized to be one-quarter the length of the activity or sequence of activities being buffered. The gray boxes of varying sizes indicate those buffers. Two types of buffer placement strategies also are illustrated. The first is buffering all purchased parts that do not lie on the ASRLT chain. Parts PPG, PPF, PPE, PPI, PPJ, PPK, PPA, and PPD all have buffers placed immediately behind them. This means that variability will be absorbed only from the external sources. There are still subassemblies (e.g., SAZ, SAC, and SAX) where internal variability can be encountered and transferred to the ASRLT chain.

A superior time buffering strategy involves placing the time buffer where a BOM leg merges with the ASRLT chain. This buffer is sized by the BOM sequence of that path. In Figure 26-24, there is a larger time buffer after SAE. That buffer is sized by calculating the longest sequence in that BOM leg—58 days. The buffer is set to one-quarter of 58 days = 15 days. This buffer is designed to protect against variability in that entire leg from passing through to the ASRLT chain to affect the delivery date of the end item.

Figure 26-25 is an example of the planning screen for all non-ASRLT parts for FPZ. Note the existence of a new column called "Buffer." That column is not populated when a buffer exists either in front of that activity or after that activity (inserted after the last part on the BOM leg).

There are obvious implications to this type of planning method for engineer to order (ETO) environments. ETO environments struggle with MRP, yet, typically, most of these companies have and use MRP systems. See Chapter 16 for a discussion of the current tools. A major milestone for these environments is to complete and enter a BOM into MRP so that purchasing and scheduling activity can occur.

An alternative strategy, as discussed in Chapter 16, is to purchase all materials to support the earliest possible start date. In environments that are hybrids between project and manufacturing, there is usually the assumption that the sooner something starts, the

## FIGURE 26-25

Planning screen for FPZ with buffer column.

| Part | Order Number | Profile | Start Date | Buffer | ASRLT | Request Date | Promise Date | Late | Quantity | Vendor | Customer | Source |
|---|---|---|---|---|---|---|---|---|---|---|---|---|
| PPG | PO-000590 | LTM | 1/19 | | 40 | 2/28 | 2/28 | | 1 | Siemens | | Driven by WO-000589, SAA |
| PPE | PO-000592 | LTM | 2/08 | 5 | 20 | 3/05 | 3/05 | | 1 | Sony | | Driven by WO-000587, SAB |
| PPF | PO-000593 | LTM | 2/14 | 4 | 15 | 3/05 | 3/05 | | 1 | SpaceTech | | Driven by WO-000587, SAB |
| PPC | PO-000594 | LTM | 2/18 | | 25 | 3/15 | 3/15 | | 1 | Boeing | | Driven by WO-000582, SAY |
| PPI | PO-000596 | LTM | 2/26 | 6 | 25 | 3/28 | 3/28 | | 1 | Raytheon | | Driven by WO-000588, SAZ |
| PPD | PO-000595 | LTM | 3/04 | 5 | 18 | 3/27 | 3/27 | | 1 | MicroTech | | Driven by WO-000583, SAD |
| PPJ | PO-000598 | LTM | 3/10 | 5 | 20 | 4/04 | 4/04 | | 1 | SpaceTech | | Driven by WO-000586, SAC |
| SAY | WO-000582 | NB | 3/04 | 11 | 45 | 4/04 | 4/04 | | 1 | | | Driven by WO-000580, SAE |
| PPK | PO-000599 | LTM | 3/13 | 5 | 20 | 4/07 | 4/07 | | 1 | Siemens | | Driven by WO-000584, SAX |
| PPB | PO-000597 | LTM | 3/20 | | 15 | 4/04 | 4/04 | | 1 | Sony | | Driven by WO-000580, SAE |
| SAZ | WO-000588 | NB | 3/28 | | 32 | 4/04 | 4/04 | | 1 | | | Driven by WO-000586, SAC |
| SAE | WO-000580 | NB | 3/20 | 15 | 58 | 4/17 | 4/17 | | 1 | | | Driven by WO-000579, FPA |
| SAC | WO-000586 | NB | 4/04 | | 35 | 4/07 | 4/07 | | 1 | | | Driven by WO-000584, SAX |
| SAX | WO-000584 | NB | 4/07 | 10 | 39 | 4/11 | 4/11 | | 1 | | | Driven by WO-000581, SAF |
| SAX | WO-000584 | NB | 4/07 | | 39 | 4/11 | 4/11 | | 1 | | | Driven by WO-000581, SAF |
| PPA | PO-000600 | LTM | 4/11 | 1 | 5 | 4/17 | 4/17 | | 1 | SpaceTech | | Driven by WO-000579, FPA |

better chance there is that it will finish early or at least on time. This assumption is most often completely false. Launching purchase orders (POs) and manufacturing orders (MOs) as early as possible exposes the company to risk associated with design changes; unnecessarily consumes cash, capacity, and materials; and chokes the manufacturing floor with unnecessary work-in-process. Employing the suggested DDMRP method staggers and chokes the release of POs and MOs in such a way that will minimize those risks while at the same time better protects and manages the critical timing needs of the end-item order.

## DECOUPLED EXPLOSION

The BOM explosion is still a critical and necessary element of planning in environments of even moderate complexity. Without exploding the BOM, there is no visibility of total requirements. This is why MRP proponents in complex environments continue to use MRP and claim that without it there is no chance at flexibility and agility.

In DDMRP, component-part requirements are still calculated by exploding down through the BOM. However, this planning is decoupled at any buffered component part (i.e., replenished, replenished override, or min-max). These parts then will explode when they reach their respective rebuild zones. Figure 26-26 shows the decoupled explosion for the earlier example Part FPA. Note that whenever a buffer position is encountered, the BOM explosion stops. The figure on the left depicts the explosion for the parent item FPA after its available stock position has been driven into the yellow zone. The middle figure represents buffered components that explode independently when they have reached their respective rebuild zones. Finally, there is the explosion for subassembly B (SAB) after its available stock equation has been driven into yellow.

**FIGURE 26-26**

Decoupled explosion example.

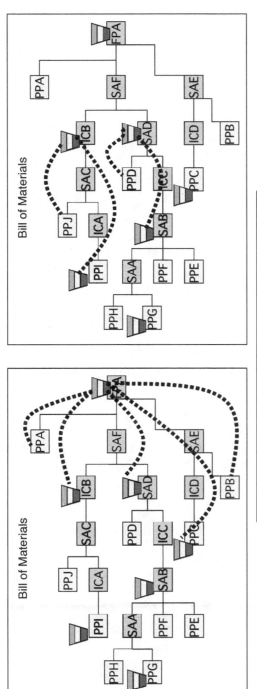

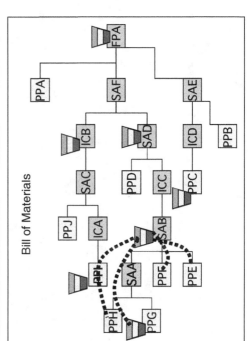

## CHAPTER 27

# Highly Visible and Collaborative Execution

The *P* in MRP stands for *planning*. As has been already discussed, material requirements planning (MRP) inherently has been a planning system and not an execution system. Conventional MRP systems lack real-time and focused visibility to actual and relative priorities associated with purchase orders (POs), transfer orders (TOs), and manufacturing orders (MOs) throughout the internal manufacturing operation and supply chain. Without an effective priority approach, conventional tools force supply chains (i.e., suppliers, manufacturing, fulfillment, and customers) to employ a rudimentary and arbitrary priority system called *priority by due date*.

## CHALLENGING PRIORITY BY DUE DATE

Common practice is that if suppliers are late, it counts against them in their performance report. If a manufacturer is consistently not able to meet customer due dates, then there will be negative business implications. These include lost opportunities and increased expedite related expenses. Companies are acutely aware of the importance of hitting due dates, especially in the current hypercompetitive market. This ripples throughout an organization, reinforcing the need to measure and act according to priority by due date. Thus priority by due date becomes the primary method of attempts to maintain expected service levels to the customer.

Consider the following question: Would you rather have your suppliers be on time or never create a shortage? This question usually elicits a very interesting reaction. Most responders blurt out, "On time, of course!" but then pause with a quizzical look on their faces. Intuitively, they begin to consider that there may be a misalignment between what companies use as a metric and what effect that metric might have. Suppliers can be consistently 100 percent on time, and the customer still has shortages. This is especially true when due dates generated out of conventional MRP systems are based on lead times that are unrealistic (either underestimated or overestimated) and possibly with quantities that are unnecessary. If a supplier hits dates that are unrealistic and quantities that are unnecessary, then there is still a significant chance that there could be shortages.

Priority by due date rarely conveys the real day-to-day inventory and materials priorities. Priorities are not static. They change as variability and volatility occur within the active life span of POs and MOs—the time from when they are opened until they are closed. This life span is called the *execution horizon*. Customers change their orders. Quality challenges occur. There can be weather- or customs-related obstacles. Engineering changes happen. Suppliers' capacity and reliability can fluctuate temporarily. The longer the execution horizon, the more volatile are priorities. This means that the company is more susceptible to adverse material synchronization issues and shortages. This variability and volatility guarantee that despite our best attempts at planning, reality will deviate from the plan. Reality has no regard for the due date assigned to the order when it is released. Usually, given the number of items that can consume capacity and time, this means that the plan underestimates requirements. This leads directly to the risk of shortages. Alternatively, the plan equally could overestimate requirements. It is not uncommon for companies to pay premium or expedite charges only to find dust on the unopened box months later.

There are other challenges associated with priority by due date. Request and promise dates change frequently. These changes often create confusion and disagreement between suppliers and customers. Suppliers could view their on-time performance as high because they delivered to their promise date, whereas customers view it much differently from their view of the request date.

Aligning a supplier's schedule with a customer's real priorities under conventional MRP approaches can be a real challenge. Frequently, a manufacturing company has several open POs to a supplier all with the same due date. Figure 27-1 is an example. Note that there are three orders all due on the same day (PO numbers 280-89, 279-84, and 276-54). With regard to these three orders, if the supplier does not have the capability to deliver all three on the same date, what is the process to decide priority? Options can include calling the buyer at the customer or choosing the order based on what the supplier perceives to be its best use of time.

If the decision is to call the buyer, can the buyer quickly convey a correct priority? In most cases with conventional MRP tools, the answer is no. Determining correct priority will require an additional amount of data analysis. If the supplier instead does what it perceives to be the best use of its time and capacity, the fact that it might pick the right priority for the customer would be completely coincidental. Complicating this is the fact that there are two other open POs that are due later (PO numbers 281-21 and 275-44). These orders could have even higher priority from the customer's perspective than the previous three, even though the due dates are later.

Figure 27-2 shows the on-hand buffer status assuming that all these POs feed replenished buffers. This information changes the picture dramatically. The on-hand buffer status now shows the two orders that likely would have been deferred until later as the highest priority.

**FIGURE 27-1**

Example of
orders prioritized
by date.

| Order # | Due Date | Customer |
|---------|----------|----------|
| PO 280-89 | 05/12 | Super Tech |
| PO 279-84 | 05/12 | Super Tech |
| PO 276-54 | 05/12 | Super Tech |
| PO 281-21 | 05/14 | Super Tech |
| PO 275-44 | 05/16 | Super Tech |

**FIGURE 27-2**

Example of
orders prioritized
by buffer status.

| Order # | OH Buffer Status | Due Date | Customer |
|---------|------------------|----------|----------|
| PO 275-44 | 3% (RED) | 5/16 | Super Tech |
| PO 281-21 | 17% (RED) | 5/14 | Super Tech |
| PO 276-54 | 27% (RED) | 5/12 | Super Tech |
| PO 280-89 | 47% (YELLOW) | 5/12 | Super Tech |
| PO 279-84 | 54% (YELLOW) | 5/12 | Super Tech |

The priority-by-due-date problem does not just affect the traditional customer-supplier relationship; it has huge implications within a manufacturer as well. When building orders to stock, there are different priorities for different orders. The shop floor usually has visibility between stock orders and orders that go directly to customers.

Figure 27-3 illustrates two examples of what a manufacturing floor might see relative to manufacturing orders. In this case, we have included at least one order that is going directly to a customer (MO number 12379). The upper chart depicts what it might look like if the MRP system gave discrete due dates for stock orders. The lower chart is what it might look like if the MRP system simply coded stock orders, "Due NOW." Most planning personnel would agree that the lower chart is more problematic for determin-

**FIGURE 27-3**

Manufacturing
orders prioritized
by date.

| Order # | Order Type | Due Date | Customer |
|---------|-----------|----------|----------|
| MO 12367 | Stock | 5/12 | Internal |
| MO 12379 | MTO | 5/12 | SuperTech |
| MO 12465 | Stock | 5/12 | Internal |
| MO 12401 | Stock | 5/14 | Internal |
| MO 12411 | Stock | 5/16 | Internal |

| Order # | Order Type | Due Date | Customer |
|---------|-----------|----------|----------|
| MO 12367 | Stock | Due Now | Internal |
| MO 12379 | MTO | 5/12 | Internal |
| MO 12465 | Stock | Due Now | Super Tech |
| MO 12401 | Stock | Due Now | Internal |
| MO 12411 | Stock | Due Now | Internal |

ing priority of stocked orders. When everything is the priority, nothing is the priority. Under these circumstances, most manufacturing personnel would focus attention on the actual customer order and wait for the expeditor to determine the most needed stock orders. Adding due dates to the picture provides some sort of sequence to the stock orders. That sequence may or may not be correct based on the actual priority of the company. However, at least it is something by which to sort. Even with discrete due dates assigned to stock orders, the actual customer order would get immediate attention by default in most manufacturing companies.

Figure 27-4 is what it would look like if buffer status were provided for the stock orders. In both cases, the views are sorted by buffer status on stock orders and place an actual customer order at the top if it is the closest order with regard to due date. Now there is clear visibility to manufacturing personnel for the relative stock order priorities.

While this information will not completely resolve the potential conflict between an actual customer order and a high-priority stock order, at least it brings more perspective to the dilemma. That perspective can be available to all potential concerned parties (i.e., operations, sales, and customer service). Included in this perspective is that replenished and replenished override buffers are strategically selected points of protection for the system. They have been carefully selected based on several business factors (see Chapter 24). Disregarding them actually may end up affecting more actual customer orders. In addition, this will cost the company lots of money in expedite-related expenses. Companies need to understand and define how to resolve these dilemmas for their unique circumstances. Without understanding and conveying buffer status priority, however, the chance of effectively resolving such dilemmas is slim at best.

In conventional MRP, any sort of visibility to or a specific answer about the real-time and relative priority of orders according to buffer status often necessitates a manual workaround or subsystem that necessitates massive daily efforts of analysis and adjustments.

## FIGURE 27-4

Manufacturing orders prioritized by buffer status.

| Order # | OH Buffer Status | Order Type | Due Date | Customer |
|---------|------------------|------------|----------|----------|
| MO 12379 | | MTO | 5/12 | SuperTech |
| MO 12401 | 12% (RED) | Stock | 5/14 | Internal |
| MO 12465 | 27% (RED) | Stock | 5/12 | Internal |
| MO 12367 | 33% (YELLOW) | Stock | 5/12 | Internal |
| MO 12411 | 41% (YELLOW) | Stock | 5/16 | Internal |

| Order # | OH Buffer Status | Order Type | Due Date | Customer |
|---------|------------------|------------|----------|----------|
| MO 12379 | | MTO | 5/12 | SuperTech |
| MO 12401 | 12% (RED) | Stock | Due Now | Internal |
| MO 12465 | 27% (RED) | Stock | Due Now | Internal |
| MO 12367 | 33% (YELLOW) | Stock | Due Now | Internal |
| MO 12411 | 41% (YELLOW) | Stock | Due Now | Internal |

**FIGURE 27-5**

DDMRP
execution alerts.

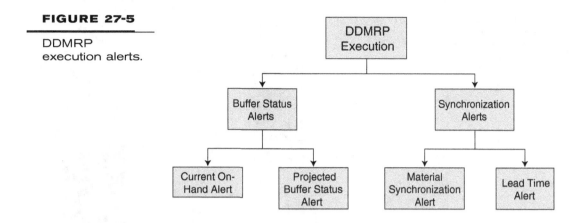

Figure 27-5 illustrates DDMRP execution alerts. There are two categories of alerts. One category, *buffer status alerts*, is focused on stocked parts. The second category, *synchronization alerts*, will cover nonstocked parts (*material synchronization alert* covers both stocked and nonstocked parts). These alerts are real-time and available online most likely through a browser pointed at a Web server with the appropriate database. Contained in subsequent sections are explanations of and samples of these alerts.

## BUFFER STATUS ALERTS

Demand-driven MRP (DDMRP) allows actual order priorities (i.e., POs, TOs, or MOs) to be conveyed effectively without additional efforts, disconnected subsystems, or other workarounds. Buffer status alerts are driven by the on-hand or projected on-hand position. Color coding gives an intuitive and easy-to-understand general reference. The percentage of buffer remaining gives a specific discrete reference for items with the same color reference. These references convey the actual and relative priority regardless of due date. Below are examples of buffer displays for geographically distributed (by location) manufactured and purchased items. There will be two types of alerts; one will focus on the immediate priority and the other on a short-range future time horizon [usually one actively synchronized replenishment lead time (ASRLT)].

Both these types of alerts will use a defined percentage of the total red zone in order to determine the color coding (general reference) of the alert. An *on-hand alert* level is set as a percentage of the total red zone. Typically, the alert is set at 50 percent of the total red zone. If a part is above this line and still in the red zone (e.g., 51 to 100 percent of red), then the alert is displayed as yellow. Under the line, the alert is displayed as red. The discrete reference is expressed as a percentage of the total red zone.

The reason why the alert is displayed in yellow when the buffer status is greater than 51 percent of the red zone is that buffer positions are set properly if the average on-hand position is in the lower half of the yellow zone. These alerts are centered on the on-hand position only. Conceptually, there is a different green zone for available versus on-hand positions. Figure 27-6 describes the conceptual connection between levels and

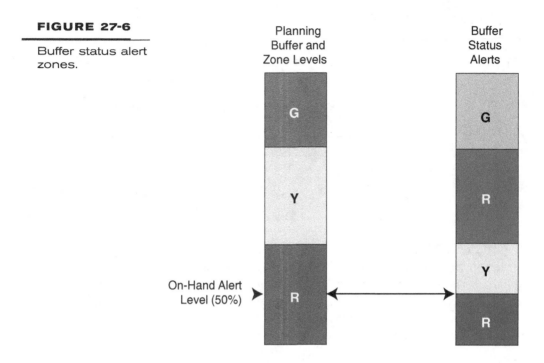

**FIGURE 27-6**

Buffer status alert zones.

zones used for planning and the ways in which the buffer status alerts translate those levels and zones for execution purposes.

As shown in the figure, the top half of the red zone from a planning perspective is actually a yellow zone position from an on-hand perspective. Figure 27-7 is another graphic depiction. If the desired distribution of on-hand inventory is in the lower half of

**FIGURE 27-7**

Buffer status alert zones—another view.

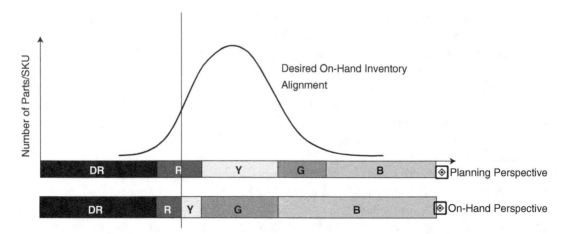

the yellow zone from a planning perspective, then it implies that when considering only an on-hand perspective, this would, in fact, represent the green zone. In this case, the entire yellow zone has been designated the on-hand green zone.

## Current On-Hand Alert

The current on-hand alert is designed to show planning and, in some cases, manufacturing personnel what replenished positions are currently in trouble from an on-hand position only. For planning and purchasing personnel, these alerts are meant to identify parts where open supply may need to be expedited. For manufacturing personnel, they provide relevant information about which manufacturing orders should take precedence. Figure 27-8 is an example of an on-hand alert (OH alert) screen sorted by manufactured parts only.

This report is filtered for stock-out and red and yellow zone on-hand status. Part FPA is in the OH alert red zone with an on-hand position of 2,000, which is 27 percent of a total red zone of 7,500. The OH alert is set to 50 percent of red (3,750). The top of green is 20,000. Available stock is at 95 percent of top of green and is in the green zone [2,000 on hand + 17,000 open supply (no demands) = 19,000]. In this case, a planner or manufacturing scheduler should consider expediting open supply. Figure 27-9 is a summary of Part FPA's current on-hand and available stock situation.

When the planner implodes FPA open supply, he or she will see the information provided in Figure 27-10. In this case, open supply consists of three different MOs (i.e., 123-72, 122-11, and 119-10). If today's date were 5/20, MO 123-72 obviously would be late with regard to its promise date. Its status shows that it is in progress at paint. Paint is the last stage in the process. In contacting the paint department, the planner/scheduler learns that the order is actually complete and is in the process of being put into stock and will be available as on hand shortly. In this case, no additional expedite needs to be considered because the addition of 6,000 units will bring the on-hand buffer status out of the yellow zone because the total of 8,000 will be above the red zone.

Part FPE is in the OH alert red zone with an on-hand position of 250. The top of green is 2,000. Available stock is at 77.5 percent of the top of the green zone and is in the green zone (250 on hand + 1,550 open supply – 250 demand = 1,550). Part FPE also should be considered for expedited open supply. Figure 27-11 is a summary of Part FPE's current on-hand and available stock situation.

### FIGURE 27-8

On-hand alert screen example (manufactured parts).

**On-Hand Alert**

| Part # | OH Buffer Status | Part Type | On-Hand Qty | Open Supply | Demand | Available Stock Status |
|--------|------------------|-----------|-------------|-------------|--------|------------------------|
| FPA | 27% (RED) | Manufactured | 2000 | 17,000 | 0 | 19,000 (GREEN) |
| FPE | 42% (RED) | Manufactured | 400 | 1,550 | 250 | 1,550 (GREEN) |
| SAE | 88% (YELLOW) | Manufactured | 100 | 0 | 60 | 10 (RED) |

**FIGURE 27-9**

Part FPA available and on-hand stock positions.

| Part: FPA | |
|---|---|
| Top of Green | 20,000 |
| Top of Yellow | 15,000 |
| Top of Red | 7,500 |
| On-Hand Alert | 3,750 |
| Available Stock | 19,000 |
| On-Hand | 2,000 (27%) |

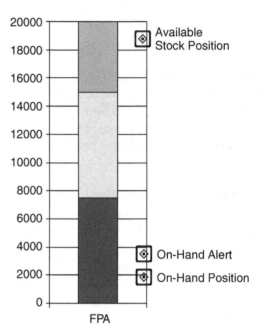

**FIGURE 27-10**

Part FPA open supply summary.

| Part #: FPA | | OH Buffer Status: 27% | | Today's Date: 5/20 |
|---|---|---|---|---|
| Order # | Quantity | Request | Promise | Status |
| MO 123-72 | 6,000 | 5/01 | 5/11 | In progress at Paint |
| MO 122-11 | 5,000 | 5/10 | 5/15 | In progress at Mill 230 |
| MO 119-10 | 6,000 | 6/01 | 6/01 | Released, awaiting start |

Part SAE is a different case. Its OH alert position is in the yellow with an on-hand position of 70. Top of green is set at 200. There is no open supply, but there is a demand of 60 units. In this case, the available stock position is in the red zone (i.e., 70 on hand + 0 open supply – 60 demand = 10). Part SAE's problem has nothing to do with open supply; there isn't any. No open supply is the problem. Thus a planning action is required to launch additional supply orders. Figure 27-12 is a summary of Part SAE's current on-hand and available stock situation.

Figure 27-13 is another example of an OH alert. In this case, the alert is for purchased parts only. The filter is set for stock-out and red and yellow zone on-hand status. Part PPA is stocked-out with demand, meaning that there is no on-hand stock and at the same time there is qualified actual demand. The source of the demand for 25 is immate-

**FIGURE 27-11**

Part FPE available and on-hand stock positions.

| Part: FPE | |
|---|---|
| Top of Green | 2,000 |
| Top of Yellow | 1,500 |
| Top of Red | 600 |
| On-Hand Alert | 300 |
| Available Stock | 1,550 |
| On-Hand | 250 (42%) |

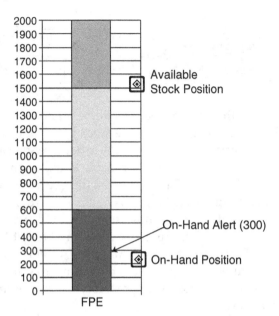

**FIGURE 27-12**

Part SAE available and on-hand stock positions.

| Part: SAE | |
|---|---|
| Top of Green | 200 |
| Top of Yellow | 160 |
| Top of Red | 80 |
| On-Hand Alert | 40 |
| Available Stock | 10 |
| On-Hand | 70 |

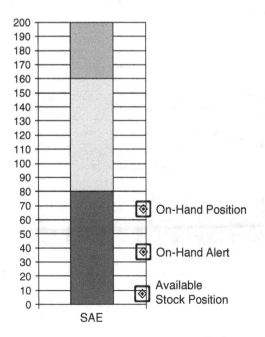

**FIGURE 27-13**

On-hand alert screen example (purchased parts).

**On-Hand Alert**

| Part # | OH Buffer Status | Part Type | On-Hand Qty | Open Supply | Demand | Available Stock Status |
|--------|------------------|-----------|-------------|-------------|--------|------------------------|
| PPA | Stock Out w/Dem | Purchased | 0 | 925 | 25 | 900 (GREEN) |
| PPJ | 47% (RED) | Purchased | 70 | 385 | 10 | 445 (GREEN) |
| PPS | 67% (YELLOW) | Purchased | 50 | 263 | 13 | 300 (GREEN) |

rial for the example but should be made available by drilling on the order activity of Part PPA. Despite this, available stock is at 90 percent and in the green, meaning that there is ample open supply. Part PPA is definitely a candidate for open-supply expedite. Figure 27-14 shows Part PPA's available and on-hand stock situation.

Figure 27-15 is the information received when a buyer implodes PPA open supply. Once again, today's date is 5/20. The order due the soonest is now late with regard to promise date. According to status, it is at the supplier's shipping facility. In this case, expedited freight might be considered to get PO 126-12 in as soon as possible.

Stocked out with demand (SOWD) is usually more devastating than simply being stocked out. Being stocked out means that there is a risk of missing opportunity. SOWD means that missing opportunity or at a minimum eroding customer confidence has become a reality.

**FIGURE 27-14**

Part SAE available and on-hand stock positions.

| Part: PPA | |
|-----------|--------|
| Top of Green | 1,000 |
| Top of Yellow | 700 |
| Top of Red | 300 |
| On-Hand Alert | 150 |
| Available Stock | 900 |
| On-Hand | 0 |

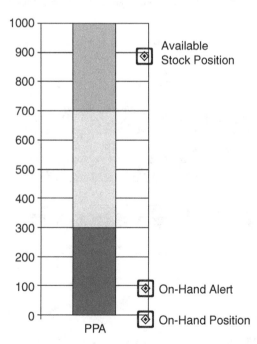

**FIGURE 27-15**

Part PPA open supply summary.

Part #: PPA                      OH Buffer Status: SOWD                              Today's Date: 5/20

| Order # | Quantity | Request | Promise | Status |
|---|---|---|---|---|
| PO 126-12 | 350 | 4/25 | 5/19 | At supplier shipping facility |
| PO 127-17 | 300 | 5/10 | 6/12 | In process at supplier |
| PO 128-27 | 275 | 5/25 | 6/15 | In process at supplier |

**FIGURE 27-16**

OH alert for Part FPJ.

On-Hand Alert: FPJ

| Part # | OH Buffer Status | Part Type | Location | On-Hand Qty | Open Supply | Demand | Available Stock Status |
|---|---|---|---|---|---|---|---|
| FPJ | 33% (RED) | Distributed | Region 1 | 100 | 675 | 25 | 750 (GREEN) |
| FPJ | 67% (YELLOW) | Distributed | Region 2 | 50 | 175 | 25 | 200 (GREEN) |
| FPJ | 33% (RED) | Manufactured | Plant | 1000 | 7000 | 500 | 7500 (RED) |

Figure 27-16 is a final example for an OH alert. In this case, the screen displays only one part. Part FPJ is a buffered part at the manufacturing plant and a distributed part in the company's regional warehouses. An additional column, labeled "Location," has been added to account for the same part but in different geographic locations.

Planners would use this report to consider expedites of open supply to all locations. Additionally, in this example, the plant's buffer position has a red OH alert. Under these conditions, it is helpful for planners to know which regional locations should get supply and which can afford to wait.

## Projected Buffer Status Alert

Projected buffer status alerts notify planning, manufacturing, and logistics personnel to situations where projected part consumption could result in an eroded buffer position prior to receipt of incoming supply orders. This is a radar screen that informs materials and planning personnel about anticipated projected on-hand alerts (particularly red zone penetrations) over the ASRLT of the part based on average daily usage, actual demand, and open supplies. Managing the projected buffer status alerts with a high degree of efficiency reduces the number of current OH alerts.

Figure 27-17 is an example of a projected buffer status alert. This particular example includes both manufactured and purchased parts. It is sorted by the severity of the alert (top being the priority). In this example, two parts are projected to stock-out in the near future. One of those parts will stock-out with demand against it, hence the acronym SOWD. The two remaining parts are not projected to stock-out, but their deepest projected on-hand dips over ASRLT are color coded and displayed as a percentage remain-

**FIGURE 27-17**

Projected buffer status alert sample.

Project Buffer Status Alert

| Part # | Projected Buffer Status | Part Type | ASRLT | Current OH | Demand Over ASRLT | ADU | ADU Over ASRLT | Open Supply |
|--------|------------------------|-----------|-------|-----------|-------------------|-----|---------------|-------------|
| SAD | Stock out in 3 days (RED) | Manufactured | 5 days | 75 | 92 | 25 | 125 | 0 |
| PPZ | SOWD in 4 days (RED) | Purchased | 10 days | 55 | 60 (RED) | 5 | 50 | 55 |
| PPL | 13% in 8 days (RED) | Purchased | 20 days | 100 | 150 | 10 | 200 | 350 |
| PPC | 75% in 3 days (YELLOW) | Purchased | 10 days | 45 | 27 | 5 | 50 | 40 |

ing of the total red zone. In this case, yellow priority items are displayed in the alert screen (PPC).

Part SAD is projected to stock-out in three days. Its current on-hand position is 75. Average daily usage (ADU) is 25. There is no open supply. This situation will require planning attention because the part is most certainly below the green level in its available stock equation. The planner will need to revisit the planning screen for this part because there is no open supply to expedite.

In determining future on-hand positions, a projected buffer status alert that uses ADU works exceptionally well in after-market parts, where there is little to no known future demand (immediate pull). This is the preferred approach in distribution environments, where the demand window tends to be very short, if not immediate. Where there are actual orders out in time (firm demand), a better option is to look at actual demand through the ASRLT in relation to supply-order receipts to determine potential negative on-hand balances and/or times of severe on-hand quantity erosion (i.e., near stock-outs). In the following examples, both types of on-hand determinations are presented.

In Figure 27-17, the column labeled "Demand Over ASRLT" represents the amount of actual demand (total customer orders) over the ASRLT. This is not limited to the qualified actual demand used when calculating available stock. It is the summation of all actual demand within ASRLT. When the summation of that actual demand over ASRLT is greater than the ADU over ASRLT, actual demand orders in relation to supply orders are used to generate the projected buffer status alert.

This occurs for Part PPZ. Note the actual demand quantity of 60 is greater than the calculated ADU quantity over ASRLT of 50. This is flagged to the planner by shading the box red. The planner now will need to drill down on Part PPZ to consider action.

Figure 27-18 represents all supply and demand activity for Part PPZ. In this case, there are three manufacturing orders (MOs 531-99, 532-10, and 532-32) that push the future on-hand balance to a stocked-out position before the next supply order receipt is due (PO 625-71). This stock-out occurs on 5/24 and actually results in a negative on-hand balance of 5. Thus Part PPZ is coded with the stock-out with demand acronym SOWD. In this case, the buyer would consider PO 625-71 for expedited status. Normally, it is more effective to expedite an order than to launch a new order in less than expected lead time.

**FIGURE 27-18**

All order activity for Part PPZ with projected on-hand position.

| Part #: PPZ | | PBS: SOWD in 6 Days | | | Today's Date: 5/20 |
|---|---|---|---|---|---|
| Order # | Quantity | Projected | Request | Promise | Status |
| MO 531-99 | −10 | 45 | 5/22 | 5/22 | Unreleased |
| MO 532-10 | −10 | 35 | 5/23 | 5/23 | Unreleased |
| MO 532-32 | −40 | −5 | 5/24 | 5/24 | Unreleased |
| PO 625-71 | 30 | 25 | 5/25 | 5/25 | In process at supplier |
| PO 626-05 | 25 | 50 | 5/29 | 5/29 | In process at supplier |

Figure 27-19 is a graphic depiction of what the projected on-hand balance looks like over the course of Part PPZ's ASRLT. The projected stock-out is the dip that occurs on 5/24. Negative on-hand position is not displayed in the graph. ADU is not being factored into the projected on-hand balance because actual demand is greater than ADU over ASRLT. To factor in both actual demand and ADU will cause "double dip" demand and unnecessarily trip the alert. Our next example with Part PPL uses the ADU to trigger a projected buffer status alert.

Part PPL has a projected buffer status alert. Its calculated ADU over ASRLT is greater than actual demand over ASLRT (200 > 150). In this case, ADU will be applied to the projected on-hand position. Figure 27-20 is the drill down on supply orders for Part PPL.

In this case, there is not a projected stock-out, but there is a projected eroded buffer position that is severely in the red zone. By using ADU, the on-hand position deteriorates to 13 percent of total red zone before the next incoming receipt. Figure 27-21 is a graphic

**FIGURE 27-19**

Projected on-hand balance for Part PPZ.

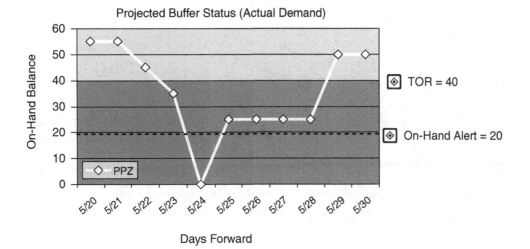

**FIGURE 27-20**

Open supply activity for Part PPL with projected on-hand position.

| Part #: PPL | | PBS: 13% in 8 Days | | | Today's Date: 5/20 |
|---|---|---|---|---|---|
| Order # | Quantity | Projected | Request | Promise | Status |
| PO 624-72 | 100 | 110 | 5/24 | 5/29 | In process at supplier |
| PO 625-11 | 100 | 170 | 5/29 | 6/29 | In process at supplier |
| PO 625-98 | 150 | 250 | 6/09 | 6/09 | In process at supplier |

depiction of this erosion, with the deepest projected point occurring on 5/28 (eight days from the current date). At this point, the buyer will need to make a decision whether to expedite or not. In any case, it would be good to call the supplier for a status update.

Part PPC is the final part on the projected buffer status alert. Its calculated ADU over ASRLT is greater than actual demand over ASLRT (50 > 27). In this case, ADU will be applied to the projected on-hand position. Figure 27-22 is the drill down on supply orders for Part PPC.

As with Part PPL, there is no predicted stock-out. However, unlike Part PPL, there appears to be no serious buffer erosion either. In most cases, parts are not brought into projected buffer status alert unless they erode through the OH alert level. This example was provided in order to further illustrate the idea that when the on-hand position is red from a planning perspective, it may not be red from an on-hand perspective.

In Figure 27-23, past-due supply orders are disregarded in determining the future on-hand position based on either ADU or actual demand. This highlights the need to adjust the promise date of the supply order to make a valid projection.

**FIGURE 27-21**

Projected on-hand balance for Part PPL.

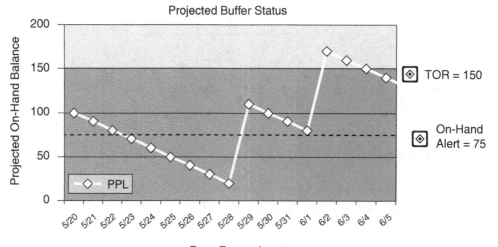

**FIGURE 27-22**

Open supply activity for Part PLC with projected on-hand position.

| Part #: PPC | | PBS: 75% in 3 Days | | | Today's Date: 5/20 |
|---|---|---|---|---|---|
| Order # | Quantity | Projected | Request | Promise | Status |
| PO 624-71 | 20 | 45 | 5/23 | 5/24 | In process at supplier |
| PO 625-36 | 20 | 45 | 5/28 | 5/28 | In process at supplier |

**FIGURE 27-23**

Projected on-hand balance for Part PPC.

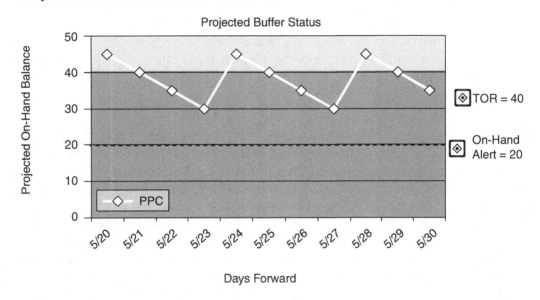

Note that all the preceding alerts still incorporate and/or use due dates in some way. This may appear to be a contradiction from the stated need to move away from priority by due date. Instead, this demonstrates that due dates, with regard to buffered items, should be a secondary factor in determining priority and/or expedite activity. In each case, the primary priority focus came from buffer status. The analysis after that identification brought due dates and other information such as status into the equation.

## SYNCHRONIZATION ALERTS

Synchronization alerts will cover non-buffered parts and in some cases buffered parts as well. Since there is no buffer for these parts, there is less slack with which companies have to work. The timing and synchronization associated with these parts require tools that promote visibility to potential problems before they happen. Synchronization alerts depend heavily on due dates. There are two types of synchronization alerts.

## Material Synchronization Alert

This alert is for all part types. *Material synchronization alerts* (MSAs) display the earliest occurrence of a negative on-hand balance within at least one ASRLT where open supply receipt is due after the required demand date. This usually occurs for a few different reasons:

- Heavy actual demand (potentially a spike) drives a buffered position to a negative on-hand position before the planned receipt of an incoming supply.
- The promise date on a supply order is pushed later in time. This also can create a negative on-hand position in the case of a buffered part. For nonbuffered parts it might mean that the part will be unavailable for a demand requirement at the time of the demand requirements release date.
- The promise date of a demand requirement is pulled in earlier in time. This can create a negative on-hand position for component parts at the time of the demand requirements release date, especially when those components parts are non-buffered parts.

Material synchronization alerts involve either an expediting action on the supply order or the reschedule of at least part of the parent order to coincide with the planned receipt of the component's supply order. Figure 27-24 is an example of an MSA screen.

The first line displays an MSA for a buffered parent component (Part SAG) and one of its buffered components (Part PPZ). The negative on-hand balance will occur in four days on the release date (5/24) of the parent order (MO 532-32). In fact, we have already seen this component part in the preceding section as one of the parts that triggered a projected buffer status alert. This alert presents the issue from a slightly different perspective. The projected buffer status alert focuses on a particular buffered part's buffer status. The MSA is designed to focus personnel on potential problems in the relationship between supply orders and demand orders. Shortages between component supply and parent-part demand create a problem with those relationships. The quantity of the negative on-hand position is displayed in the "Shortage" column.

Figure 27-25 shows the previous order activity example for Part PPZ.

In this case, there are a few possible solutions:

- The buyer can attempt to expedite the supply order (PO 625-71) or a portion of the supply order. It may be possible that the supplier could ship part of the PO today in order to be received in time to cover the projected shortfall.

**FIGURE 27-24**

MSA example.

**Material Synchronization Alert**                                             Today's Date: 5/20

| Demand Order # | Part # | Release Date | Qty | Order Type | Shortage | Supply Order # | Part # | Order Type | Qty | Promise Date |
|---|---|---|---|---|---|---|---|---|---|---|
| MO 532-32 | SAG | 5/24 | 40 | Replenished | 5 | PO 625-71 | PPZ | Replenished | 30 | 5/25 |
| MO 531-47 | FPS | 5/28 | 60 | NB | 60 | PO 611-54 | PPY | NB | 60 | 6/02 |

**FIGURE 27-25**

All order activity for Part PPZ with projected on-hand position.

| Part #: PPZ | | PBS: SOWD in 6 Days | | | Today's Date: 5/20 |
|---|---|---|---|---|---|
| Order # | Quantity | Projected | Request | Promise | Status |
| MO 531-99 | −10 | 45 | 5/22 | 5/22 | Unreleased |
| MO 532-10 | −10 | 35 | 5/23 | 5/23 | Unreleased |
| MO 532-32 | −40 | −5 | 5/24 | 5/24 | Unreleased |
| PO 625-71 | 30 | 25 | 5/25 | 5/25 | In process at supplier |
| PO 626-05 | 25 | 50 | 5/29 | 5/29 | In process at supplier |

- A planner/scheduler could push the release of the parent order back. In this case, the parent is buffered. The planner/scheduler should check the buffer status of the parent item before pushing the date back. If the buffer is in an acceptable position (from both available stock and on-hand perspectives), then pushing the order back one to two days may be a good option. This is one of the benefits to having a buffer at the parent—the absorption of variability from both supply and demand.
- A planner/scheduler could amend the quantity of the demand order to a lower amount (five fewer pieces). In this case, this might work because the parent is a buffered item. As in the preceding case, the planner/scheduler should check the buffer status of the parent item before doing this. If the buffer is in an acceptable position (from both available stock and on-hand perspectives), then the quantity change can be done with little to no risk to the system. Once again, this is one of the benefits to having buffers—the absorption of variability from both supply and demand.
- A planner/schedule could look to substitute an alternate material if this is an option.

Regardless of what action is taken, there needs to be a defined process to resolve MSAs. Attempts to rectify the MSA can involve changes to either component orders or parent demand requirements or possibly some combination of both. Frequently, these components will be under the control of different personnel and/or departments. The MSA is meant to focus the relevant personnel to collaborate and find the best potential solution together.

Continuing with the example in Figure 27-24, there is an additional MSA. In this case, the MSA is between a nonbuffered parent (Part FPS) and a non-buffered component (Part PPY). The "NB" in the "Order Type" column stands for non-buffered. The synchronization problem is to happen eight days from today. It occurs because the component's supply-order promise date is now five days behind the planned release of the parent demand order. The quantity shortage is for the entire order; both parts are non-buffered. In this case, there will be a different, albeit more limited, set of options because neither of the parts is buffered.

- The buyer can attempt to expedite supply to synchronize with the release date of the parent order.
- The planner can push the release date of the parent order back in time to synchronize with the incoming supply's promise date. Attempts then could be made to expedite the MO in an attempt to meet its promise date. Sales or customer service may need to be alerted because this order is going directly to a customer.

Whatever the resolution, the good news is that the problem is known well in advance of it actually occurring. In fact, MSAs should be triggered instantly or in "what if" mode when promise dates change or are proposed to change.

## Lead-Time Alerts

Figure 27-26 is a graphic depiction of the lead-time-managed (LTM) alert concept. A timeline from late January to early June is at the top. The lead time of each part/order is displayed as a horizontal bar. The size of the bar is in proportion to the length of the lead time. The last third of the bar is stratified by equal zones of green, yellow, and red. The end of the bar is set to the promise date on the timeline. If the current date is 5/02, then only one part (PPD) will generate a lead-time alert. On 5/02, Part PPD's current lead-time alert status is yellow.

Lead-time alerts are used to prompt personnel to check up on the status of critical non-stocked parts before those parts become an issue. Lead-time-managed (LTM) items are tracked, and at a defined point in the part's lead time, personnel are prompted for follow-up. Typically, this point is two-thirds of the way through the lead time of the part, as

**FIGURE 27-26**

Lead-time alerts concept.

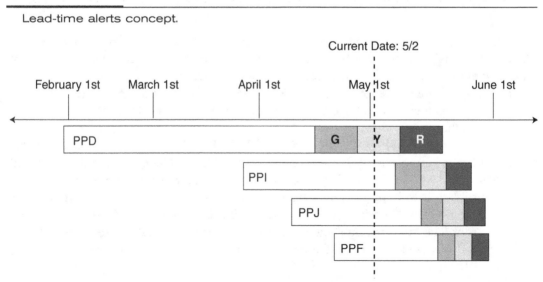

**FIGURE 27-27**

Lead-time alert concept showing progression from 5/02 to 5/20.

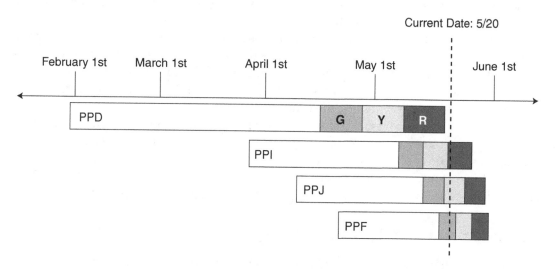

measured from the promise date. The final third then is divided into zones of typically equal proportions.

Figure 27-27 shows how time progression changes the status of the various LTM parts used in the example. As each zone is crossed, an alert is triggered with a color that corresponds to the zone boundary that was just crossed. If an action is not taken, then the alert remains active or intensifies. This intensification is expressed as an icon that conveys urgency.

The action that can clear the alert can be an entered note against the order generating the alert. This note could contain any type of status about the order. Regardless of whether the news is bad or good, it is news nonetheless. When a note is entered, a symbol that represents "followed up on" is displayed. This is displayed as a checkmark in Figure 27-28.

This figure provides an example of a possible lead-time alert screen that corresponds with Figure 27-7. There are four orders (each a different part number) that have triggered a lead-time alert. The "Days Left" column represents the number of days

**FIGURE 27-28**

Lead-time alert example.

**Lead Time Alerts**                                                      Today's Date: 5/20

| Status | Order # | Days Left | Part Type | Part # | ASRLT | Request Date | Promise Date |
|--------|---------|-----------|-----------|--------|-------|--------------|--------------|
| ! | PO 4532 | Late | Purchased | PPD | 105 | 5/15 | 5/19 |
| ! | PO 5120 | 6 | Purchased | PPI | 63 | 5/26 | 5/26 |
| ✔ | PO 5214 | 10 | Purchased | PPJ | 45 | 5/24 | 5/30 |
| ✔ | PO 5290 | 12 | Purchased | PPF | 36 | 6/01 | 6/01 |

remaining until the promise date of the order. An exclamation point in the status column means that the part has entered the respective displayed color zone and there has been no note/action against that order. The checkmark means that there has been an action or note against the order since its entry into a zone.

In this case, there are two orders that need attention and two orders that have already been statused. PO 4532 is late because the promise date is one day prior to the current date. A late order also represents another zone (dark red) that triggers an alert. PO 5120 has entered its red zone, has six days remaining until its promise date, and there has been no note or action against the order.

A note or action against an order could be either the assignment of a follow-up date (temporary resolution) or the assignment of a final confirmed date and decision (could be sooner, on time, or later). Regardless of what the note or action is, it prompts and brings visibility to what might happen ahead of time. This proactive effort often prevents potential problems resulting in better due-date performance for these types of components.

## EXECUTION COLLABORATION

Finally, the mechanisms are described in which all of the preceding alerts are used to generate collaboration and integration of various areas within a company and even across a supply chain. This requires various front-office tools to work hand in hand with these execution alerts. The use of e-mail, contact lists, calendar events, and reminders, as well as notes and even instant messaging, can enhance the communication, visibility, and ultimately the collaboration in a relatively short period of time. Figure 27-29 shows the stages of execution action once an alert has been triggered and what the appropriate electronic tool might be to accomplish each stage.

Figure 27-30 is an example of the MSA between Parts FPS and PPY. In this case, the promise date of the component part is four days after the release date of the parent item.

For this example, assume that the initial response to this MSA is to first approach the supplier to determine if it might be able to adjust its promise date to meet the synchronization requirement. For this example, the synchronization alert was triggered by the sales order's promise date being pulled in four days. The promise-date change moved

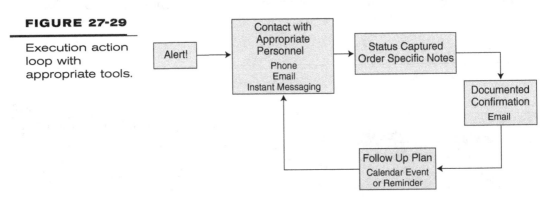

**FIGURE 27-29**

Execution action loop with appropriate tools.

**FIGURE 27-30**

Material synchronization alert for Part FPS.

| Material Synchronization Alert | | | | | | | | | | Today's Date: 5/20 |
|---|---|---|---|---|---|---|---|---|---|---|
| Demand Order # | Part # | Release Date | Qty | Order Type | Shortage | Supply Order # | Part # | Order Type | Qty | Promise Date |
| MO531-47 | FPS | 5/28 | 60 | NB | 60 | PO 611-54 | PPY | NB | 60 | 6/02 |

the release date backward four days. Thus this MSA is not the supplier's fault, but still, the request will be to ask the supplier to adjust if possible (see Figure 27-31).

- *Step 1.* The first thing the buyer will do is to call the supplier. In this case, the supplier agrees to adjust its promise date to meet the synchronization requirement.
- *Step 2.* The buyer now will enter a note that the promise date is now 5/28. The buyer makes the change to the promise date and notes that as well. The material synchronization alerts will be cleared as a result of this action.
- *Step 3.* The buyer launches an e-mail to the customer-service representative who approved the new promise date. In the event of a critical part, perhaps additional people at higher levels in both organizations are copied. In this case, the buyer copies his or her boss. Delivery and read receipts are requested. This establishes a documented and archived trail of communication.
- *Step 4.* The buyer initiates a calendar reminder—the subject of which is the PO number for the morning of 5/25. On the morning of 5/25, the buyer will receive

**FIGURE 27-31**

Execution actions in relation to MSA for Part FPS.

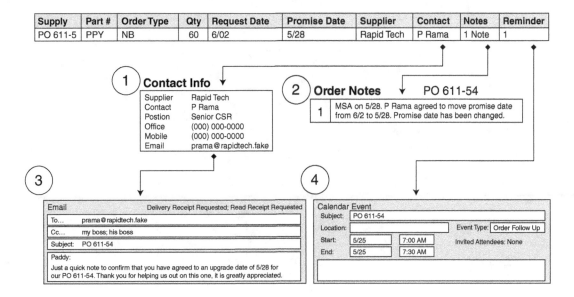

an order-specific reminder through his or her calendar system and, perhaps, mobile device.

Figure 27-31 represents the preceding chain of events. At the top of the figure is the order activity and workbench. "Contact," "Notes," and "Reminder" all can be linked electronically to appropriate electronic tools and databases.

## CHAPTER 28

# Demand-Driven Material Requirements Planning (DDMRP) Performance Reporting and Analytics

**P**erformance reporting can be the subject of another whole book. There are many nuances to what executives and decision makers want to see for decision making. Figure 28-1 shows a grid of potential inventory performance reports to consider. These reports can be extremely narrow in focus or encompass all part numbers.

Below are three examples based on different combinations of the preceding categories. The first example (Figure 28-2) is a trend report for the last 90 days that displays over top of green (OTOG) dollars for all parts.

According to this report, total OTOG dollars have been declining steadily over the reported period. Total reduction in OTOG dollars is over 50 percent for that time period ($527,323 to $250,455).

The second example (Figure 28-3) is a snapshot report for the last 180 days for stock-outs and stock-outs with demand (SOWD) for all suppliers.

According to the figure, over the last 180 days, Ace Supply has had a total of 60 stock-out occurrences, 45 of which were with demand. Rapid Tech has had more total stock-outs but significantly fewer with demand. Some could claim that SOWD is not a fair way to measure supplier performance because the supplier has no control over the company's demand. SOWD has less to do with the particular supplier and more to do to provide focus on where the organization might be more exposed with its total purchased-parts strategy. Possibly these purchased parts have more quality challenges or are involved in parent end items that are experiencing high sales growth. Seeing a snapshot weighted by SOWD provides an inward-looking metric rather than determining the specific level of supplier non-conformance to buffer status. These particular parts may need to be reviewed for manual adjustment or different buffer profiles.

The total number of stock-outs is certainly an issue to bring up with suppliers. This assumes that supply orders according to the available stock equation for the replenished

## FIGURE 28-1

DDMRP performance-reporting combinations.

| Primary Category | Secondary Category | Date Range | Trend or Snapshot |
|---|---|---|---|
| Part/SKU # | OTOG Zone (OH) | | |
| Planner | OTOG Dollars | | |
| Location | Green Zone (OH) | | |
| Supplier | Yellow Zone (OH) | | |
| All Parts/SKU | Red Zone (OH) | User Defined | User Defined |
| Buffer Profile | All Zones (OH) | | |
| Part Type | Stock-Out and SWOD | | |
| Part/SKU Family | Execution Alerts | | |
| | On-Time Delivery/Fill Rate | | |
| | OTOG Zone (AS) | | |
| | Green Zone (AS) | | |
| | Yellow Zone (AS) | | |
| OH = On-Hand | Red Zone (AS) | | |
| AS = Available Stock | All Zones (AS) | | |

## FIGURE 28-2

Example of a trending report.

**Example 1**

| Primary Category | Second Category | Date Range | Trend or Snapshot |
|---|---|---|---|
| All Parts/SKU | OTOG Dollars | Last 90 Days | Trend |

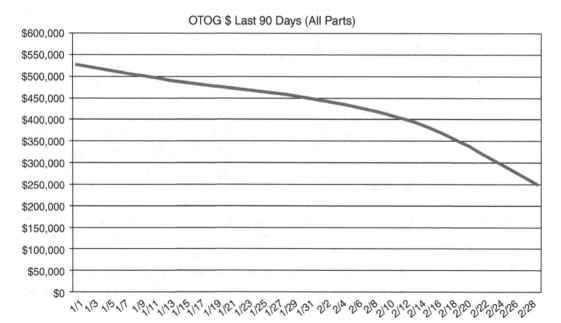

OTOG $ Last 90 Days (All Parts)

**FIGURE 28-3**

Example of a snapshot report.

**Example 2**

| Primary Category | Second Category | Date Range | Trend or Snapshot |
|---|---|---|---|
| Supplier | Stock-Out and SWOD | Last 180 Days | Snapshot |

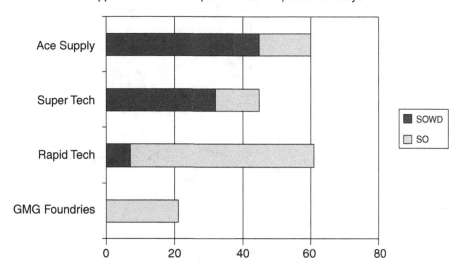

buffers are being generated at the appropriate times. In many cases, companies have changed the nature of their relationship with their suppliers by providing them with the up-to-date buffer status for items with open purchase orders. See Chapter 29 for additional information.

The third example (Figure 28-4) is a trend report over the last 60 days for zone performance for the 169 parts that a particular planner controls. The report shows the trended zonal distribution (i.e., OTOG, green, yellow, red, and stock-out) of the 169 parts over the last 60 days.

According to this report, the planner's performance is going in the right direction. The planner is minimizing both the OTOG and stock-out situations while moving the majority of the parts to the yellow zone.

These reports are just an example of what is possible given the DDMRP report. Important to note is that planners, buyers, and suppliers have real visibility on the true priorities for the company. Most important is the overall bottom-line improvement for the company. The actual format of the operations performance report is less important. Next are the case studies of two companies that have been early adopters of these concepts and their results.

**FIGURE 28-4**

Trended zone performance report over time range.

Example 3

| Primary Category | Second Category | Date Range | Trend or Snapshot |
|---|---|---|---|
| Planner | All Zones | Last 60 Days | Trend |

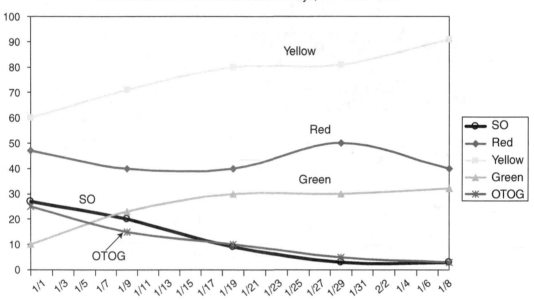

Kirk's OH Zone Performance Last 60 Days, 169 Total Parts

# OREGON FREEZE DRY RESULTS

Oregon Freeze Dry is the largest diversified food freeze dryer in the world. Prior to implementing DDMRP, it used traditional MRP with standard minimum batch practices. By implementing DDMRP with no additional capital expenditure, overhead, or other improvement initiatives, Oregon Freeze Dry reported the following gains:

**Mountain House Division:**
- Sales increased 20 percent.
- Customer fill rate improved from 79 to 99.6 percent.
- Inventory was reduced by 60 percent.

**Industrial Ingredient Division:**
- Make-to-order lead time was reduced by 60 percent.
- On-time delivery was 100 percent.
- Inventory was reduced by 20 percent.

- Raw materials:
  - No stock-outs
  - Inventory reduction of $2.5+ million

## LETOURNEAU TECHNOLOGIES RESULTS

The LeTourneau Technologies, Inc. (LTI), companies include some of the world's leading innovators in manufacturing, design, and implementation of systems and equipment for mining, oil and gas drilling, offshore oil drilling, power control and distribution, and forestry. LTI has two main manufacturing facilities (Longview, TX, and Houston, TX) that are similar in terms of capability, product complexity, and size. The performance reports in Figure 28-5 demonstrate dramatic differences in performance between the two comparable campuses of Longview and Houston. The type of manufacturing is very similar in terms of both complexity and scale. The difference was how each was managed. Longview used DDMRP tactics, whereas Houston used traditional MRP tactics.

Beginning in 2005, the market began to expand rapidly for all LTI business segments. This boom-bust cycle was not a new phenomenon for LTI. Typically, LTI's inventory and expenses would rise at a similar rate as revenue; service levels deteriorated at a proportionate rate. However, in the 2005 boom, the Longview facility embraced the new business rules of DDMRP in conjunction with its partial implementation of drum buffer rope (DBR). This facility was able to dramatically control inventory and expenses while maintaining excellent service levels. This is noted in the graph on the left in Figure 28-5.

All boom markets eventually end. In this case, the markets began to cool off in 2008. 2009 brought a significant decline in revenue. When the boom times were over, DDMRP at Longview minimized the company's exposure to inventory liabilities. No matter what

**FIGURE 28-5**

Total revenue (TR) versus inventory (INV)—Longview and Houston facilities.

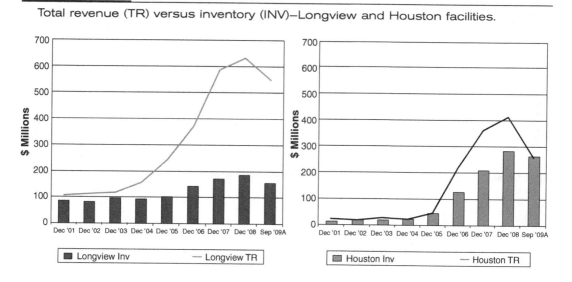

kind of economic times a company finds itself in, good inventory practices that minimize inventory exposure while maintaining service levels provide sustainable financial success for the company.

Figure 28-5 shows total revenue versus inventory from 2001–2009 from both the Longview and Houston sites. Note that beginning in 2005 there was dramatic growth at both sites. In Longview, revenue grew by a factor of greater than 300 percent (over $400 million). Over that same period, inventory rose only by 80 percent (about $80 million). In Houston's case, however, inventory ended up growing at nearly the same rate as revenue. There is about a six- to nine-month lag, but it is pacing at the same rate. The lag was due to the policy of building to a forecast. When the future is projected from the past, the boom ends, and the future looks nothing like the past, the end result is obvious.

When the market began to turn down at the beginning of 2008, LTI Houston was burdened with significant inventory liability. Owing to the nature of forecasting, there is a real risk that the inventory actually could grow beyond revenue in the short run without a massive course correction in the form of purchase order and manufacturing order cancellations and/or delay. Figure 28-5 shows that this actually occurred in Houston in 2009. This exposure is a classic effect of traditional planning environments.

Important to note is that the people in the Houston facility were smart, professional manufacturing personnel. They did not have the tools and business rules at their facility to replicate what happened at Longview. Figure 28-5 is not an indictment of the people. Rather, it is an effective illustration that the generally accepted rules for planning material represents a huge liability in the volatile and variable manufacturing environments that tend to be today's rule rather than exception.

## CHAPTER 29

# DDMRP Future

The demand-driven material requirements planning (DDMRP) solution laid out in Chapters 22 through 28 is a blueprint for a long-overdue overhaul of MRP rules and tools so that manufacturers and supply chains can remain competitive in the twenty-first century and beyond. Early adopters have already paved the way for mainstream usage. Referring back to the first edition of this book, Orlicky suggested the following research opportunities that still fit today!

## RESEARCH OPPORTUNITIES

There is ample opportunity and a genuine need for further research into MRP and related areas. Thus far, the subject has received minimum or no attention—except for questions of order quantity—in published literature, research papers, and academic curricula. Operations research in general consistently has shown a predilection for inventing techniques (algorithms), but practicing managers have little interest in the results of research that produces mathematically elegant solutions to trivial or nonexistent problems.

Those who represent the industry point of view have had occasion to wish for the researcher to address what needs to be researched as opposed to what is eminently researchable and for the educator to teach what needs to be taught as opposed to what is eminently teachable.

An improvement on both sides of this question can only come from better cooperation and dialogue between academia and industry. In the future, the inhabitants of the real world must take the initiative in articulating and communicating their problems to researchers and educators, and they should actively support valid research. Researchers, on the other hand, should make an attempt to validate their research targets before actually proceeding with projects. This is not at all difficult to do—the researcher will find industry people helpful and cooperative in answering his or her question: "Is this one of your more pressing problems, and should I be working on it?" The future of an academic institution can well be at stake if concepts are not taught and researched that actually benefit real-world companies.

In the area of MRP, the following are suggested as affording opportunity for productive research:

1. Theory:
   - Links between the MRP systems across a supply chain, including the frequency of communication across the supply chain
2. Justification:
   - Applicability of MRP
   - Costs of an informal system
3. System design:
   - Design criteria for different business environments
   - Bill-of-material (BOM) modularization
   - Alternatives in the treatment of optional product-feature data
4. System implementation and use:
   - Analysis of implementation problems
   - Master production schedule (MPS) development and management
   - Operational aspects of MRP system use
5. Education:
   - Curricula design and teaching tools

Links between MRP systems across the supply chain need to be better defined and the systems better integrated. We need better techniques for answering the crucial question of whether the holistic material requirements plan actually can be met, as far as its timing is concerned, on an order-by-order basis.

Applicability of the MRP method is a subject that calls for exploration. What are the criteria of advantageousness as contrasted with applicability of the application of MRP methods in a specific case? Costs of an informal system are unquestionably high, but no one seems to know how high. They seem to be undermeasured and underappreciated. Conventional accounting methods are ill suited for capturing costs in this category. The ledger does not include accounts for costs of confusion, unnecessary handling of excessive material, mistakes made by operating management owing to a lack of valid information, inefficiencies of component staging and expediting, missed schedules, time spent by supervisory personnel on chasing parts, machine teardowns owing to rush work, the amount of dust collecting on unopened crates of expedited materials, and inventory write-offs attributable to poor planning—the list seems endless. Note that to the extent that the informal system is being maintained parallel with and because of a formal system that does not function satisfactorily, the costs of an informal system are avoidable. Is this a serious problem? It is indeed despite the fact that its dimensions are hidden under a basket. Field research into this problem would produce startling results.

What is the optimal design of an MRP system for a given environment? Can factors bearing on elements of design be isolated and quantified? BOM modularization (see Chapter 11) and other techniques of BOM structuring thus far have received only the scantest attention in literature. The information is contained in anonymous system documentation of scattered MRP system users. Principles and ground rules for the guidance of prospective MRP system users have yet to be formulated.

This subject invites and requires research. The most intriguing question is this: Can the logic of BOM structuring be formally described, captured, and programmed for a computer to execute? Can software be developed that would analyze and restructure a BOM file correctly and in optimal fashion for purposes of MRP in a specific case?

Alternatives in the treatment of optional product-feature data (see Chapter 11) need to be explored, evaluated, and documented in literature. This problem area is related to BOM structuring, as well as to product design and the organization of product specifications data. Phenomena of nested options (i.e., option within option within option) and sub-options invite inquiry.

Analysis of MRP system implementation problems, that is to say, management and operating problems encountered in implementing (installing) MRP systems, would serve a very useful purpose because it would act to reduce both system implementation cost and the rate of system failure. It is known that the most serious obstacles to MRP system success lie outside the (technical) system boundaries. The problems must be sought not in computer hardware and software but in people, their attitudes, habits, and knowledge level—the *thoughtware* within a system.

Continued MPS development and management (see Chapter 12) represent an area of rich research "pay dirt." The MRP system acts as a mirror being held up to the MPS—it illuminates its quality by translating it into specific consequences in terms of material, lead time, and capacity availability. Once an MRP system is installed, it tends to force management to reexamine the entire process of master scheduling, including the procedures of schedule development, maintenance, and revision. Management of the manufacturing operation through the MPS becomes feasible—an opportunity and a responsibility. Much more needs to be learned about this area. How should MPSs be developed, and what should they consist of in different business environments? How should sales and operations plans (SOP) be developed in the new demand-driven world? What are the organizational implications? What are the limits of MPS changeability? What techniques should be employed to ensure and safeguard the integrity of shop priorities? How does an integrated demand-driven materials and finite-capacity scheduling system produce a realistic but easy-to-understand picture of both capacity and material limitations.

Operational aspects of MRP system use, despite their importance, have not been investigated, maintained, and documented adequately in literature. An entire generation of MRP users has lost connection with the latent capabilities of MRP systems. How does an inventory planner, capacity planner, master scheduler, and manager make better use of the tool represented by an MRP system? What are the limits of information that the system can provide? What is the set of situations that a planner potentially may face, and what is the correct response to each one of those situations? In what respects are MRP systems found lacking from the user's point of view? Answers to many of these and similar questions may be found, case by case, in the field. Their analysis, classification, and compilation would provide a much needed generalized guide to the use of MRP systems.

Curricula design and teaching tools are, at this time, still largely undeveloped. Teaching operations management has become almost extinct in the United States. However, to have sustainable success, every company can only exploit its unique operational capability that provides value to its customers at a profit to itself. This makes the subjects of operations management, production management, and inventory management demand the inclusion of the subject of MRP more important than ever. Where the subject is included, it tends to be given cursory treatment. This area is wide open to creative research and development. Much work is needed in curriculum redesign; development of teaching materials, classroom examples, and exercises; writing of case studies; and the construction of computer-based simulators suitable for student use. Here academic/industry cooperation would foster rapid progress.

The problem remains the same today as it was when Orlicky wrote the first edition of this book—the problem was not one of planning but of replanning. Replanning at the rate changes were taking place proved to be impossible and extremely disruptive. Orlicky, at that time, perceived the solution to lie in stabilizing the MPS, in making the marketing people live with the original forecast, and in freezing the design of the product—ending the volatility of the environment and stopping the turbulence. Needless to say, that never happened. Orlicky went on to write the following prophesy:

> From today's vantage point, it is clear that such thinking is invalid and the perception false. The manufacturing business environment, in most cases, is inherently unstable and turbulent. Change is the rule. Change, in fact, is "the name of the game." The solution lies not in methods to stabilize and freeze, but rather in an enhancement of the ability to accept change and to respond to it promptly and correctly—and do it routinely, as a matter of course. In the future, instability and propensity to change will, if anything, increase. An ability to accept change fosters more change.
>
> The future should bring further enhancements in the ability to respond to change through expanded use of computer-based methods of time phasing, i.e., material requirements planning and time-phased order point, and through a linkage of such systems between vendor, manufacturer, and distributor. Both the shop and the vendor will need to be reeducated to the volatility of need and to formal, routine methods of conveying the information, in contrast with traditional expediting. Expediting, a symptom of an inability to maintain a valid plan and the most inefficient method of gathering and conveying information (this is what expediting really is), is on the wane. Formal systems are bound to displace informal systems because they can be vastly more efficient—and we now have formal systems that can do the job.

What is clear is that early adopters such as the two described in Chapter 28 are experiencing huge gains with demand-driven concepts and rudimentary software capability. Figure 29-1 is a comparison and summary chart for DDMRP versus traditional MRP.

## PREDICTION OF THE FUTURE

Planning systems since the 1920s evolved from the core perspective of inventory. This is no longer true. Now demand must be at the core. Thus DDMRP is not the next evolution in inventory management; it is a revolutionary shift in perspective and tactics. It is disruptive methodology that will drive disruptive technology. DDMRP is not just a different planning approach with regard to a single company. Inherent in its rules and tools is a level of common sense and visibility that easily translates beyond the confines of a single entity. For progressive and/or highly integrated supply chains, DDMRP represents a common set of rules and tools that make a supply chain more efficient, competitive, and profitable for all players. DDMRP enables a win-win-win across the supply chain.

In order to capitalize on this potential, companies will need to collaborate and share data and information. This will be possible only if each link of the supply chain understands the benefit to itself. Cloud computing would appear to be a valid vehicle to support this vision. The technological direction should be on-demand and highly visible connections between people, among groups, and throughout the suppliers and customer network. The Internet revolutionized business and started this process. During the next decade, it will be interesting to watch how the cloud and even social networking might enable business opportunity and supply-chain integration.

## SUCCESS LEVERAGING TECHNOLOGY

There are six questions to evaluate and ensure the true impact of any new technology on a business—including DDMRP.

1. What is the power of the new technology?
2. What current limitation or barriers does the new technology eliminate or vastly reduce?
3. What usage rules, patterns, and behaviors exist today that consider the limitation?
4. What rules, patterns, and behaviors need to be changed to get the benefits of the new technology?
5. What is the application of the new technology that will enable this change without causing resistance?
6. How do we build, capitalize, and sustain the business?[1]

---

[1] Goldratt, 2002; Goldratt, with Schragenheim and Ptak, 2000; Schragenheim, 2010, North River Press, New Barrington MA.

# FIGURE 29-1

Traditional MRP versus DDMRP.

| | Typical MRP Attributes | | Demand-Driven MRP Attributes | | DDMRP Effects |
|---|---|---|---|---|---|
| **Planning Attributes** | *MRP uses a forecast or master production schedule as an input to calculate parent and component level part net requirements.* | ▲ | DDMRP uses buffer profiles in combination with part traits to set initial buffer size levels. These buffer sizes are dynamically resized based on actual demand. Buffer levels are replenished as actual demand forces buffers into their respective rebuild zones. Planned adjustments are used to "flex" buffers up or down. | ▲ | DDMRP eliminates the need for a detailed or complex forecast. Planned adjustments to buffer levels are used for known or planned events/circumstances. |
| | *MRP pegs down the ENTIRE Bill of Material to the lowest component part level whenever available stock is less than exploded demand.* | ▲ | Pegging is decoupled at any buffered component part. | ▲ | Larger BOM environments are often stratified into independently planned and managed horizons separated by buffered positions. This prevents or dampens "nervousness." |
| | *Manufacturing Orders are frequently released to the shop floor without consideration of component part availability.* | ▲ | Material Synchronization and Lead Time Alerts are designed to warn planners of shortfalls when incoming supply will not be in time for parent demand order release. | ▲ | Planner can take appropriate action and eliminate excess and/or idle WIP. |
| | *Limited future demand qualification. Limited early warning indicators of potential stock outs or demand spikes.* | ▲ | An Order Spike Horizon in combination with an Order Spike Threshold qualifies spike quantities over the ASRLT of the part. The qualified spike is then added to the available stock equation and is compensated for in advance. | ▲ | Reduces the materials and capacity implications of large orders and/or limited visibility. Allows stock positions to be minimized since spike protection does not have to be "built in." |
| | *Lead time for parent part is either the manufacturing lead time (MLT) or the cumulative lead time (CLT) for the parent item.* | ▲ | DDMRP uses ASR Lead Time, which is the longest unprotected/coupled sequence in the bill of material whenever that lead time exceeds manufacturing lead time. | ▲ | Creates a realistic lead time for customer promise and/or buffer sizing. Enables effective lead time compression activities by highlighting the longest unprotected path. |
| **Stock Management Attributes** | *Fixed reorder quantity, order points, and safety stock that typically do not adjust to actual market demand or seasonality.* | ▲ | Buffer levels are dynamically adjusted as the part specific traits change according to actual performance over a rolling time horizon. Planned adjustments "flex" buffers for product phase in/out and seasonality. | ▲ | DDMRP adapts to changes in actual demand and planned/strategic changes. |
| | *Past due requirements and orders to replenish safety stock can be coded as "Due Now."* | ▲ | All orders get an assigned a realistic due date based up on ASR lead times. | ▲ | Creates a realistic lead time for customer promise and/or buffer sizing. Enables effective lead time compression activities by highlighting the longest unprotected path. |
| | *Priority of orders is managed by due date (if not Due Now).* | ▲ | All DDMRP buffered parts are managed using highly visible zone indicators including the percentage encroachment into the buffer. This gives you a general reference (color) and discrete reference (%).* | ▲ | Planning and Materials personnel are able to quickly identify which parts need attention and what the real-time priorities are. |
| | *Once orders are launched, visibility to those orders is essentially lost until the due date of the order when it is either present or late.* | ▲ | DDMRP gives special consideration to some critical non-stocked parts called LTM Parts. These parts are given visibility and color coded priority for pre-expediting activities through Lead Time Alerts. | ▲ | Better synchronizes key non-stocked components with demand orders and reduces schedule surprise and slides due to critical part shortages. |

The power of DDMRP is that the conflict between planning (predictability) and flexibility (responsiveness) is finally broken. Now it is possible for a company to plan materials effectively while increasing its responsiveness to the market. This brings in the notion of agility into the realm of the achievable rather than the impossible. This enables a company to sense changing customer demand and then adapt planning and production while pulling from suppliers—all in real time. DDMRP enables a company to stop pushing excess inventory and start pulling in more profits. However, the new rules as described in Part 4 of this book are necessary. Yet to be answered is how this supporting technology will be developed and enabled without causing resistance.

In order to help foster and focus the required research and curriculum development, the Demand Driven Institute has been founded. We invite all interested parties to help in this endeavor. More information can be obtained at **www.demanddriveninstitute.com**. We look forward to your feedback on this edition and your input for future editions.

# Joseph Orlicky's Contributions to Material Requirements Planning

*By Gene Thomas*

**B**ased on recent acquaintances with Carol Ptak and Chad Smith leading to an intense discovery of mutual past experiences regarding the eras of MRPs, I feel honored to contribute to recollections about Joe Orlicky's contributions to the advent of material requirements planning (MRP) revelations.

I was an IBM marketing representative in Des Moines, IO, during the late 1950s with a specialized group of heavy manufacturing customers, including John Deere, Maytag, John Morrell, Pella, and Fischer Controls. I was subsequently assigned as a five-state district manufacturing specialist with coverage of our IBM plant in Rochester, MN, to foster IBM's leadership with manufacturing applications. During this coverage in the early 1960s, I recognized the core requirement to develop standardized newly offered disk-oriented software to build and maintain the basic manufacturing files for bills of material and routings related to inventory masters. I invented and published in 1961 the specifications for the canned software package BOMP (Bill-of-Material Processor).

My coverage territory included Milwaukee, where I first met Joe Orlicky when he was the production control manager at the Kearney & Trecker machine-tool plant and helped implement a shop-floor data-collection system similar to the one I initiated in Rochester. Then Joe went to J. I. Case in Racine, WI, again as production control manager, but with the experienced and obstinate attitude that MRP weekly/monthly regeneration cycles were insufficient to plan and control large operations with configured product lines. With the introduction of disk hardware (IBM 305 RAMAC), Joe's staff initiated a homegrown "net change" MRP system with the limited file size of 5 millibytes—even using the file addresses as part numbers to facilitate explosion processing. They had only enough processing power to maintain BOM sublinkages, so no where-used linkages were automatically maintained. This explosion process of both BOMs and routings was termed *GNETS* (Gross-to-Net, Time Series). The IBM plants had been processing GNETS for several years using very large-scale sequential tape systems (no disks yet) by passing the entire single-level BOM and routings files at every level. Since they had to preprocess the BOM batch files each time to maintain low-level and continuity coding, net-change functionality had never been implemented before.

Meanwhile, in parallel with Case's development, I was promoted to marketing manager manufacturing in Milwaukee with direct control of the IBM and customer resources to develop the BOMP software (with where-used and low-level coding) as a canned application package. Then Joe and the Case staff upgraded their IBM hardware and used the BOMP package as a generalized file-organization program.

We enlisted the assistant of one of our Milwaukee customer's vice presidents of manufacturing to write the GNETS RPG netting, lot-sizing, and off-setting logic embedded within the BOMP package. It was used as the *LAMP* (Labor and Material Planning), the first canned MRP package, and was implemented by many Milwaukee manufacturers. It was later published as a marketing package by the IBM Manufacturing Industry Group that I managed. The Production Information Control System–Requirements Planning System (PICS-RPS) package then followed, which turned into MAPICS and COPICS.

"Papa" Joe Orlicky later joined IBM as a manufacturing specialist working closely with the Ollie Wight and George Plossl consulting firm. The work developed in Milwaukee, supported by superlative guidance from Joe and his staff, has remained the basis for the core MRP architecture. The publishing of Joe's book has served as stellar support for the basis of a practitioner's real-life experience with pragmatic, sometimes dogmatic resolve! The GNETS kernel has remained intact, with many add-ons over time and hardware/software improvements. I ended my highly revered relationship with Joe working directly with him while overseeing the IBM plant system consolidations in 1969 (reference the May 5, 1967 *LIFE Magazine* article on IBM's internal frustrating warfare of business).

The future of more things to evolve based on this solid foundation that Joe continued to espouse would include more mass customization with front-to-backend configuration processed in real time (on demand). There will be developing systems power to generate pegged (re)demand requirements to be exploded down to strategic componentry. It is the process of constant rescheduling to balance any/all key constraints (material, labor, and cost), because they disrupt manufacturing execution, that will arrive as hardware and application software continue to evolve.

*Gene Thomas, Founder Emeritus*
http://mrp-to-lean.blogspot.com/

# Definitions: APICS Terms and Their Place in DDMRP

| Known and Accepted Terms from the *APICS Dictionary*, 12th Ed. | The Term's Place in DDMRP |
|---|---|
| **actual demand:** Actual demand is composed of customer orders | After setting initial buffer levels, actual demand (not forecasted demand) is used to generate new supply requirements as well as adjust buffer levels. |
| **agility:** The ability to successfully manufacture and market a broad range of low-cost, high-quality products and services with short lead times and varying volumes that provide enhanced value to customers through customization. Agility merges the four distinctive competencies of cost, quality, dependability, and flexibility. | One objective of a DDMRP system. |
| **bill-of-material explosion:** The process of determining component identities, quantities per assembly, and other parent/component relationship data for a parent item. Explosion may be single level, indented, or summarized. | Explosion still occurs in DDMRP, but it is a decoupled explosion (see definition later in this appendix). |
| **buffer:** 1) A quantity of materials awaiting further processing. It can refer to raw materials, semifinished stores or hold points, or a work backlog that is purposely maintained behind a work center. 2) In the theory of constraints, buffers can be time or material and support throughput and/or due date performance. Buffers can be maintained at the constraint, convergent points (with a constraint part), divergent points, and shipping points. | In DDMRP, buffers represent any stocked position. Whether those buffers are strategically replenished, replenished over-ride, or min-max will determine how they are sized and managed. |

| Known and Accepted Terms from the *APICS Dictionary*, 12th Ed. | The Term's Place in DDMRP |
|---|---|
| **control point:** In the theory of constraints, strategic locations in the logical product structure for a product or family that simplify the planning, scheduling, and control functions. Control points include gating operations, convergent points, divergent points, constraints, and shipping points. Detailed scheduling instructions are planned, implemented, and monitored at these locations. Other work centers are instructed to "work if they have work; otherwise, be prepared for work." In this manner, materials flow rapidly through the facility without detailed work center scheduling and control. | Control points are important considerations in the strategic inventory positioning process. |
| **cumulative lead time:** The longest planned length of time to accomplish the activity in question. It is found by reviewing the lead time for each bill of material path below the item; whichever path adds up to the greatest number defines cumulative lead time. Syn: aggregate lead time, combined lead time, composite lead time, critical path lead time, stacked lead time. *See* planning horizon, planning time fence. | Cumulative lead time still can be a factor in DDMRP if, after determining the proper inventory positioning, there are now stocked components in a parent item's bill of material. |
| **decoupling:** Creating independence between supply and use of material. Commonly denotes providing inventory between operations so that fluctuations in the production rate of the supplying operation do not constrain production or use rates of the next operation. | Decoupling is a critical aspect of DDMRP. Decoupling is used to mitigate the increased variability, volatility, and required variety in the "new normal." |
| **decoupling inventory:** An amount of inventory kept between entities in a manufacturing or distribution network to create independence between processes or entities. The objective of decoupling inventory is to disconnect the rate of use from the rate of supply of the item. *See* buffer. | Decoupling inventory is a critical aspect of DDMRP. Stock buffers decouple explosion. Actively synchronized replenishment lead time (ASRLT) is a decoupled cumulative lead time. |
| **decoupling points:** The locations in the product structure or distribution network where inventory is placed to create independence between processes or entities. Selection of decoupling points is a strategic decision that determines customer lead times and inventory investment. *See* control points. | Decoupling points are critical in DDMRP. These strategic points are identified in the strategic inventory positioning step. |
| **demand chain management:** A supply chain inventory management approach that concentrates on demand pull rather than supplier push inventory models. | DDMRP is a demand-chain management approach. |

## Known and Accepted Terms
## from the *APICS Dictionary*, 12th Ed.

## The Term's Place in DDMRP

**demand lead time:** The amount of time potential customers are willing to wait for the delivery of a good or a service. Syn: customer tolerance time.

Demand lead time or customer tolerance time factors heavily into inventory positioning strategy.

**demand uncertainty:** The uncertainty or variability in demand as measured by the standard deviation, mean absolute deviation (MAD), or variance of forecast errors.

Demand uncertainty is a factor for inventory positioning and buffer profile selection.

**divergent point:** An operation in a production process in which a single material/component enters and, after processing, can then be routed to a number of different downstream operations.

Divergent points factor in positioning inventory.

**drum:** In the theory of constraints, the constraint is viewed as a drum, and nonconstraints are like soldiers in an army who march in unison to the drumbeat; the resources in a plant should perform in unison with the drumbeat set by the constraint.

Drums are control points that represent places around which schedules are frequently synchronized; thus they need to be protected. Drums factor into inventory positioning.

**effective date:** The date on which a component or an operation is to be added or removed from a bill of material or an assembly process. The effective dates are used in the explosion process to create demands for the correct items. Normally, bills of material and routing systems provide for an effectivity start date and stop date, signifying the start or stop of a particular relationship. Effectivity control also may be by serial number rather than date. Syn: effectivity, effectivity date.

Effective dates will be used in conjunction with planned adjustments for buffered parts.

**flexibility:** The ability of a supply chain to mitigate, or neutralize, the risks of demand forecast variability, supply continuity variability, cycle time plus lead-time uncertainty, and transit time plus customs-clearance time uncertainty during periods of increasing or diminishing volume.

As with agility, flexibility is a primary objective of DDMRP. The way in which DDMRP mitigates or neutralizes certain factors represents a departure from traditional MRP tactics (see Chapter 22).

**hub-and-spoke systems:** In warehousing, a system that has a hub (or center point) where sorting or transfers occur, and the spokes are outlets serving the destinations related to the hub.

As with divergent points, a hub-and-spoke system is the basis for DDMRP tactics in the distribution network.

**implosion:** The process of determining the where-used relationship for a given component. Implosion can be single-level (showing only the parents on the next higher level) or multilevel (showing the ultimate top-level parent). *See* where-used list.

Implosion is used heavily in determining inventory positions. Implosion is used in combination with matrix BOMs and ASRLT.

| Known and Accepted Terms from the *APICS Dictionary*, 12th Ed. | The Term's Place in DDMRP |
|---|---|
| **inventory velocity:** The speed with which inventory passes through an organization or supply chain at a given point in time as measured by inventory turnover. | Inventory velocity is important so long as there is not a high degree of stock-outs and/or stock-outs with demand. |
| **law of variability:** The more that variability exists in a process, the less productive that process will be. | This is an important basic assumption in DDMRP. The law of variability is not limited to a resource. In fact, the more resources that work in combination and experience variability, the more is the productivity loss of the total system. This is a key reason for decoupling. |
| **manufacturing lead time:** The total time required to manufacture an item, exclusive of lower level purchasing lead time. For make-to-order products, it is the length of time between the release of an order to the production process and shipment to the final customer. For make-to-stock products, it is the length of time between the release of an order to the production process and receipt into inventory. Included here are order preparation time, queue time, setup time, run time, move time, inspection time, and put-away time. Syn: manufacturing cycle, production cycle, production lead time. | Manufacturing lead time (MLT) is still used as primary input for ASRLT. Whenever a parent part has a component that is not stocked, however, MLT will not be the lead time to calculate the parent's buffer level (for stocked parts) or determine a due date (for nonstocked parts). |
| **matrix bill of material:** A chart made up from the bills of material for a number of products in the same or similar families. It is arranged in a matrix with components in columns and parents in rows (or vice versa) so that requirements for common components can be summarized conveniently. | Matrix BOM used in combination with ASRLT becomes crucial in the inventory positioning equation. |
| **min-max system:** A type of order point replenishment system where the "min" (minimum) is the order point, and the "max" (maximum) is the "order up to" inventory level. The order quantity is variable and is the result of the max minus available and on-order inventory. An order is recommended when the sum of the available and on-order inventory is at or below the min. | Min-max parts will be nonstrategic and readily available stocked parts. |
| **natural variations:** These variations in measurements are caused by environmental elements and cannot be removed. | Acceptance of some variability and the effects it can have on planning and execution is a basic assumption in DDMRP. |

| Known and Accepted Terms from the *APICS Dictionary*, 12th Ed. | The Term's Place in DDMRP |
|---|---|
| **on-hand balance:** The quantity shown in the inventory records as being physically in stock. | Part of the available stock equation. |
| **on-order stock:** The total of all outstanding replenishment orders. The on-order balance increases when a new order is released, and it decreases when material is received against an order or when an order is canceled. | Part of the available stock equation—also referred to as open supply. |
| **pre-expediting:** The function of following up on open orders before the scheduled delivery date, to ensure the timely delivery of materials in the specified quantity. | Pre-expediting is accomplished through lead-time-managed designations and the various execution alerts in DDMRP. |
| **protective capacity:** The resource capacity needed to protect system throughput—ensuring that some capacity above the capacity required to exploit the constraint is available to catch up when disruptions inevitably occur. Nonconstraint resources need protective capacity to rebuild the bank in front of the constraint or capacity-constrained resource (CCR) and/or on the shipping dock before throughput is lost and to empty the space buffer when it fills. | Protective capacity is a kind of buffer that can reduce the inventory requirements of a system. It is a factor for inventory positioning and even buffer profile selection. |
| **pull signal:** Any signal that indicates when to produce or transport items in a pull replenishment system. For example, in just-in-time production control systems, a kanban card is used as the pull signal to replenish parts to the using operation. *See* pull system. | The concept of a pull signal is crucial to DDMRP. In DDMRP, the signal is electronic and takes into account the factors within the available stock equation. |
| **purchasing lead time:** The total lead time required to obtain a purchased item. Included here are order preparation and release time; supplier lead time; transportation time; and receiving, inspection, and putaway time. | Purchasing lead time (PLT) is used in DDMRP to size buffers of purchased parts and generate request dates. |
| **pull system:** 1) In material control, the withdrawal of inventory as demanded by the using operations. Material is not issued until a signal comes from the user.... 3) In distribution, a system for replenishing field warehouse inventories where replenishment decisions are made at the field warehouse itself, not at the central warehouse or plant. | DDMRP is a dynamic pull system with a high degree of visibility and relative priority for parts and products. |

| Known and Accepted Terms from the *APICS Dictionary*, 12th Ed. | The Term's Place in DDMRP |
|---|---|
| **replenishment lead time:** The total period of time that elapses from the moment it is determined that a product should be reordered until the product is back on the shelf available for use. | Replenishment lead times should be measured and used to adjust manufacturing lead time (MLT) and production lead time (PLT). |
| **risk pooling:** A method often associated with the management of inventory risk. Manufacturers and retailers that experience high variability in demand for their products can pool together common inventory components associated with a broad family of products to buffer the overall burden of having to deploy inventory for each discrete product. | The statistical reasons behind risk pooling are a key reason for the hub-and-spoke tactic of DDMRP distribution tactics as well as the placement of inventory at a divergent point. |
| **where-used list:** A listing of every parent item that calls for a given component, and the respective quantity required, from a bill-of-material file. *See* implosion. | Where-used lists still can be used, understanding that they contain a limited slice of information. For inventory positioning, matrix BOMs in combination with ASRLT should be used. |

# New Terms in Demand-Driven Material Requirements Planning

**actively synchronized replenishment:** A set of demand-driven inventory and materials positioning, planning, and execution techniques.

**actively synchronized replenishment (ASR) equilibrium:** The point of indifference between the investment in adding stock positions (typically component parts) and the lead-time inventory compression resulting from that investment (typically parent items).

**ADU alert:** An alert indicating a significant change in ADU within a defined set of parameters (quantity and time).

**ADU alert horizon:** A defined shorter rolling range within the broader rolling horizon used to calculate ADU.

**ADU alert threshold:** A defined level of change in ADU that triggers the alert within the ADU alert horizon.

**ADU-based recalculation:** A process of dynamically adjusting strategically replenished buffers incorporating a rolling horizon.

**artificial batch:** Any batch that is not equal to actual demand.

**ASR lead time (ASRLT):** A qualified cumulative lead time defined as the longest unprotected/unbuffered sequence in a bill of material.

**available stock:** A planning calculation to determine the planning status of a stocked item. The equation is on-hand + on-order stock (also referred to as *open supply*) – unfulfilled qualified actual demand. If the available stock equation yields a position in the green zone of a buffer, then typically no planning action is required.

**average daily usage (ADU):** Average usage of a part, component, or good on a daily basis.

**buffer penetration:** The amount of remaining buffer, typically expressed as a percentage.

**buffer profile:** A globally managed group of parts with similar lead time, variability, control, and order management characteristics.

**buffer zone:** A stratification layer within a stock buffer. Typically, buffer zones are color-coded with red, yellow, and green assignments.

**consumption-based supply generation:** The process of creating supply orders based on the available stock equation and decoupled explosion.

**current on-hand alert:** An execution alert generated by on-hand penetration into a buffer. Typically, on-hand alert will be set for red zone penetrations and out-of-stock situations.

**decoupled explosion:** The cessation of bill of material explosion at any buffered/stocked position.

**demand-driven material requirements planning (DDMRP):** An innovative method to plan materials that enables a company to build more closely to actual market requirements.

**dynamic buffers:** Buffer levels that are adjusted either automatically or manually based on changes to key part traits.

**execution horizon:** The life cycle of orders from the time the order is created and/or released to the time it is closed.

**green zone:** The top layer of a replenished and replenished override buffer. If available stock is in this zone, then no additional supply is created.

**lead-time alert:** An alert/warning generated by an LTM part. An alert will be triggered whenever the part enters a different time zone from its buffer. Green is the first alert to be encountered, follow by yellow and then red.

**lead-time alert zone:** The zone associated with the percentage of lead time that provides the definition for lead-time alerts. The LTM alert zone has three equal sections color-coded green, yellow, and red.

**lead-time-managed (LTM) part:** A critical nonstocked part that will have special attention paid to it over its execution horizon. Typically, LTM parts are critical, long-lead-time components that do not have sufficient volume to justify stocking. A portion of the lead time of the part (typically 33 percent) will have a three-zoned warning applied to it. That portion is typically divided into three equal sections.

**market potential lead time:** The lead time that will allow an increase in price or the capture of additional business either through existing or new customer channels.

**material synchronization alert:** An alert generated by the earliest occurrence of a negative on-hand balance within at least one ASRLT where open supply receipt is due after required demand date.

**nonbuffered part:** All parts that are not stocked.

**occurrence-based recalculation:** A method to adjust buffers based on the number and severity of specific occurrences in predefined fixed interval.

**on-hand alert level:** The percentage of the red zone used by buffer status alerts in order to determine a yellow or red color designation.

**order spike horizon:** A defined future time frame used to qualify order spikes in combination with an order spike threshold. Typically, order spike horizon is set to one ASRLT.

**order spike threshold:** A defined amount used to qualify order spikes in combinations with an order spike horizon. Typically, the order spike threshold will be expressed as a percentage of the total red zone (or min value) of a part's buffer.

**over top of green (OTOG):** A situation in which either available stock or on-hand stock is over the top of defined green zone, indicating an excessive inventory position.

**planned adjustments:** Manipulations to the buffer equation that affect inventory positions by raising or lowering buffer levels and their corresponding zones at certain points in time. Planned adjustments are often based on certain strategic, historical, and business intelligence factors.

**projected buffer status alert:** An alert generated by projected on-hand positions over a part's ASRLT based on on-hand, open supply, and either actual demand or ADU.

**qualified actual demand:** The demand portion of the available stock equation comprised of qualified order spikes, past-due demand, and demand due today.

**qualified order spike:** A quantity of combined daily actual demand within the order spike horizon and over the order spike threshold.

**ramp-down adjustment:** Manipulations to the buffer equation that affect inventory positions, lowering buffer levels and their corresponding zones at certain points in time. Ramp-down adjustments typically are used in part deletion.

**ramp-up adjustment:** Manipulations to the buffer equation that affect inventory positions, raising buffer levels and their corresponding zones at certain points in time. Ramp-up adjustments typically are used for part introduction.

**red zone:** The lowest-level zone in a replenished and replenished override part buffer. The zone is color-coded red to connote a serious situation. The red zone is the summation of red zone safety and red zone base.

**red zone base:** The portion of the red zone sized by lead-time factors.

**red zone safety:** The portion of the red zone sized by variability factors.

**relative priority:** The priority between orders filtering by zone color (general reference) and buffer penetration (discrete reference).

**replenished override part:** A strategically determined and positioned part using a static (buffer zones are manually defined) three-zoned buffer for planning and execution. Planned adjustments, however, can be used with these buffers.

**replenished part:** A strategically determined and managed part using a dynamic three-zoned buffer for planning and execution. Buffer zones are calculated using buffer profiles, ADU, and ASRLT.

**seasonality adjustment:** Manipulations to the buffer equation that affect inventory positions by adjusting buffers to follow seasonal patterns.

**significant minimum order quantity:** A minimum order quantity that sets the green zone of a buffer.

**strategic inventory positioning:** The process of determining where to put inventory that will best protect the system against various forms of variability to best meet market needs and leverage working capital.

**thoughtware:** The analysis and process employed to define the relevant factors and dependencies in an organization or system in order to construct appropriate business rules and operating strategies that maximize velocity, visibility, and equity. Within the DDRMP framework, thoughtware is commonly referred to with regard to applying the inventory positioning factors.

**top of green (TOG):** The quantity of the top level of the green zone. TOG is calculated by the sum of red, yellow, and green zones.

**top of red (TOR):** The quantity of the top level of the red zone.

**top of yellow (TOY):** The quantity of the top level of the yellow zone. TOY is calculated by the sum of the red and yellow zones.

**yellow zone:** The middle layer of the buffer level coded with yellow to convey a sense of warning. The yellow zone is the rebuild zone for replenished and replenished override buffers.

# To My Best Recollection: The Eras of Material Requirements Planning with Packaged Software

*By Gene Thomas*

**G**ene Thomas, Founder Emeritus, has been an independent consultant and developer of product management objects. He was formerly with IBM for 15 years in manufacturing software development and marketing activities, including sales manager, Milwaukee, and branch manager, Baltimore, where he supported development of the pioneering Black & Decker multiplant, net-change, automatic engineering change inventory balancing MRP system in 1967.

Gene invented IBM's Bill-of-Material Processor (BOMP) package in 1961; initiated the first two MRP software packages, LAMP in 1964 and PICS in 1966; authored IBM's first online "dock to stock" and central-dispatching systems at its Rochester plant in 1960; and directed the first installation of IBM's U.S. Kraus/CLASS/CAPOSS series of finite-capacity scheduling systems in 1968.

This appendix bullets a number of timelined events over the previous 60-some years. Since most of these are personal recollections from a 79-year-old semiretired founder emeritus, I would like to elicit comments, reinterpretations, and corrections from the many professional friends exposed to these events during their influences over this body of work.

Many thanks,
Gene Thomas, Founder Emeritus
Comments are welcome at http://mrp-to-lean.blogspot.com/.

- *First mechanization: punched cards or sequential tape.* In the early 1950s, the only mechanization generally available to manufacturers was either punched-card processing or very large-scale tape-oriented computers (no disks yet). In either case, it was a sequentially processed procedure, punching up bill-of-material (BOM) decks or maintaining tape files from punched cards. Processing involved copying (reproduction of punched-card BOM decks), sorting, and summarizing level-by-level explosions. There often were excessive numbers of levels designed to delineate componentry at multiple identifiable and inventorial levels.

Then, in the mid-1950s, as computing began to appear, the structuring and maintenance of single-level punched-card BOM decks often were supported by the various assembly and fabrication drawing componentry part-numbering schemes. These were normally stored in tub files and pulled manually to support the reproduction and sorting to summarize each of the time-bucketed componentry requirements. A sequencing technique often was used to ensure that no cards were missing in the BOM and routing decks by using a collator to compare preceding and succeeding sequence numbers.

- *Shop paper ditto mats and ledger cards.* Before the 1950s, most manufacturers were manually supporting their shop work-order paperwork with graphic duplication procedures (ditto mats) for bills and routings along with drawing copies. Part-numbered nomenclatures were rampant with all kinds of part number significance schemes such as a drawing-size prefix and revision-level suffix. Inventory balances on hand (BOHs) were mostly posted to ledger cards, sometimes with both sold/released allocations and stock room transfers to work-in-progress from fabrication completions and purchases.

  In a locally managed Motorola plant in Rome, Italy, they actually had a manual posting station manned at the end of each stock room aisle but also punched-card standard costs to four decimal positions in lira! Manually built-up standard costs then were used to "charge off" the dollar values from work-in-progress to finished goods to summarize cost of sales (COS).

- *IBM's computer-intensive leadership.* In the mid-1950s, manufacturing education and training classes for IBM customers at Endicott, NY, were held in the training center covering requirements planning techniques for IBMers and customers using unit record equipment and large sequential tape-oriented computers at the IBM plants. There was always much contention from IBM customers that the computer costs were subsidized internally and generally out of reach of normal manufacturers!

  I remember a couple of instructors (but I wish I could remember their names) explaining that the volume of componentry generated by an explosion process was a "google-plex"—a google to the google power!

- *BOM shortcuts.* During the later 1950s, requirements planning used punched-card reproductions of inventory BOH-summarized ledger cards and exploded BOM quick decks (presummarized for each model part number across all levels of a product BOM) for each time bucket of the future planning horizon, normally, monthly.

  Early John Deere plants attempted to maintain a "100-machine load" quick deck (presummarized by component part number for use at all levels for 100 of the top-level models or families) as a shortcut but suffered a major problem with inaccurate maintenance changes from the single-level engineering drawings. Separate cost accounting "buildup and charge-off" quick decks (the term *back-flush* hadn't been commissioned yet!) were even in worse shape because they were maintained by accounting personnel who were not very communicative with engineering changes.

- *BOM explosion techniques.* Then, by upgrading computers to single-level sequential magnetic tape systems storing the BOMs, they were used for level-by-level explosions requiring a complete pass of all the BOMs for each level and a subsequent merging of the inventory netting files by low-level code. Several technical improvements were developed for the IBM plants to segment the inventory netting tapes by low-level code to reduce the passing time, but it then also required newly maintained codes to be regenerated prior to any requirements planning explosion cycle.

    I understood that the Collins Radio Company in Cedar Rapids, IA, installed 7-7070 tape systems (for separate processing runs for each level) in line to try to continuously process requirements changes, the first inkling of an interest in simulating a net-change process!

- *Bottom-up quick-deck explosions.* Maytag, in Newton, IA, pioneered the use of a 650 tape-drum system in the late 1950s in which the limited assembly model schedule (provided by Joe Dorzweiler, production control manager) was capable of being stored on the 2,000-word drum (like a revolving hard drive that also contained the programming instructions). It projected the future time series of several months of weekly bucketed schedules. However, the BOMs again had to be maintained on tape in inventoried component part number sequence in a bottom-up summarized where-used quick-deck format, thus, again, requiring excruciatingly cumbersome maintenance.

- *BOM representations.* As could be observed, a plethora of computerized BOMs were being used in the late 1950s for a myriad of siloed organizational interests: (1) single-level drawings, (2) summarized top-down quick decks, (3) single-level and (4) summarized bottom-up where-used sortations, (5) indented explosions and (6) implosions, and (7) special-use explosion costing BOMs with (8) single-level labor routings.

    The aerospace and Department of Defense–oriented engineering/manufacturers were even more complicated, having to foster a project orientation and traceability functionality to the BOM–routing maintenance requirements.

- *Introduction of random-access disk processing with where-used.* About this time, some of my John Deere customers were overloading the medium-sized 650 tape drum systems and requesting larger drums to cover more models and a master production schedule (MPS) for their growing families. But with the announcement of the IBM 305 and 650 RAMAC in the late 1950s, I initiated the idea of a disk-oriented packaged software program, Bill-of-Material Generator (BONG), to maintain the where-used quick-deck tape files as a by-product from maintenance of the conventional single-level engineering drawing documentation.

This package then could be generalized to cover most of the perceived engineering interests, including a simultaneously maintained where-used feature (the first manufacturing relational database other than that used for straight sequentially part-numbered inventory).

- *BOMP development basis for standardization of manufacturing applications.* I was then assigned to an internal IBM plant sales oversight function (IBM sales divisions were trying to help our plant implement far-reaching applications for effective customer demonstration—"How do you use your own equipment?"), which also involved assistance to our Midwest sales force and manufacturing customers.

  Over the next three years from 1960 at the IBM Rochester, MN, plant, I initiated and participated in four major manufacturing-oriented projects:

  1. Invented and published the IBM internal TIE (Technical Information Exchange Project 3319, September 1961 for BOMP Bill-of-Material Processor) specifications as a basis for a manufacturing application packaged program.

  2. Recruited John Schleier from John Deere as MIS manager and initiated the Rochester plant's control center centralized dispatching implementation with punched-card data collection and sequenced authorization of operation queues.

  3. Initiated the plant's "dock-to-stock" online receiving, warehouse locating, shortage filling, and picking systems, both subsequently documented as marketing brochures.

  4. Assisted the implementation of Joe Orlicky's design of the first net-change MRP system at the J. I. Case–Claussen Works, Racine.

- *Milwaukee IBM development of BOMP.* Would you believe that it took me over two years (carrying around my "worm chart" BOMP database chaining specifications) to convince IBM development management, along with several customer sites, and our Milwaukee sales office staff that we should begin the BOMP programming development managed from the field sales staff? Everyone thought that the engineering function was much too complicated to be standardized across marketplace product lines.

  Things finally came together with my assignment as the manufacturing sales manager in Milwaukee, where we had the resources under our own command. We then were assisted with the enthusiastic customer support offered in the early 1960s by Allis Chalmers (Doc Rue, MIS, and Bob Blair, later with Cummins Engine), Cutler-Hammer (Oscar Reak, president), Milwaukee Faucets (Sandy Kohn, vice president for manufacturing), and Vollrath (Terry Kohler, to-be president). J.I. Case-Claussen Works in Racine (Joe Orlicky, production control manager, Jack Chobanion, production control, and Fred Brani and Harold Jones, MIS) was already under way with our oversight (IBM System Engineer Ted Musial) as well as Graco-Minneapolis (Gene Laguban, MIS, and later at Allen-

Bradley, TLA and Oshkosh Truck), so they implemented their own customized RAMAC BOM maintenance system, but without the processing burden of the where-used cross-linkage functionality.

- *Joe Orlicky's net-change influence at J. I. Case.* Restricted by disk capacity restraints, Case used the six-character RAMAC file address for its part numbering and direct file linkages. They stored a combination of a 20-day build schedule (six days frozen plus a year of monthly buckets) to maintain both a forecasted MPS horizon and also the shorter-term configured customer order final assembly schedule (FAS). A configured 15-character generic coded model numbering scheme (features and options) was used to select preconfigured modular option BOMs during the explosion routines, which later showed up in our logic of the IBM PICS MRP-RPS and the subsequent MAPICS package.
- *BOMP became the basis for packaged manufacturing software.* With the resources of the Milwaukee sales office and the contributing customer staffs, we programmed the Bill-of-Material Processor (BOMP, then fortuitously changed from being called "BONG") in 4 kB of assembly language packaged for use initially on the smaller 1401 disk system, recognizing that the contributing customers also would use the package on larger equipment. Cutler-Hammer even modified the where-used chaining to use four anchors balanced to reduce the chain chasing for deletions within long where-used linkages on their larger 1410 hardware.

    Many additional implementations of the BOMP package were developed in the Milwaukee area until we figured that we could advance the concept of packaged manufacturing software use to cover the requirements planning functionality.
- *Further development of manufacturing applications by Milwaukee companies.* Following from the customized J. I. Case implementation, many additional functions were developed around BOMP from the Milwaukee IBM staff's influence during the mid-1960s. Jim Burlingame, vice president manufacturing at Twin Disc in Racine (and later the president of APICS national) fostered an MRP explosion program with multilevel pegging of specific sold customer requirements intermixed with the forecasted lot-sized planned orders. He also built on top of this current accounting procedures that "charged off" the builtup "standard" costs as finished to-order products were shipped.

    These transactions were too late to be used for netting work-in-process, so they fenced up the stocked componentry BOHs and then hard-issued them to assembly work-in-process balances. Then they charged off the assembly work order BOM componentry on shop floor completions, reducing the work-in-process componentry part balances to support subsequent netting during the daily MRP net-change processing.

    We were all chagrined about the inaccuracy of the originally manually maintained BOMs in those days and had to wait for successful coordination of the different files from engineering, accounting, and production control silos over another decade. In the meantime, though, this accuracy-building cultural issue was coined in Racine as *backflushing* (subject to initially fostering "crappy" results) and began a learning curve leading up to today whereby the accuracy issues are normally taken for granted owing to the success of BOMP maintenance!

- *Development of the first packaged MRP software products.* With the assistance of a small customer, the vice president for manufacturing of Milwaukee Faucets (Sandy Kohn), I participated by writing the netting, lot-sizing, and time-series off-setting logic in RPG for the standard BOMP summarized explosion routines.

  In those days, most data-processing (tabulation managers) departments reported to financial management, and product structuring was characterized as designed to maintain inventory valuations. There were excessive levels in the BOMs to accommodate identification for counting inventory such as cut-length part numbering, subassembly drawings, semifinished raw materials, and so on. Lot sizing at these levels complicated explosion netting and time-bucket off-setting.

  The problems were exacerbated with the common use, then, of order-point replenishment procedures executed independently from the BOHs at every level. Level-by-level explosion processing gave the opportunity to gate lot sizing in synchronization with levels above and below to save excessively generated safety time and stocking levels.

  We published the programs as the first packaged MRP system called Labor and Material Planning 520-1231-0 (LAMP). Kohn's company was still using punched-card equipment to maintain an inventory BOH stock status, so we used a bank's 1401 disk system to maintain and store the BOMs. Kohn then generated gross requirements for subsequent netting during a punched-card stock status inventory report back at the plant.

- *Early design of MRP level-by-level, time-series, explosion logic.* As can be observed during this early era of packaging manufacturing (MRP) software functionality, progress steps were always contingent on computer technology advancements. An MRP explosion process, then commonly referred to as Gross-to-Net Explosions, Time Series (GNETS), was very computing-intensive with an extremely high volume of generated transactions against a massive database (BOM, routings, and inventory/order status).

  From the original Milwaukee area experiences with design direction, we structured the GNETS processes to accommodate the currently developing computing power as well as projected functionality. As a result, most of the original designs were structured for regenerative explosions with monthly time bucketing for long monthly processing runs—normally over weekends!

  "Papa" Joe Orlicky's influence at J. I. Case, Racine, was a very significant step up in functionality but required much more sophistication in the GNETS logic. They followed some of the logic prevalent in the automotive industry, where requirements were maintained to cumulative build schedules of a model year. Orlicky retained some of the cumulative logic (especially involving behind-schedule bucketed control balances) but broke the regenerative cycles down to be performed as often as weekly or even daily as required. They also maintained both a monthly forecast horizon for about a year and a separate shorter-term actual final assembly schedule (FAS) supported by a configured channel order backlog.

From these experiences, we architected the GNETS process in such a fashion that it could be used for both regeneration and net-change cycles, with an eye to future real-time processing (this still has not been universally delivered). The forecasted and made-to-stock (MTS) marketplace wouldn't soon press the need for these shorter processing cycles, but the churning of the make-to-order (MTO) product lines with configuration requirements would provide most of the pressure for advancement of the state of the art.

Most of those critical of MTO lean MRP functionality still don't realize that they should turn off the master scheduling (MPS) of forecasted planning with suspect planning BOMs and work only with the configured sold order horizon! The GNETS logic would remain similar for real-time cycles, but the computer database upgrading would become enormous.

- *A J. I. Case testimonial re: Use of BOMP as a generic file organizational tool.* In an IBM Application Brief Publication of implementations of the late 1960s entitled, "The Production Information and Control System at J. I. Case Company with Ernie Roeseler—Burlington Plant" (GK20-0364-0 3/70), featuring the implementation history of the BOMP RPS system, a byline by Orlicky's team was included as "The bill of material processor program represents support by IBM in the area of manufacturing application and implementation. It provides the support to organize, maintain, and reorganize the four basic manufacturing data files—part number master file, product structure file, standard routing file, and work center file. Case has found, however, that the program can be used very effectively in other application areas as well as in manufacturing. Their views are expressed in the quotation from a letter to IBM. '. . . This is one of the most powerful tools . . . for the purpose of organizing direct access files. We are presently using it not only in areas for which it was designed, such as the parts master, routing, etc., but . . . also . . . in payroll, and we will be using it in our order entry and accounting areas. . . . It is misnamed and . . . should be listed and supported as a file organization program rather than as a bill of material processor. . . .'"

- *Internal IBM world trade training on BOMP and LAMP.* In late 1964, I traveled to London, Paris, Sindelfingen, Milan, and Stockholm to present the BOMP and LAMP software to the World Trade IBMers covering the European marketplace. During this trip, I met with Werner Kraus and Gunter Evers, authors of the German finite scheduling/sequencing software called Kraus—later to be upgraded to CLASS and CAPOSS.

As we discussed the differences in marketplace approaches, we envisioned the merging of finite scheduling and MRP into a single system, but it might take

10 to 15 years for the cultures to coincide! They related the German situation as a scheduling intensity where any production schedule changes would be added only to the end of the horizon because all plants were overcommitted anyway. I related our U.S. situation as that of a sales manager walking through the shop floor with a case of beer, "suggesting" schedule accommodations to cover his favorite customer's changes!

It has been 50 years, and neither of us is where we prognosticated! My Baltimore IBM staff installed Kraus (the first APS—Advanced Planning System in German) at the Koppers Piston Ring plant, supported with the data-collection dispatching through a control center later in 1968. They were processing an average of 34 machining operations per shop order in a randomly pathed, functionally organized shop-floor layout!

- *Publishing of MRP-PICS application software.* Subsequently, as manufacturing industry staff manager in Chicago in the mid-1960s, I initiated the development of repackaging of the systems as Production and Inventory Control System–Requirements Planning System E20-0280-2 (PICS-RPS), subsequently managed by Pat Reilly and Will Couch of IBM's development staff, and we published it as an application brochure for use internationally (the third edition copyrighted in 1968). It later became the basis for IPICS, RICS, MAPICS, and COPICS.
- *Database evolution for manufacturing applications.* Another Racine company with Department of Defense traceability requirements added modifications to the BOMP logic where Bob Haddox, MIS manager of Sunbeam, built support for a "chain-of-chains" database traceability pegging capability.

This architecture was used later at Black & Decker by Haddox for calculated net-change componentry use-up balancing, again supported by my staff at the Baltimore IBM branch. It later showed up in the architecture of the MAC-PAC lot-control and traceability functionality.

- *Influential professionals contributing to manufacturing software.* Several newly formed MRP software firms and consultancies were trained and originated later during the 1960s using the BOMP architecture and backgrounds: Dick Lilly's Software International (SI) using the BOMP tree logic for MRP and general ledger structuring,; Dick Ling's Arista, which became XCS and Glovia, and who later develop the sales and operations planning (S&OP) concept (see Chapter 20 in this book for his latest thoughts on S&OP); Tom Nies's Cincom system that used a speedier file organization procedure for the item masters; Romey Everdell, Nick Edwards, Tricia Moody, and Woody Chamberlain of Rath & Strong with fully pegged requirements planning (PIOS) for aerospace traceability; Dale Colosky's Computer Strategies (CSI) first second-tier automotive support with cumulative-repetitive, orderless processing; Gerry Roch user's architect of M2M; and Jerry Bowman (of my Milwaukee staff) as founder of fourth-shift MRP, the first MRP product written specifically for the personal computer.

Books, consultancies, and publications were added by highly acclaimed George Plossl and Ollie Wight in 1968, with Ray Lankford and Chris Gray's standard system specs; Dave Garwood's BOM consulting in the 1970s; Joe Orlicky's net-change MRP in 1972; the first BOM handbook by Dick Bourke later in 1975; and Terry Schultz (from my Milwaukee staff) partnering with Dave Buker Education and subsequently as founder of the Forum Consultancy.

Subsequently, Andersen Consulting (now Accenture) purchased the RPG package through MRM, the Milwaukee firm I helped to start when I left IBM in 1969. It was reworked and patented under project managers Bill Darnton and Jeff Rappaport as MAC-PAC RPG and then revised again as MAC-PAC DOD from work done by the Minneapolis Comserv group emanating from Milwaukee Allen-Bradley experience.

All subsequent MRP systems used various offshoots of BOMP functionality as their base file organization—the where-used linkage was really the first randomized "relational" support for disk database architecture.

- *Manufacturing "fraternity" of influential practitioners.* As a result of these influential relationships among people who frequently communicated professionally, during an engagement to Cummins Engine in June 1971, I arranged a "fraternity" meeting to cover advance MRP techniques using large-scale disk systems continuously processing GNETS replanning explosions.

The invited attendees included a number of those related to BOMP-MRP development during the 1960s, including Jim Burlingame and Bill Wassweiler, Twin Disc; Jack Chobanian, J. I. Case; Larry Colla, Cutler-Hammer; Bob Haddox, Bob McKain, and Carl Euker, Black & Decker; Dave Hargrove, Pat Murray, and Jim Fortier, Baxter Labs; Herb Friedman, Thomas Friedman Associates; Ted Musial, Tom Long, Ray Fritch, Sam Huffman, Herb Pereyra, Cliff Smith, and Ed Souders, IBM; Gene Laguban, Allen Bradley, Duane Segebarth, and Bud Vogel, John Deere; key management of Cummins, including Bob Blair, Gale Shirk, and Earl Hahn, S&DP; Hal Smitson, plant manager, and Bob Dice, John Wertz, Leo Underwood, and Doug Taylor, production control.

The participants viewed Cummins' pioneering conversion from large-scale processed tape systems to disk-oriented relational database (IMS DB-2) architecture. They were able to gain a decade advantage over their competitor, Caterpillar, which was still stuck with monstrous shadow file functionality against legacy batch processing!

In retrospect, there have been many occasions where this "fraternity" has cross-communicated, competed, and supported inventive conceptual activities that all have contributed to the ERAs of MRP to ERP, especially the dearth of practitioner feedback to keep the IT folks germane—it has been a fun ride!

# Index

*References to figures are in italics.*